T0205915

Traditional and Indigenous Knowledge for the Modern Era

A Natural and Applied Science Perspective

Traditional and Indigenous Knowledge for the Modern Era

A Natural and Applied Science Perspective

Edited by

David R. Katerere, Wendy Applequist,
Oluwaseyi M. Aboyade, and Chamunorwa Togo

CRC Press is an imprint of the
Taylor & Francis Group, an informa business

CRC Press
Taylor & Francis Group
6000 Broken Sound Parkway NW, Suite 300
Boca Raton, FL 33487-2742

First issued in paperback 2021

ISBN-13: 978-1-03-208941-6 (pbk)
ISBN-13: 978-1-138-03429-7 (hbk)

Publisher's Note
The publisher has gone to great lengths to ensure the quality of this reprint but points out that some imperfections in the original copies may be apparent.

Visit the Taylor & Francis Web site at
http://www.taylorandfrancis.com

and the CRC Press Web site at
http://www.crcpress.com

Contents

Foreword

Traditional and Indigenous Knowledge for the Modern Era: A Natural and Applied Science Perspective

Biodiversity underpins life on earth, and traditional knowledge along with medicinal plants have contributed to our health and well-being, and have underpinned the development of modern medicine. Yet at the dawn of this new millennium, one of the most pressing challenges of our time is the continuing, and at times irreversible, loss of biodiversity and its associated precious knowledge on our planet.

The Convention for Biological Diversity (CBD) enacted after the 1992 Rio Earth Summit has played a central role in sensitizing the world to the plight of biodiversity loss and in asserting the sovereignty of countries over their bio-resources. Today the CBD has 193 parties (or governments) as members.

This seminal moment has represented a dramatic step forward in the conservation of biological diversity, the sustainable use of its components, and the fair and equitable sharing of benefits arising from the use of genetic resources. Subsequent to this major effort, the Nagoya Protocol has gone that extra mile in further enshrining the need to reconsider Access and Benefit Sharing.

To keep reminding the world of the urgency of preserving our biodiversity, the United Nations has declared the period 2011–2020 to be the Decade on Biodiversity. Nowhere is the need for conservation and sustainable utilization of biodiversity greater than in sub-Saharan Africa, whose biodiversity wealth is uniquely important from a global conservation viewpoint.

The African continent is home to around 60,000 plant species, of which at least 35,000 are found nowhere else. Africa's biodiversity wealth is, unfortunately, not uniformly distributed. Countries including Madagascar, the islands of the southwestern Indian Ocean, and South Africa, have been classified as "megadiverse" countries. The world's 17 most biologically diverse countries, including megadiverse African countries, together account for nearly 70% of global species diversity.

Despite its enormous natural wealth, sub-Saharan Africa faces daunting conservation challenges. Its flora and fauna are under threat, and climate change is exacerbating that. Biodiversity loss, in Africa and in other parts of the world, has a significant impact on economic growth and social development.

For rural citizens, loss of biodiversity has the effect of removing key sources of food, fuel, and medicines, and even adversely affecting sectors like tourism and the pharmaceutical industry, which would be hurt from a reduction in the availability of medicinal plants. New knowledge, about conservation and whole plant utilization, is needed, not just to strengthen the conservation effort but to harness this unique patrimony of natural resources to foster economic development, reduce poverty, protect the environment, and more importantly, ensure our own survival on this planet.

Seen in this context, this volume, *Traditional and Indigenous Knowledge for the Modern Era: A Natural and Applied Science Perspective*, edited by David Katerere, Wendy Applequist, Oluwaseyi Aboyade, and Chamunorwa Togo, adds new contributions for celebrating traditional knowledge and understanding how it has supported our well-being, providing us with food and medicine, and ensuring other ecosystem services.

This book will further advance our understanding of the increasingly crucial role that traditional and indigenous knowledge play in the economic, cultural, medical, and social spheres of our lives. It brings on board contributions from several continents, and it is a welcome addition in terms of safeguarding this precious knowledge for humanity as a whole.

Sustainable utilization and management of plant genetic resources is a topic of contemporary significance. By marshalling the latest evidence and cutting-edge as well as age-old knowledge, this volume should find broad appeal among academics, scientists, farmers, policymakers, and all those who are committed to reducing traditional knowledge and biodiversity loss on our planet and promoting new leads for the development of new drugs, foods, and other products and services that will sustain our well-being.

Ameenah Gurib-Fakim, PhD, DSc
Former President of Mauritius and Founder of
The Ameenah Gurib Fakim Foundation

Preface

The idea for this book was planted at one of the Innov8 events, panel discussion platforms held by The Innovation Hub Management Company in Pretoria. On that cool July evening in 2016, three of the four editors had been called to participate as part of a panel on "Indigenous Food and Nutrition". It was a lively discussion, with contributions from the different panellists and audience members about the importance of "going back to the food of our forebears". The irony was that none of that food was served at the buffet dinner afterwards!

But the evening had served its purpose and challenged us to ask two simple questions, "How relevant is traditional and indigenous knowledge in modern times? Does that knowledge, in fact, still exist?"

We hope the answers to those questions (and others) can be found in this book. The book intends to illustrate the continuing relevance of age-old "knowledges" from around the world in solving some of the world's current and complex challenges, be they global warming, health and wellness, or food. Thus, the book addresses topics which are relevant to the natural and social sciences and the humanities, topics that illustrate the inclusive utility and sheer breadth of traditional and indigenous knowledge.

We have assembled a veritable list of subject experts who utilize a variety of original research, reviews, and global case studies to convincingly argue the continued relevance of traditional and indigenous knowledge. We hope that this book will be a valuable resource to scholars and policymakers in the field and that it will also be a useful introduction to this area of research for newcomers and the casual reader.

We are indebted to the authors of the chapters for the different and new perspectives they bring to this book. As editors, it was a great delight and honour to work with them all. We thoroughly enjoyed editing the chapters while learning many new philosophies and hitherto hidden knowledges of different cultures of the world. We hope that you, the reader, will find these well-researched contributions, written in highly accessible language, worth your while.

We would also like to thank our editor, Randy Brehm, and her team for their patient guidance, without which this project would not have been a success. Thokozani Skhosana is duly acknowledged for proofreading and last-minute formatting of various chapters.

David R. Katerere
Wendy Applequist
Oluwaseyi M. Aboyade
Chamunorwa A. Togo

Editors

David R. Katerere, PhD, trained as a pharmacist in Zimbabwe before completing a PhD at University of Strathclyde in pharmaceutical science (Natural Products Chemistry). After a stint as a postdoctoral fellow at the University of Pretoria (2001–2003), Dr. Katerere briefly joined Farmovs-Parexel CRO Pty Ltd as chief bioanalyst before working at the SA Medical Research Council. Dr. Katerere has a keen interest in medicinal plant research and botanical medicine. He has published extensively and is co-editor of the book *Ethnoveterinary Botanical Medicine.* He comes from a family of traditional healers, with grandparents on both sides having been herbalists and/or traditional birth attendants. Dr. Katerere is currently a full professor of pharmaceutical science at Tshwane University of Technology, where he teaches and conducts research. He is also a consulting partner with Pharmakine Pty Ltd and Afrobotanix.

Wendy Applequist, PhD, trained as a plant systematist and earned a PhD from Iowa State University. Dr. Applequist has worked at the Missouri Botanical Garden since 2000; she is currently an associate curator in the Garden's William L. Brown Center and editor-in-chief of *Novon.* She manages natural products discovery collections, collaborates with researchers studying medicinal plants, and conducts taxonomic research on the flora of Madagascar. Dr. Applequist is the author of a book on the morphological identification of medicinal plants commonly used in North America.

Oluwaseyi M. Aboyade, PhD, has spent almost two decades researching the therapeutic benefits of indigenous plants across Africa. Dr. Aboyade earned her under graduate and master's degrees in biochemistry and pharmacognosy from universities in Nigeria before moving to South Africa to complete her PhD in botany with a focus on ethnobotany at the University of Fort Hare. Since then, her research has focused on traditional herbal medicine and non-communicable diseases. Presently, Dr. Aboyade coordinates several research projects funded by the MRC, NRF, and TIA in these fields at Tshwane University of Technology. She has also launched a start-up company which specializes in food innovation using indigenous knowledge.

Chamunorwa Togo, PhD, has extensive research experience and background in biotechnology, ranging from the industrialization of indigenous fermented foods to environmental biosensors. Dr. Togo earned his undergraduate and master's degrees at University of Zimbabwe, after which he completed PhD studies at Rhodes University, South Africa, and a master's degree in business leadership at UNISA. A combination of Dr. Togo's upbringing in the rural areas of Chivhu (Zimbabwe) and knowledge gained in scientific research ignited his interest in demonstrating the complexity and advanced nature of indigenous knowledge practices. Dr. Togo lectured at Zimbabwe

Open University, Midlands State University, and the University of Witwatersrand before moving into a consultancy at Aurecon Training Academy. Currently he is a general manager for the bio-economy at The Innovation Hub Management Company (TIHMC), an innovation agency of Gauteng Province, where he works with a team driving commercialization of technologies and services in the bio-economy sector.

Contributors

Roseanna Avento
Development Services
University of Eastern Finland
Kuopio, Finland

Sechaba Bareetseng
NEPAD Southern Africa Network for Biosciences (SANBio)
CSIR Biosciences
Pretoria, South Africa

Callistus Bvenura
Department of Chemistry
Tshwane University of Technology
Pretoria, South Africa

Ereck Chakauya
NEPAD Southern Africa Network for Biosciences (SANBio)
CSIR Biosciences
Pretoria, South Africa

Weiyang Chen
Department of Pharmaceutical Sciences
Tshwane University of Technology
Pretoria, South Africa

Felix D. Dakora
Department of Chemistry
Tshwane University of Technology
Pretoria, South Africa

Trish Flaster
Botanical Liaisons, LLC
Boulder, Colorado

Edson Gandiwa
School of Wildlife, Ecology and Conservation
Chinhoyi University of Technology
Chinhoyi, Zimbabwe

Estonce T. Gwata
School of Agriculture
University of Venda
Thohoyandou, South Africa

Philip F. Iya
IKS Centre
North West University
Potchefstroom, South Africa

Shakkie Kativu
Tropical Resource Ecology Programme
University of Zimbabwe
Harare, Zimbabwe

Olga L. Kupika
School of Wildlife, Ecology and Conservation
Chinhoyi University of Technology
Chinhoyi, Zimbabwe
and
Institute of Cooperate Citizenship
University of South Africa
Pretoria, South Africa

Mothusiotsile E. Maditsi
IKS Centre
North West University
Potchefstroom, South Africa

Lilian Mukandiwa
Queensland Alliance for Agriculture and Food Innovation (QAAFI)
The University of Queensland
St Lucia, Queensland

Godwell Nhamo
Institute of Corporate Citizenship
University of South Africa
Pretoria, South Africa

Norman Z. Nyazema
Department of Pharmacy
University of Limpopo
Mankweng, South Africa

David Picking
Natural Products Institute
The University of the West Indies
Kingston, Jamaica

Donald R. Sibanda
Department of Agriculture and Animal Health, School of Agriculture and Life Sciences
University of South Africa
Pretoria, South Africa

Darshan Shankar
Trans-Disciplinary University
Bangalore, India

Soul Shava
Department of Science and Technology Education
University of South Africa
Pretoria, South Africa

Naotoshi Shibahara
Division of Kampo Diagnostics, Institute of Natural Medicine
University of Toyama
Toyama, Japan

Renée A. Street
South African Medical Research Council, Environment and Health Research Unit & Department of Occupational and Environmental Health
University of KwaZulu-Natal
Durban, South Africa

Zvikomborero Tangawamira
BioFISA 2, NEPAD Southern Africa Network for Biosciences (SANBio), CSIR Biosciences,
Pretoria, South Africa

Kaizer Thembo
Complementary Medicines Unit, South African Health Products Regulatory Authority (SAHPRA), CSIR Campus
Pretoria, South Africa

Ina Vandebroek
Institute of Economic Botany
The New York Botanical Garden
The Bronx, New York

CHAPTER 1

Indigenous/Tribal Knowledges – Definition and Relevance in the Modern Era

Soul Shava

CONTENTS

ABSTRACT

Background Indigenous knowledges are generated from interactions at the nexus between indigenous peoples and their lived environment over time. They are the cumulative body of diverse intergenerational and transgenerational knowledges that have sustained indigenous communities, enabling them to adapt to and thrive within their ever-changing environment.

Relevance Indigenous knowledges are transdisciplinary, cutting across all sectors of society. Policy makers should therefore pay attention to the possible role that indigenous knowledges can play within their different sectors, including their contribution to the identity and representation of indigenous peoples and their application (potential roles) in education, science, and technology.

INTRODUCTION

This chapter introduces indigenous knowledges. In writing about indigenous knowledges as an indigenous person and indigenous scholar, I draw upon my own and other indigenous people's experiences and research on indigenous people and their knowledges, which comprise the growing body of literature on indigenous knowledges that runs counter to colonial/Western/Euro-Americentric misrepresentations of indigenous peoples as people without their own knowledges and histories or people with primitive, irrelevant, and/or invalid knowledges. This chapter attempts to provide a definition of indigenous knowledges, explores indigenous knowledge systems, discusses indigenous knowledges in the context of the global geopolitics of knowledges, looks into innovative applications of indigenous knowledges in the 21st century, and emphasizes why indigenous people should view indigenous knowledges as living knowledges and a continuous decolonizing and transgressive process towards attaining epistemological pluralism and equity in education, research, knowledge generation, and "formal" knowledge representation in socio-economic contexts.

DEFINING INDIGENOUS KNOWLEDGES

The knowledges of indigenous peoples have been variously labelled as indigenous knowledge (IK), indigenous knowledge systems (IKS), tribal knowledge, traditional knowledge, traditional ecological knowledge (TEK) (Inglis, 1993), traditional environmental knowledge (Johnson, 1992), endogenous knowledge (Hountondji, 1997; Crossman, 2002), local knowledge, folk knowledge, and aboriginal knowledge. These synonyms refer to the epistemologies (indigenous ways of knowing and doing things) of indigenous peoples relating to all aspects of their life. In essence, it is not the terminology that matters; it is the description and application.

Indigenous knowledges are characterized by some key aspects, namely, people, place, culture, language, knowledge, practices, and dynamism (Shava, 2013).

People

Indigenous peoples (first peoples, aboriginal peoples) are the generators (source) of indigenous knowledges, imbuing them with indigenous discourses emanating from experiences in their lived world contexts.

The term "people" is a racialized concept in the politics of identity, where indigenous people are made to occupy the margins of humanity. Unfortunately, the emphasis on the identity of indigenous people in global discourses has been on minority groups of society (UNDESA, 2009, 2016, 2017; Minority Rights Group International [MRG], 2006, 2007, 2008, 2011, 2012, 2016). However, in Africa, indigenous peoples are the majority and have largely been non-dominant in the political sphere due to European colonization. This sidelining of the majority indigenous peoples is a global political process that continues to marginalize their concerns, issues, and interests. For instance, even in the post-independence era in most African states, the knowledge and languages of indigenous peoples are largely marginalized and excluded in mainstream education and socio-economic contexts due to the continuing hegemony (coloniality) of Western/Euro-Americentric epistemologies and social systems.

Place (Spatio-Temporal Context)

Place speaks to origin, history, and ancestry. Indigeneity (indigenousness) has a strong connection to land, memory, and history. Indigenous people live on the land (local landscapes comprising physical landforms, riverscapes, climates, and associated ecosystems), which they have interacted with and imbued with meaning from human experiences over time, transforming it into socio-ecological landscapes. They are part of (not separate from) the natural world. Dotted across these landscapes are named aspects (biophysical, cultural, spiritual, utilitarian, etc.) that indigenous people identify with (memory), giving them a sense of place, of belonging. Linked to the land are the issues of connectedness to place, indigenous rights, and access to indigenous/local resources.

Due to their direct reliance on the land, most indigenous peoples do not consider themselves above nature. They instead acknowledge that there is a relational link between themselves and the land (Odora Hoppers, 2005) in which a negative impact on the land also affects them (directly and indirectly). They therefore emphasize living in harmony with nature, as opposed to the dominance over nature and its consideration as merely an economic resource, which is promoted by Western/Euro-Americentric capitalistic perspectives.

We have to acknowledge that because indigenous knowledges are generated in different places/locales by different groups of indigenous people, they cannot be conflated into a singular collective entity under the commonly used unifying term "indigenous knowledge". They are instead polycentric and plurally definable **heterogeneous bodies of knowledge** or **indigenous knowledges** that are the products of the diverse contexts from which they evolve (Shava, 2013). We can therefore, broadly speaking, talk about African (Southern African, West African, East African, Central African, North African), Australian, New Zealand, South American, North American/Ameri-Indian, and Asian (Chinese, Japanese, Indian) knowledges, among others. However, while indigenous knowledges are localized, they can be applicable and transferable to other contexts (for example, indigenous medicines that are made into pharmaceutical drugs for particular diseases or ailments can be used globally), and there is some interchange of knowledge between indigenous peoples and other peoples.

Language

What characterizes a group of indigenous people is their distinctive language. Language serves as a means of communication within the group, and more importantly, language is the main medium through which indigenous knowledge is represented and transmitted within and across generations.

The continuing marginalized role of indigenous languages in formal education contexts is an issue of concern for indigenous peoples in Africa and across the globe. European languages, the languages of the colonizers, continue to dominate as the formal languages in socio-economic and political contexts and in contexts of teaching and learning in formal education thereby impacting on indigenous learners' epistemological access (Shava & Manyike, 2018).

Culture and Practice

Each group of indigenous people has a distinctive culture. Their indigenous knowledge is "embedded in their culture and embodied in their practices" (Shava, 2013: 384). Indigenous knowledges are therefore an embodiment of indigenous people's cultures, practices, identities, and histories (the knowledge-experience-practice nexus).

Some similarities amongst indigenous cultures and practices have been observed across the globe; these include attachment to the land, reciprocity, relationality/kinship (Gelfand, 1981; Odora Hoppers, 2005; Goduka, 2005; Goduka et al., 2013), communality (the "self" embedded in community), and respect for the elderly.

Knowledge

Knowledge is valuable inasmuch as it is beneficial and applicable to human survival and livelihood sustenance. Indigenous knowledge is the cumulative body of knowledge (understandings, meanings, interpretations/explanations, experiences, philosophies, pedagogies, methodologies, practices, skills, and strategies) of an indigenous people, in its diverse representations, that has developed through their interactions with the land (lived environment) and is sustained over time through culture and practices. The knowledges of indigenous people are holistic, signifying the interdependence and interconnectedness of the human (social, cultural, economic, spiritual) and biophysical (land/nature) aspects, and they span all areas of their life (practical, physical, health, material, cultural, spiritual, ethical/moral). Indigenous knowledges are not homogenous within a group of indigenous people (Dei, 2008). They have both explicit (tangible) and tacit (intangible/implicit) elements as well as generalized/common/public and specialized/sacred knowledge aspects. They are generated from experiences on the land (in the lived environment). Indigenous knowledges are transmitted within and across generations orally through stories, myths, poetry, songs, ceremonies, rituals, taboos, proverbs, and idioms; visually (e.g. through observation); in "written" form (e.g. San/bushman paintings, Zulu love letters (beaded bangles which convey messages), and other cultural art forms and decorations); spiritually through dreams and visions; and practically through sharing between peers, demonstrations, apprenticeship (into specialized trades), work and play (indigenous games), through gendered roles, and from the elders. Indigenous knowledges are lived and practiced knowledges that constantly circulate in interactions within indigenous communities of practice.

Dynamism

While indigenous knowledges embody the historical experiences of indigenous peoples over time, they also enable the indigenous community to adapt to and cope with emerging complex situations from within and without the lived context. Dei (2008: 6) points to the "adaptability, vitality, and agency" of indigenous knowledges. They are therefore dynamic rather than static knowledges (Dei et al., 2002; Pottier et al., 2003; Masuku Van Damme & Neluvhalani, 2004; Dei, 2008, 2010; Shava, 2013) that are living and constantly being renewed (formed, reformed, and transformed).

INDIGENOUS KNOWLEDGE SYSTEMS

While indigenous knowledges are generally holistic in nature and not structured into disciplines, because they span several Western/Euro-Americentric knowledge disciplines, they can be organized according to their functionalities into "systems of knowledge" relating to different aspects of life, such as the biophysical (natural), technological, and social spheres (Odora Hoppers, 2005). These indigenous knowledge systems include:

1. *Indigenous (traditional) ecological knowledge* – This is indigenous peoples' knowledge of local species diversity (knowledge and indigenous classifications of plants, animals, fungi, etc., and their habits and habitats) and ecology (indigenous classifications of local ecosystems and their interrelationships) (Campbell et al., 1991; Gagdil et al., 1993; Berkes et al., 1994; Campbell, 1997; Dold & Cocks, 1999; Nygren, 1999; Byers et al., 2001; Murombedzi, 2003; Cocks, 2006a, 2006b; Brown et al., 2011; O'Donoghue et al., 2013; Shava & Schudel, 2013). Indigenous knowledges express the interrelationship, interconnectedness, and interdependence between indigenous communities and their lived environments (the human/nature interface).
2. *Indigenous agricultural knowledge* – This includes knowledge of agrobiodiversity (e.g. crops and animal diversity, including indigenous underutilized crops and indigenous/traditional animal breeds); indigenous/traditional agricultural practices (shifting cultivation, fallow system, multicropping, the commons, conservation of vegetation, and agroforestry), and indigenous soil taxonomy (Shava et al., 2009; O'Donoghue et al., 2013).
3. *Indigenous meteorology* – This is the knowledge indigenous peoples have of weather (weather forecasting) and climate. Indigenous meteorology incorporates the indigenous ways of interpreting and predicting weather and their application in decision-making processes.

4. *Indigenous cuisine* – This includes food preparation, preservation, food diversity, and nutrition (balanced diet) (Arnold et al., 1985; Gadaga et al., 1999; Shava, 2000a).
5. *Indigenous cosmetics* – This pertains to indigenous beauty creams, ochre, indigenous pastes, and indigenous oils (Van Wyk & Gericke, 2000; Morekhure et al., 2017; Dlova et al., 2013).
6. *Indigenous technologies* – These include indigenous metallurgy (metal mining, smelting, and processing/blacksmithing), textiles, craftware (e.g. wood carving, weaving/basketry), architecture (buildings [e.g. huts], stone masonry, indigenous designs, etc.), and indigenous food processing (Childs, 1991; Fox & Young, 1982; Childs & Killick, 1993; Shava & O'Donoghue, 2014).
7. *Indigenous/traditional medicine* – This includes indigenous herbal medicines, ethnoveterinary medicines, indigenous/traditional pharmacology, and traditional medicinal practice, including bone-setting and obstetrics, midwifery, and circumcision (Watt & Breyer-Brandwijk, 1962; Gelfand et al., 1993; Hutchings et al., 1996; Marck, 1997; Dold & Cocks, 2001; Cocks & Dold, 2006; Luseba & Van De Mwere, 2006; Katerere & Luseba, 2010; MacGaw, 2008; Shava, 2011). Traditional medicines are derived from natural products.
8. *Indigenous governance systems* – These include the indigenous/traditional institutions and systems of governing people, the land and natural resources (including mineral resources, environmental management, and resource [biodiversity and water] conservation and utilization). Indigenous governance structures and systems include indigenous laws, norms, and collective decision-making processes that enable effective organization of indigenous societies (Porter, 1998; Kuokkanen, 2011; von der Porten & de Loë, 2013).
9. *Indigenous culture* – This includes indigenous/traditional music, dance, poetry, art, craft, painting, etc. (Fortune & Hodza, 1974; Odora Hoppers, 2005; Shava, 2015).
10. *Indigenous philosophies* – Indigenous philosophies are embedded in proverbs, idioms, taboos, stories, and principles which convey meanings relating to human/human and human/nature interactions (e.g. "Ubuntu"/"Hunhu" – African indigenous humanity) (Mayr, 1912; Chimhundu, 1980; Letseka, 2000; Shava, 2000b; Odora Hoppers, 2005; Goduka, 2000, 2005; Chivaura, 2007; Dei, 2008; Chigidi, 2009; Letseka, 2011; Mandova, 2013; Makaudze & Shoko, 2015).
11. *Indigenous spirituality/religions* – This includes indigenous religious rituals and norms such as respect (not worship) of ancestors, respect of others, sensitivity, humility, morality, compassion, and gentleness (Goduka, 2000; Dei, 2002). Such spiritual values are conveyed through culture, myths, proverbs, songs, folk tales, mediums/ancestors, elders, etc.

KEY CHALLENGES FOR INDIGENOUS KNOWLEDGES

The Crisis of Representation: Indigenous Knowledges and the Geopolitics of Knowledge – the West and the Rest

One main challenge to indigenous knowledges is the continuing domination, subjugation, marginalization, primitivization, devaluation, decontextualization invalidation, rejection, exclusion in/by Western/Euro-Americentric discourses and socio-economic spheres (Dei, 2002a; Shava, 2008, 2013). Dei (in Dei, Hall & Rosenberg, 2002) made the following observation about education processes encountered in the North American context while teaching in the Canadian education system:

> I hear my students – especially though not exclusively those from minoritized groups – ask me why certain experiences and histories count more than others when "valid" academic knowledge is being produced and validated. I hear students lament the effort it is taking for educators to recognize the powerful linkages between identity, schooling, and knowledge production. But more importantly, I hear my students worry how indigenous knowledges are being marginalised in the academy, and about the impact that the ranking of knowledges may well have on the prospects for educational transformation and social change.
>
> **(2002, p. xi–xii)**

In discussing indigenous knowledges, we need to be conscious of the constant power/knowledge relationships that continue to undermine, marginalize, and exclude the knowledges of indigenous peoples across the globe (Dei, 2004; Shava, 2008, 2013).

Indigenous scholars question why Western/Euro-Americentric epistemologies have dominated what constitutes in/valid knowledge and why indigenous peoples' histories, experiences, and innovations have been, and continue to be, excluded from education discourses and mainstream knowledge (Shiva, 1993; Semali & Kincheloe, 1999; Dei, 2002; Odora Hoppers, 2005). Western rationality/philosophy/scientificity, portrayed as universal, decontextualized, and apolitical, has often been tagged as the benchmark criteria by which the knowledge of other cultures should be evaluated (Shiva, 1993; Dei, 2002; Cornell, 2009; Shava, 2008). Indigenous/traditional knowledges of indigenous peoples, on the other hand, have been frequently portrayed/labelled as primitive, closed, pragmatic, utilitarian, value laden, indexical, contextual, and spiritual. However, this portrayal of indigenous overlooks the fact that Western knowledge disciplines and their associated discourses have been imposed on indigenous peoples through processes of colonization and that Western knowledge disciplines have grown by cannibalizing other knowledge systems (Nader, 1996; Hountondji, 2002). Knowledge is considered valid when contextualized in the knower's subject position and location (Dei & James, 1998). Indigenous epistemologies have their own inherent scientific validity and should not be validated on the basis of Western/Euro-Americentric epistemologies.

Santos (2007, 2016) argues that, in the ecologies (geopolitics) of knowledges, Western thinking is abyssal thinking by its granting modern science the exclusionary monopoly of universal distinction between true and false while excluding and making indigenous knowledges invisible. Indigenous knowledges, which are on the other (invisible) side of the abyssal line are considered to be not real knowledge. They are considered to be beliefs, opinions, intuitions, and subjective understandings, which at the most may become objects or raw material for scientific inquiry. In so doing, Western/Euro-Americentric thinking commits epistemicide on indigenous knowledges.

The crisis of representation is a clarion call for indigenous scholars and researchers to continually research and generate publications on indigenous knowledges and the need for curriculum transformation in academia to open up to plural epistemologies and to embrace indigenous knowledges.

IK Appropriation

IK appropriation refers to the processes of subversive extraction of indigenous knowledge forms (cannibalization – see Hountondji, 2002) and resources for economic benefit (biopiracy – see Shiva, 1997) without due acknowledgement of the indigenous source and compensation of the indigenous originators of the knowledge forms (theft, usually under the guise of labelling IKS as common/general knowledge).

Indigenous knowledges have been disembodied, decontextualized, appropriated, and subsumed into Western knowledge and natural and social science disciplines (e.g. medicine, pharmacy, botany, anthropology) without due credit to the knowledge sources (Hountondji, 2002; Shava, 2008).

With regards to indigenous resources, equitable access and benefit sharing of intellectual products derived from indigenous knowledges as defined in the Convention on Biological Diversity (CBD) (United Nations, 1992) is still a key challenge facing many indigenous communities (Wynberg et al., 2009; Chikombero & Luseba, 2010). This extends to shared/joint ownership of the intellectual products and projects with indigenous peoples. However, an emphasis only on sharing economic benefits overlooks the contributions of indigenous peoples in research processes. This speaks to "hit and run" short-term research approaches on indigenous peoples (Smith, 1999) and the continuing "ethical" research practices of non-disclosure of indigenous participants under the guise of protecting their identity, namely, intellectual appropriation. Indigenous peoples and scholars/researchers need to ask the following questions: Who does what research on indigenous peoples? Who decides what research should be done on indigenous peoples and on indigenous lands? Who has authority over indigenous research outputs? How are indigenous people represented in research outputs? It is important for indigenous people to own their research, to be in control of research processes, and to write their own research stories/publications. This calls for indigenous people to engage in transgressive research by locating themselves in research through owning and growing their own research projects, establishing research ethics review processes driven and authorized by indigenous peoples, promoting research methodologies that are developed by indigenous peoples, and requesting external researchers to do research that acknowledges indigenous peoples as co-researchers and co-authors/contributors.

Another key, and often overlooked, aspect of colonial appropriation is indigenous peoples' sovereignty, that is, ownership of land, and their rights to this land and access to associated resources. Indigenous peoples have had their land appropriated by colonial settlers through forced mass evictions that have in many cases been accompanied by bloody colonial massacres that have almost exterminated indigenous populations. The struggles by indigenous people to reclaim their ancestral lands are a global phenomenon.

CONCLUSION: WHY INDIGENOUS KNOWLEDGES?

Indigenous peoples across the globe have suffered the colonial onslaught of the imposition of European histories on the global south, where histories of the global south have been histories of European conquest (and being called people without a history), the dominance of colonially derived European languages (as languages of instruction and commerce), the imposition of European socio-economic structures and systems, and the hegemony of Western/Euro-Americentric education (as the "formal" education). This misrepresentation of indigenous peoples undermines and overlooks the fact that indigenous people have their own histories, functional knowledges, and socio-economic systems.

Indigenous knowledges are in an ongoing process of indigenous resistance (Semali & Kincheloe, 1999; Hill, 2009; Dei, 2008, 2012; Shava, 2016). They represent a process of decolonization that continuously challenges the coloniality (the continuing forms of colonial hegemony in the present period, especially in knowledge systems) of Western/Euro-Americentric/hegemonic epistemologies (worldviews, knowledge systems, and knowledge generation processes and practices). They dismantle the epistemic privilege of Western/Euro-Americentric knowledge and its mythical claim as universal, unlocated, disembodied, neutral knowledge by drawing attention to the non-hierarchical **plurality of knowledges and their localized origins** (pluriversality) (Shiva, 1993; Dei, 2008; Shava, 2008, 2013).

Indigenous knowledges present alternative ways of looking at and understanding our world, that is, they present alternative epistemologies or worldviews. From an indigenous peoples' perspective, we need to move indigenous peoples and indigenous issues from the margins to the center of postmodern/postcolonial history (waThiong'o, 1993). This requires education and research that prioritize and respond to our contextual issues and the problems that occupy us (Dei, Hall & Rosenberg, 2002). This calls for a re-evaluation of our education and research goals and our attitudes towards transgressive pedagogies, theoretical perspectives, and research methods aligned to the knowledge systems of indigenous peoples (hooks, 1994; Smith, 1999; Kovach, 2009; Dei, 2002, 2010, 2012; Chilisa, 2012). This further requires us to self-transform our locus of enunciation (where we speak from). We must move from speaking from a Western/Euro-Americentric perspective to speaking from our own indigenous knowledges, cultures, and contexts – what Ngugi waThiong'o refers to as the "decolonising the mind" (waThiong'o, 1981).

Indigenous knowledges are alternative knowledges that refuse to fit into the destructive, artificial, Euro-Americentric Cartesian duality between man and nature (a duality that posits humankind above nature). This Cartesian foundation of modern/Western science has led to the reckless destruction of nature in an attempt to dominate and harness the environment for the economic benefit of humankind. Indigenous knowledges are built on a holistic and relational foundation that emphasizes the continuity between humans and nature (humans as part of the natural world) in which an effect on one aspect has a concomitant/domino effect on the other.

Indigenous scholars point out the problems with Western/Euro-Americentric capitalistic development, which is unsustainable, and how such (Western/

Euro-Americentric) epistemologies cannot necessarily provide the solutions to the global crises they have created (Shiva, 1993; Shava, 2008, 2013). In this regard, indigenous knowledges possess the power to transform the arena of knowledge discourses and practices that currently favor Western/Euro-Americentric knowledge systems. They usher forth alternative solutions to problems created by Western modernity, which Western/Euro-Americentric knowledge is failing to solve. These solutions include alternative and sustainable forms of agriculture, alternative forms of medicine with possibly fewer side effects, alternative and sustainable forms of living, alternative forms of community, alternative forms of food that are healthy, etc. However, positioning indigenous knowledges in mainstream knowledge should not be considered as the extreme notion of replacing existing Western/Euro-Americentric knowledges. Instead, they open up the knowledge arena to plural knowledges (pluriversalism as opposed to universalism) and provide multiple perspectives of our world (waThiong'o, 1993; Shava, 2008, 2013, 2016).

IK agency should therefore be prioritized through the affirmation of our indigenous scholarly and transformative presence in academic contexts across diverse disciplines, including biodiversity conservation (and use), politics, economy, healthcare, society, and community development contexts as our locus of enunciation (epistemic location).

REFERENCES

Arnold T.H., Wells M.J., Wehmeyer A.S. 1985. Khoisan food plants: taxa with potential for future economic exploitation. In: Wickens G.E., Goodin J.R., Field D.V. (eds) *Plants for Arid Lands*. Dordrecht: Springer. pp. 69–86.

Berkes, F., Folke, C. and Gadgil, M. 1994. Traditional ecological knowledge, biodiversity, resilience and sustainability. In: Perrings, C.A., Maler, K.-G., Folke, C., Holling, C.S. and Jansson, B.-O. (eds). *Biodiversity Conservation: Problems and Policies*. Dordrecht: Springer. pp. 269–287.

Brown, R., Brown, J., Readron, K. and Merrill, C. 2011. Understanding STEM: current perceptions. *Technology and Engineering Teacher* 70(6): 5–9.

Byers, B.A., Cunliffe, R.N. and Hudak, A.T. 2001. Linking the conservation of culture and nature: a case study of sacred forests in Zimbabwe. *Human Ecology* 29(2): 187–218.

Campbell, B.M. 1997. The use of wild fruits in Zimbabwe. *Economic Botany* 41(3): 375–385.

Campbell, B.M., Clarke, J.M. and Gumbo, D.J. 1991. Traditional agroforestry practices in Zimbabwe. *Agroforestry Systems* 14: 99–111.

Chigidi, W.L. 2009. Shona taboos: the language of manufacturing fears for sustainable development. *The Journal of Pan African Studies* 3(1): 174–188.

Chikomborero, M. and Luseba, D. 2010. Logistical and legal considerations in ethnoveterinary research. In Katerere, D. and Luseba, D. (eds). *Ethno-Veterinary Botanical Medicine: Herbal Medicines for Animal Health*. Boca Raton: CRC Press. pp. 25–40.

Childs, S.T. 1991. Style, technology, and iron smelting furnaces in Bantu-speaking Africa. *Journal of Anthropological Archaeology* 10: 332–339.

Childs, T. and Killick, D. 1993. Indigenous African metallurgy: nature and culture. *Annual Review of Anthropology* 22: 317–337.

Chilisa, B. 2012. *Indigenous Research Methodologies*. Los Angeles: Sage Publications.

Chimhundu, H. 1980. Shumo, tsumo and socialisation. *Zambezia VIII*(i): 38–51.

Chivaura, V.G. 2007. Hunhu/Ubuntu: a sustainable approach to endogenous development, biocultural diversity and protection of the environment in Africa. In Haverkort, B. and Rist, S. (eds). *Endogenous Development and Biocultural Diversity: The Interplay of Worldviews and Locality.* Leusden: ETC-COMPAS Centre for Development and Environment. pp. 229–240.

Cocks, M. 2006a. Biocultural diversity: moving beyond the realm of 'indigenous' and 'local' people. *Human Ecology* 34(2): 185–200.

Cocks, M. 2006b. Wild Plant Resources and Cultural Practices in Rural and Urban Households in South Africa: Implications for Bio-cultural Diversity Conservation. Doctoral thesis, Wageningen University, Wageningen, The Netherlands.

Cocks, M.L. and Dold, A.P. 2006. Cultural significance of biodiversity: the role of medicinal plants in urban African cultural practices in the Eastern Cape, South Africa. *Journal of Ethnobiology* 26(1): 60–81.

Cornell, R. 2009. *Southern Theory: The Global Dynamics of Knowledge in Social Science.* Cambridge: Polity Press.

Crossman, P. 2002. *Teaching Endogenous Knowledge in South Africa: Issues, Approaches and Aids.* Pretoria: Centre for Indigenous Knowledge, Department of Anthropology, University of Pretoria.

Dei, G.J.S. 2002a. *Rethinking the role of indigenous knowledges in the academy.* NALL Working Paper No. 58.

Dei, G.J.S. 2002b. *Spiritual knowing and transformative learning.* NALL Working Paper No. 59.

Dei, G.J.S. 2008. Indigenous knowledge studies and the next generation: pedagogical possibilities for anticolonial education. *Australian Journal of Indigenous Education* 37(Supplementary): 5–13.

Dei, G.J.S. 2010. *Teaching Africa: Towards a Transgressive Pedagogy.* Dordrecht: Springer.

Dei, G.J.S. 2012. Indigenous anti-colonial knowledge as 'heritage knowledge' for promoting Black/African education in diasporic contexts. *Decolonization: Indigeneity, Education and Society* 1(1): 102–119.

Dei, G.J.S., Hall, B.L. and Rosenberg, D.G. (eds). 2002. *Indigenous Knowledges in Global Contexts: Multiple Readings of Our World.* Toronto: Buffalo, published in association with University of Toronto Press.

Dei, G.J.S. and James, I.M. 1998. 'Becoming black': African-Canadian youth and the politics of negotiating racial and racialized identities. *Race, Ethnicity and Education* 1(1): 91–108.

Dlova, N.C., Nevondo, F.T., Mwangi, E.M., Summers, B., Tsoka-Gwegweni, J., Martincigh, B.S. and Mulholland, D.A. 2013. Chemical analysis and in vitro UV-protection characteristics of clays traditionally used for sun protection in South Africa. *Photodermatology, Photoimmunology and Photomedicine* 29(3): 164–169.

Dold, A.P. and Cocks, M.L. 1999. Preliminary list of Xhosa plant names from the Eastern Cape, South Africa. *Bothalia* 29(2): 267–292.

Dold, A.P. and Cocks, M.L. 2001. Traditional veterinary medicine in the Alice district of the Eastern Cape Province, South Africa. *South African Journal of Science* 97(9&10): 375–379.

Fortune, G. and Hodza, A.C. 1974. Shona praise-poetry. *Bulletin of the School of Oriental and African Studies, University of London* 37(1): 65–75.

Fox, F.W. and Young, E.N. 1982. *Food from the Veld: Edible Wild Plants of Southern Africa.* Craighall: Delta Book (Pty) Ltd.

Gadaga, T.H., Mtukumira, A.N., Narvhus, J.A. and Feresu, S.B. 1999. A review of traditional fermented food and beverages of Zimbabwe. *International Journal of Food Microbiology* 53: 1–11.

Gadgil, M., Berkes, F. and Folke, C. 1993. Indigenous knowledge for biodiversity conservation. *Ambio* 22(2/3): 151–153.

Gelfand, M. 1981. *Ukama: Reflections on Shona and Western Cultures in Zimbabwe.* Harare (Salisbury): Mambo Press.

Gelfand, M., Mavi, S., Drummond, R.B. and Ndemera, B. 1993. *The Traditional Medical Practitioner in Zimbabwe: His Principles and Pharmacopoeia.* Gweru: Mambo Press.

Goduka, I.N. 2000. African/indigenous philosophies: legitimizing spiritually centred wisdoms within the academy. In Higgs, P., Vakalisa, N.C.G., Mda, T.V. and Assie-Lumumba, N.T. (eds). *African Voices in Education.* Cape Town: Juta. pp. 63–83.

Goduka, N. 2005. Eziko: Siphekasisophula: Nguni foundations for educating/researching for sustainable development. *South African Journal of Higher Education* 19(3): 467–481.

Goduka, N., Madolo, Y., Rozani, C., Notsi, I. and Talen, V. 2013. Creating spaces for eZiko-SiphekaSisophula theoretical framework for teaching and researching in higher education: a philosophical exposition. *Indilinga African Journal of Indigenous Knowledge Systems* 12(1): 1–12.

Hill, G. 2009. *500 Years of Indigenous Resistance.* Oakland: PM Press.

hooks, b. 1994. *Teaching to Transgress: Education as the Practice of Freedom.* New York, London: Routledge.

Hountondji, P. (ed). 1997. *Endogenous Knowledge: Research Trails.* Dakar: CODESRIA.

Hountondji, P.J. 2002. Knowledge appropriation in a post-colonial context. In Odora Hoppers, C. (ed). *Indigenous Knowledge and the Integration of Knowledge Systems: Towards a Philosophy of Articulation.* Claremont: New Africa Books. pp. 23–38.

Hutchings, A., Scott, A.H., Lewis, G. and Cunningham, A. 1996 *Zulu medicinal plants: an inventory.* Scottsville: University of Natal Press.

Inglis, J.T. 1993. *Traditional ecological knowledge: Concepts and cases.* Ottawa: International Development Research Centre.

Johnson, M. (ed). 1992. *Lore: Capturing Traditional Environmental Knowledge.* Ottawa: Dene Cultural Institute and International Development Research Centre.

Katerere, D. and Luseba, D. (eds). *Ethno-Veterinary Botanical Medicine: Herbal Medicines for Animal Health.* Boca Raton: CRC Press.

Kovach, M. 2009. *Indigenous Methodologies: Characteristics, Conversations, and Contexts.* Toronto, Buffalo, London: University of Toronto Press.

Kuokkanen, R. 2011. Indigenous economies, theories of subsistence, and women: exploring the social economy model for indigenous governance. *American Indian Quarterly* 35(2): 215–240.

Letseka, M. 2000. African philosophy and educational discourse. In Higgs, P., Vakalisa, N.C.G., Mda, T.V. and Assie-Lumumba, NT (eds). *African Voices in Education.* Cape Town: Juta. pp. 179–193.

Letseka, M. 2011. In defence of Ubuntu. *Studies in Philosophy and Education* 31: 47–60.

Luseba, D. and Van Der Mwere, D. 2006. Ethnoveterinary medicine practices among Tsonga speaking people of South Africa. *Onderstepoort Journal of Veterinary Research* 73(2): 115–122.

MacGaw, L.J. 2008. Ethnoveterinary use of southern African plants and scientific evaluation of their medicinal properties. *Journal of Ethnopharmacology* 119(3): 559–574.

Makaudze, G. and Shoko, H. 2015. The reconceptualization of Shona and Venda taboos: towards an Afrocentric discourse. *The Journal of Pan African Studies* 8(2): 261–275.

Mandova, E. 2013. The shona proverb as an expression of Unhu/Ubuntu. *Matatu – Journal of African Culture and Society* 41(1): 357–368.

Marck, J. (1997). Aspects of male circumcision in subequatorial African culture history. *Health Transition Review* 7(sup.): 337–360.

Masuku Van Damme, L.S. and Neluvhalani, E.F. (2004). Indigenous knowledge in environmental educational processes: perspectives on a growing research arena. *Environmental Education Research* 10(3): 353–370.

Mayr, Fr. (1912). Zulu proverbs. *Anthropos* Bd. 7(H.4): 957–963.

Minority Rights Group International. 2006. *State of the World's Minorities and Indigenous Peoples*. London: Minority Rights Group International.

Minority Rights Group International. 2007. *State of the World's Minorities and Indigenous Peoples*. London: Minority Rights Group International.

Minority Rights Group International. 2008. *State of the World's Minorities and Indigenous Peoples*. London: Minority Rights Group International.

Minority Rights Group International. 2011. *State of the World's Minorities and Indigenous Peoples*. London: Minority Rights Group International.

Minority Rights Group International. 2012. *State of the World's Minorities and Indigenous Peoples*. London: Minority Rights Group International.

Minority Rights Group International. 2016. *State of the World's Minorities and Indigenous Peoples*. London: Minority Rights Group International.

Morekhure-Mphahlele, R., Focke, W.W. and Grote, W. 2017. Characterisation of vumba and ubumba clays used for cosmetic purposes. *South African Journal of Science* 113(3/4): 1–5.

Murombedzi, J.C. 2003. *Pre-Colonial and Colonial Conservation in Southern Africa and Their Legacy Today*. Washington, D.C.: World Conservation Union (IUCN).

Nader, L. 1996. *Naked Science: Anthropological Inquiry into Boundaries, Power and Knowledge*. New York: Routledge.

Nygren, A. 1999. Local knowledge in the environment-development discourse. *Critique of Anthropology* 19(3): 267–288.

O'Donoghue, R., Shava, S. and Zazu, C. (eds). 2013. *African heritage knowledge in the context of social innovation*. United Nations University – Institute of Advanced Studies (UNU_IAS), Tokyo. http://www.ias.unu.edu/resource_centre/UNU_Booklet_MB2013_FINAL_Links_v12.pdf.

Odora Hoppers, C. 2005. *Culture, indigenous knowledge and development: The role of the university*. Centre for Education Policy Development. Occasional Paper No. 5.

Porter, R.B. 1998. Decolonizing indigenous governance: observations on restoring greater faith and legitimacy in the government of Seneca Nation. *Kansas Journal of Law & Public Policy* 8(2): 97–141.

Pottier, J., Bicker, A. and Sillitoe, P. 2003. *Negotiating Local Knowledge: Power and Identity in Development*. London: Pluto Press.

Santos, B.D.S. 2007. Beyond abyssal thinking: from global lines to ecologies of knowledges. *Review* (Fernand Braudel Center) 30(1): 45–89.

Santos, B.D.S. 2016. *Epistemologies of the South: Justice against Epistemicide*. New York: Routledge.

Semali, L.M. and Kincheloe, J.L. (eds). 1999. *What is Indigenous Knowledge? Voices from the Academy*. New York & London: Falmer.

Shava, S. 2000a. The use of indigenous plants as food by a rural community in the Eastern Cape: an educational exploration. Unpublished master's (Education) thesis, Rhodes University, Grahamstown, South Africa.

Shava, S. 2000b. *Tales of Indigenous Trees of Zimbabwe*. Howick: Share-Net.

Shava, S. 2008. Indigenous knowledges: a genealogy of representation and applications in developing contexts of environmental education and development in southern Africa. Unpublished doctor of philosophy thesis, Rhodes University, Grahamstown, South Africa.

Shava, S. 2013. The representation of indigenous knowledges. In Stevenson, R.B., Brody, M., Dillon, J. and Wals, A. (eds). *International Handbook of Research on Environmental Education*. New York: Routledge. pp. 384–393.

Shava, S. 2015. African aesthetic, the. In Shujaa, M.J. and Shujaa, K.L. (eds). *The SAGE Encyclopedia of African Cultural Heritage in North America*. Thousand Oaks, London, New Delhi: SAGE Publications, Inc. p. 992.

Shava, S. 2016. The application/role of indigenous knowledges in transforming the formal education curriculum for contextual and epistemological relevance: cases from southern Africa. In Msila, V.T. and Gumbo, M.T. (eds). *Africanising the Curriculum: Indigenous Perspectives and Theories*. Stellenbosch: Sun Press. pp. 121–139.

Shava, S. and Manyike, T.V. 2018. The decolonial role of African indigenous languages and indigenous knowledges in formal education processes. *Indilinga: African Journal of Indigenous Knowledge Systems* 17(1): 36–52.

Shava, S. and O'Donoghue, R. 2014. *Teaching Indigenous Knowledge and Technology. Natural Sciences and Technology Grades 4–6*. Grahamstown: Fundisa for Change Programme, Environmental Learning Research Centre, Rhodes University.

Shava, S., O'Donoghue, R., Krasny, M. and Zazu, C. 2009. Traditional food crops as a source of community resilience in Zimbabwe. *International Journal of African Renaissance Studies – Multi-, Inter- and Transdisciplinarity* 4(1): 31–48.

Shava, S. and Schudel, I. 2013. *Teaching Biodiversity. Life Sciences Grades 10–12*. Fundisa for Change Programme, Environmental Learning Research Centre, Rhodes University, Grahamstown, South Africa.

Shiva, V. 1993. *Monocultures of the Mind: Perspectives on Biodiversity and Biotechnology*. London & New York: Zed Books.

Shiva, V. 1997. *Biopiracy: The Plunder of Nature and Knowledge*. Cambridge, MA: South End Press.

Smith, L.T. 1999. *Decolonizing Methodologies: Research and Indigenous Peoples*. London, New York: Zed Books Ltd.

UNDESA. 2009. *State of the World's Indigenous Peoples, Volume 1*. New York: United Nations.

UNDESA. 2016. *State of the World's Indigenous Peoples, Volume 2: Indigenous Peoples' Access to Health Services*. New York: United Nations.

UNDESA. 2017. *State of the World's Indigenous Peoples, Volume 3: Education*. New York: United Nations.

United Nations. 1992. *Convention on Biological Diversity*. Geneva: Interim Secretariat for the Convention on Biological Diversity.

Van Wyk, B.E. and Gericke, N. 2000. *People's Plants: A Guide to Useful Plants of Southern Africa*. Pretoria: Briza Publications.

Von der Porten, S. and de Loë, R. 2013. Water governance and indigenous governance: towards a synthesis. *Indigenous Policy Journal* 23(4): 1–12.

waThiong'o, N. 1981. *Decolonising the Mind: The Politics of Language in African Literature*. Harare: Zimbabwe Publishing House (Pvt.) Ltd.

waThiong'o, N. 1993. *Moving the Centre: The Struggle for Cultural Freedoms*. Nairobi: East African Educational Publishers.

Watt, J.M. and Breyer-Brandwijk, M.G. 1962. *The Medicinal and Poisonous Plants of Southern and Eastern Africa: Being an Account of Their Medicinal and Other Buses, Chemical Composition, Pharmacological Effects and Toxicology in Man and Animal*. Edinburgh & London: E.S. Livingstone.

Wyneberg, R., Schroeder, D. and Chenells, R. (eds). 2009. *Indigenous Peoples, Consent and Benefit Sharing: Lesson from the San-Hoodia Case*. Dordrecht, Heidelberg, London, New York: Springer.

CHAPTER 2

How Research Funding Can Drive the Commercialization of IK-Based Technologies: The Case of SANBio

Ereck Chakauya, Zvikomborero Tangawamira, and Sechaba Bareetseng

CONTENTS

ABSTRACT

Background Science, technology, and innovation (STI) policies should be aligned with economic development. There is increasing recognition among African governments that STI should be central to the economic development agenda. This is evident from the adoption by the African Union (AU) of the Science and Technology Consolidated Plan of Action (CPA) in 2005 and the Science Technology and Innovation Strategy for Africa (STISA) 2014–2024 in 2014. The Southern Africa Network for Biosciences (SANBio) derives its mandate from the CPA and was set up in 2005 as a platform to address and find means to resolve key problems in health, nutrition, agriculture, and environment in Southern Africa utilizing bioscientific innovations which incorporate indigenous knowledge and genetic resources.

Relevance Funding can be used to incentivize research which results in commercial outputs. In this chapter, SANBio showcases some of the projects they have funded in the past decade and draws the lessons learnt from both successful projects and those that failed.

INTRODUCTION

Africa's population was almost 1.3 billion in 2017 and is expected to double by 2050 (United Nations Department of Economic and Social Affairs Population Division, 2017). More than 60% of the population is below 25 years of age. It is envisaged that if the continent continues building on the developmental outcomes of better institutions and policies on education, labour productivity, and health, the demographic dividend can account for 11–15% of GDP (Ahmed et al., 2016). However, the future of the continent is largely dependent on how much it embraces science, technology, and innovation (STI) to provide solutions to the fundamental issues of poverty, unemployment, health, food security, and effects of climate change. It has been reported that Africa bears 71% of global burden of diseases (mainly HIV/AIDS, TB, and Malaria) (Mathers, 2017; Murray et al., 2012) but only 4% of the global healthcare workforce (Eyal et al., 2016). Moreover, poverty in the continent has been rising, with 42.3% of the population in sub-Saharan Africa (SSA) living on $1.90 or less per day, a principal factor of widespread hunger. Unfortunately, most of these people are caught in the poverty cycle, and the figure can only get worse. Food security is one of the biggest challenges facing the continent, especially with the effects of climate change increasingly becoming a threat (Brown et al., 2007; El Mokhtar et al., 2019).

CONTINENTAL STRATEGIES TO ENHANCE STI POLICIES

In 2014, the Assembly of Heads of State and Governments of the AU adopted Agenda 2063: *The Africa We Want* and Science, Technology, and Innovation Strategy for Africa (STISA) 2014–2024. Agenda 2063 is a continental blueprint designed to guide Africa's development processes until 2050, with STI as a critical enabler to achieving its goals. On the other hand, STISA 2024, the First Ten-Year Implementation Plan (2014–2023) for Agenda 2063, calls upon African countries to enhance their technical competencies, invest in research infrastructure, promote innovation and entrepreneurship, and facilitate the policy environment in order to stimulate diversified economic growth. Evidence-based understanding of STI systems is important for policy formulation on research and innovation. These policies will in turn create evidence-based solutions and innovations that will help to bring about poverty reduction, job creation, sustainable livelihoods, and the improved well-being of citizens (Heher, 2006).

African political leadership is increasingly aware and supportive of STI as critical drivers of economic development and competitiveness. This is demonstrated in their active advocacy for research and development (R&D) and innovation as precursors to significant growth in African economies and prosperity. In the past five years, the continent has recorded an average economic growth of just about 3.4%, which is slightly above the world average of 3%. In fact, the continent boasts of having some countries with some of the fastest-growing economies in the world e.g. Ethiopia and Rwanda (though admittedly from a low base). To accelerate the transition of African economies from factor-based/resource-based to knowledge-based, African countries must be committed to invest about 1% of Gross Domestic Expenditure on R&D (GERD) as a percentage of GDP in R&D by 2023 (UNECA, 2016). If this is achieved, it would be a giant leap in STI support in the continent. However, questions are now being asked on the absorptive capacity of all these financial resources and how this links to economic growth.

It is no secret that a lot of knowledge products have been produced in the continent, including peer-reviewed publications and patents. However, the continent has failed to commercialize its globally acknowledged indigenous knowledge and the resources linked to it for the benefit of its citizens. One can argue that generally, commercialization of knowledge outputs has always been an afterthought for intellectual property from institutions of higher learning. More specifically, converting traditional knowledge into products of commercial value is controversial and raises emotions in most African countries (Odei, 2007). Valorization of indigenous knowledge has been one of the most undocumented failed projects of the continent. Yet this might be a way of conserving and developing such knowledge to the benefit of communities (Campbell & Vainio-Mattila, 2003; McAfee, 1999). In modern times, what cannot be exploited for benefit (commercial or otherwise) is usually discarded (see also Chapter 14).

THE HISTORY OF THE SOUTHERN AFRICA NETWORK FOR BIOSCIENCES (SANBIO)

The CPA, endorsed by the AU Summit of Heads of States and Governments in 2005, formed the basis for implementing the New Partnership for Africa's Development (NEPAD) Science and Technology programmes (AU Secretariat, 2005). The CPA sought to strengthen existing capacity. Part of the strategy of capacity strengthening included the initiation of Centres of Excellence at national, regional, and continental levels. It focuses on three pillars: capacity building, knowledge production, and technology innovation. CPA programme implementation has rolled out five cluster R&D flagship programmes in the different pillars.

The programmes include (1) Biodiversity, Biotechnology, and Indigenous Knowledge; (2) Energy, Water, and Desertification; (3) Material Sciences, Manufacturing, Laser Technologies, and Post-Harvest Technologies; (4) ICT and Space Sciences; and (5) Mathematical Sciences.

NEPAD's African Biosciences Initiative (ABI) is a programme under Cluster 1 of the CPA and is directed at facilitating the establishment of state-of-the-art R&D facilities that can enable institutions to pool resources to address common biosciences challenges affecting the people of Africa. The focus areas include human health, agriculture productivity, sustainable water resource use management, biodiversity management, and sound environmental management.

The CPA considered networking to be an essential ingredient for stimulating and expanding capacity and collaborative research efforts among organizations involved in biosciences in Africa. Accordingly, its first six years of implementation focused on creating networks.

The Southern Africa Network for Biosciences (SANBio) is a platform to address and find means to resolve key problems in health, nutrition, agriculture, and the environment in Southern Africa utilizing bioscientific innovations. SANBio was established alongside four other networks and incorporates 13 of the 16 countries in the Southern African Development Community (SADC) region. It funds and coordinates research and innovation which aims to incentivize researchers to commercialize research based on traditional and indigenous African knowledge to solve problems in the life and biological sciences, primarily medical and veterinary, and nutrition.

NEPAD SANBio was formed in 2005. It coordinates research programmes and funding in the region from a hub based at the Council for Scientific and Industrial Research (CSIR) in South Africa. With the advent of the AU Science, Technology, and Innovation strategy for Africa 2024, NEPAD SANBio has responded to the call for all NEPAD Centres of Excellence to take responsibility for the implementation and effective coordination of regional programmes in the priority areas. Adopted at the 23rd Heads of State and Governments Summit in 2014, STISA-2024 is a continental strategy which aims to respond to the demand for science, technology, and innovation to impact across critical sectors of development. The vision for this framework is to accelerate Africa's transition to an innovation-led and knowledge-based economy.

SANBio programmes are therefore focused on the priorities identified by the CPA.

Under priority area 1, "Eradicate hunger and ensure food and nutritional security", SANBio is implementing regional projects in ensuring nutritional security of the people of Southern Africa. In this regard, we are addressing the value addition of indigenous and neglected foods and food processing. Under nutritional security, we funded projects addressing the following:

- alternative sources of proteins, micronutrients, and carbohydrates, e.g. promotion of nutrient-rich foods targeting women and children
- food processing, e.g. improved food handling and storage, and preservation technologies

- value addition of indigenous and neglected foods
- indigenous coping strategies to natural and man-made disasters, e.g. climate change, HIV/AIDS, and nutritional education
- innovation in animal production (e.g. curating indigenous breeds), animal health (developing therapeutics, vaccines, and diagnostics), developing innovative agricultural products (e.g. biocontrol, biosafety and fertilisers), and supporting aquaculture projects in communities and schools

Under priority area 2, "Prevention and control of diseases", SANBio funds research and development in bioprospecting for remedies, diagnostics, e.g. point-of-care or surveillance, food allergies, and genomics.

SANBIO'S ROLE IN ENABLING INDIGENOUS KNOWLEDGE-BASED BIO-INNOVATION

Africa is rich in natural resources of plant species that are coupled to the cultures and traditions of local and indigenous communities. This unique mix can be used to innovate and develop a wide range of products and services in food and health for socioeconomic benefit. It has been documented that the way in which genetic resources are accessed and how the benefits arising from their utilization are shared can create incentives for conservation and sustainable use of the biodiversity and creation of a fairer and more equitable economy to support sustainable development (Campbell & Vainio-Mattila, 2003).

The understanding of biodiversity often comes from traditional knowledge of local and indigenous communities which has been handed down over generations. Access and Benefit Sharing (ABS) refers to the way in which genetic resources may be accessed and how the benefits that result from their use are shared between the people or countries using the resources (users) and the people or countries that provide them (providers) (de Jonge, 2011). Essentially, the rights of indigenous and local communities are considered during the ABS negotiations. The Nagoya Protocol on Access to Genetic Resources and the Fair and Equitable Sharing of Benefits Arising from Their Utilisation provides the framework to grant the local and indigenous communities the rights to decide about benefit from the utilization of their traditional knowledge and associated plant genetic resources in R&D and commercial activities (Buck & Hamilton, 2011). Depending on a country's legislation on ABS, indigenous and local communities are often the ones that act as providers of traditional knowledge and/or associated plant genetic resources and enter into ABS agreements with users. The users in this case are the R&D institutions and commercial companies.

Yet in many African countries the concept of ABS is still relatively new and often poorly understood, even though the majority of African countries ratified the Nagoya Protocol on ABS, including 14 countries from the SADC region. The R&D institutions and commercial companies, including the local and indigenous communities, are not or are only insufficiently aware of the concept of ABS and the related national and international frameworks defining their rights and roles in the interplay between users and providers (de Jonge, 2011).

Realizing the pitfalls of working with plant genetic resources without understanding the international protocols involved, SANBio decided to play a part in training and capacity building in this area. As part of the SADC IKS technical working group, we have designed guidelines for Intellectual Property and Access and Benefit Sharing for the Southern African region. Based on this experience, we are providing the necessary ABS support and tools to collaborative projects in food that utilize traditional knowledge and associated plant genetic resources. We will now discuss these projects and some of the lessons learnt beyond the implementation of ABS principles and protocols. We will also look at lessons learnt from a general sustainability and commercialization point of view.

PROJECT 1 – FROM TRADITIONAL KNOWLEDGE TO PRODUCTS: A JOURNEY WITH TRADITIONAL HEALTH PRACTITIONERS

Background: The majority of plant-based traditional medicines that are being used by Traditional Health Practitioners (THPs) are often based on anecdotal evidence, without scientific data that demonstrate safety, quality, and efficacy. The purpose of the project was therefore (1) to assist the selected THPs in South Africa in transforming their plant-based traditional medicines into safe, high-quality, and standardized products through scientific research and product development, and (2) to equip, through training, the THPs with relevant knowledge and skills for starting their own businesses.

Intervention: The project is championed by the SANBio Hub at the CSIR in Pretoria together with The Innovation Hub and South African Bureau of Standards (SABS). The project trains THPs in business development and commercialization. The training incorporates product formulation and branding.

Outcome: The first group of five THPs has already been selected, and three products have been completed and reviewed by each of the THPs concerned.

One of the THPs on the programme with a traditional medicine called Prijap won the business plan competition at The Innovation Hub, South Africa, in 2017/2018; as a result, this THP has received extensive incubation and mentoring support and has been provided with a laboratory and equipment to scale up the production of the Prijap traditional medicine.

Lessons learnt: This project is a clear demonstration of how co-development can be achieved between traditional knowledge practitioners and biomedical scientists for value addition of indigenous resources. There is a need to introduce training around regulatory compliance, especially because medical claims are being made.

PROJECT 2 – RESURRECTING THE DEMAND FOR HERBAL TEAS IN ZIMBABWE

Background: Resurrection bush (*Myrothamnus flabellifolia*/*Mupfandichimuka* [Shona]/*Umafavuke* [Ndebele]) is a woody shrub with tough branches (Engelhardt et al., 2016). For most of the year, it looks like an upright bundle of reddish brown sticks, no more than 30–50 cm high. It is called resurrection bush for the speed with which apparently dead leaves revive when the rains come. The resurrection bush is widespread in Zimbabwe and grows especially in shallow soil over rock, crevices, and rocky hillsides where few other plants survive – in full sun, usually at altitudes between 900 and 1,200 m.

A tea is made from its leaves and twigs traditionally used to treat, among other ailments, colds, kidney problems, asthma, backaches, and headaches (Bhebhe et al., 2015). The volume of resurrection bush drunk by herbal tea drinkers in Zimbabwe is small and is even smaller in South Africa. However, at present, the domestic volumes are lower than it is believed they could be if there were a more premium product aimed at higher-end and tourist markets. In addition, the project believes it is possible to enter the international market with the required safety and toxicology data in place.

Intervention: The project is aimed at commercializing resurrection bush tea in Zimbabwe by working with the communities (and there is also an attempt to propagate the plant in order to guarantee security of its supply). The commercialization of the tea will increase the volume of resurrection bush currently being harvested by rural collectors (mostly women in ecological zones 4 and 5 of Zimbabwe) and will create jobs and increase annual income for the rural economy without detracting from other sources of income. The need to secure the production levels of the plant has increased R&D in the physiology and agronomy of the plant for purposes of domestication. This R&D has attracted the attention of a big Swiss company that for the past 12 years has tried to domesticate resurrection bush.

Outcome: To date, the project has demonstrated safety and an absence of toxic effects with acute and chronic use of resurrection bush tea. Different prototypes of blended teas have been produced and user-tested, and this testing will be followed by product upscaling. The project has attracted the attention of governments, non-governmental organizations, and the private sector. The government of Zimbabwe is promoting value addition of indigenous resources, and this project is a flagship for private sector and community partnerships.

Lessons learnt: Community–Private Partnerships (CPP) can succeed if there is mutual respect and all stakeholders in the community are involved. Information being gathered in this project will feed into the national ABS guidelines with traditional Zimbabwe communities. This sharing of information may help to inform future policy on indigenous resources and knowledge in Zimbabwe. Thus, this project is a good example of a "living lab".

PROJECT 3 – IMPROVING THE QUALITY OF MARULA SEED OIL IN BOTSWANA

Background: Marula (*Sclerocarya birrea* subsp. *caffra*) trees are in high concentrations in several areas of Botswana. With the advent of the natural products revolution, the benefits of African seed oils to human skin, hair, and nails are becoming common knowledge (Vermaak et al., 2011). The chemical composition of marula oil lends itself well to the treatment of stretch marks, scars, blemishes, and uneven skin tone. Its high antioxidative stability makes it perfectly suited for the prevention of the ageing of the skin (Komane et al., 2015). Marula processing has, however, been inefficient due to the lack of suitable technology to extract marula kernels to replace the traditional means of hand-and-stone extraction.

Intervention: Blue Pride, a small family-owned enterprise, has been able to formulate a simple but efficient strategy for becoming a dominant producer of marula oil in Botswana and has developed marula decorticating machinery with an engineering firm in South Africa to optimize the process.

Outcome: Through this project, Blue Pride has mechanized the marula kernel extraction process and the oil extraction process to multiply current marula oil output and thereby deepen the social and economic impact that commercialization of marula oil production has on rural communities in Botswana. The technology enables the company to reach new production levels at a fraction of the current cost (in both time and money).

Lessons learnt: By sourcing wild marula from the communities and placing part of the processing equipment in rural villages, interventions such as these can benefit rural economies. Breaking into the global market is a challenge for raw materials which quickly deteriorate with transport, so trying to formulate the finished product as close to the source as possible is highly recommended.

PROJECT 4 – HEALTHY SMA²RT SNACKS FROM CLIMATE-SMART CROPS

Background: Traditionally rural communities in South Africa, Botswana, and Lesotho prepared nutritious snacks and meals from indigenous cereal, legume, fruit, and vegetable crops. Traditional vegetables are an important contributor to food security (Ebert, 2014). However, with urbanization and adoption of Western-type foods, consumption of such diets has declined in favour of convenient, easy-to-prepare, or ready-to-eat products. Snack foods available to consumers, while tasty, are often high in fat, sugar, and salt, and are poor in protein, micronutrients, and dietary fibre. The need for healthy snacks exists and this led to the innovation around safe, market-ready, acceptable African, ready-to-eat/use, trendy (SMA²RT) and healthy snack foods.

Intervention: SMA²RT foods formulate snacks using crops which are resistant to climate change and currently neglected (e.g. sorghum, millets, melons, and cowpeas). The project involves the use of existing products of sorghum baked biscuits to develop entrepreneurship skills in students from the University of Pretoria and the National University of Lesotho. The goal was the commercialization of the traditional motoho (non-alcoholic sorghum) drink with a commercial partner. An instant fermented/soured cereal–melon composite powder will be developed in Botswana to assist in the production of instant melon porridge.

Outcome: To date, a microbakery enterprise (Healthily Baked Product Pty Ltd), wholly owned by three female graduates of the National University of Lesotho, has been established.

Lessons learnt: It is important for universities to work with private partners in the development and commercialization of technologies. Many African universities do not work within the ecosystem of industry and meeting industry needs both in student training and product development. In an effort to strengthen the collaboration between academia and the private sector, several agreements have been signed between the partners to support commercialization of the products developed and to provide entrepreneurship training for the microbusiness owners.

PROJECT 5 – SUSTAINABLE BROILER CHICKEN FEED WITH *MORINGA OLEIFERA*

Background: The prices of broiler chicken feed have been escalating, thereby reducing the viability of the poultry industry in the SADC region. Additionally, high mortality rates of 5–10%, morbidity, decreased preference for broiler meat due to higher fat content, and lack of appealing colour have been reported.

Intervention: This project seeks to produce low-cost *M. oleifera* broiler feed that results in reduced mortality and morbidity rates, tasty meat of low fat content, longer shelf life, and appealing golden brown colour. The project will improve chicken nutrition through the utilization of *M. oleifera* feed within the SADC region.

The project involves developing and testing the feed prototype, protecting the intellectual property (IP), developing the business model (establishing the feed distribution channels and distribution contracts signed with distribution agents), and commissioning a feed manufacturing plant for the commencement of feed production and product registration.

Outcome: The expected economic benefits of the project will be reduced costs to small-scale farmers who are in poultry production in Zimbabwe. The social benefits derived from using *M. oleifera* feed include improved diets and employment generation. The feed derives some ingredients from the *M. oleifera* tree. Moringa trees also provide ecological benefits such as being wind breaks, providing improved water quality, and improving air quality by carbon sequestration (Le Houerou, 2000).

Lessons learnt: While it is good to be ambitious and think big, a step by step approach to product development ensures technical and business success. With products that use plant genetic resources, thinking about the supply chain is the first step otherwise all the work goes to waste.

PROJECT 6 – IMPROVING NUTRITION, MILK PRODUCTION, AND HEALTH IN DAIRY GOATS IN SWAZILAND

Background: The demand for livestock products in Africa has outpaced domestic production, rendering the continent heavily reliant on the importation of basic livestock products (Rakotoarisoa et al., 2012). In Swaziland, for example, dairy product consumption exceeds domestic production by over 50 million litres. Production is hampered by under-nutrition of dairy animals, and undernourished animals are immunocompromised and hence become susceptible to parasitic infections (e.g. *Haemonchus contortus*). Control of parasites through the use of veterinary drugs is expensive and out of reach for many smallholder farmers, and is ultimately unsustainable.

Intervention: The project involves manufacturing and commercializing a brand of nutritious and medicinal feed pellets made using the nutrient-rich and anti-parasitic *Melia azedarach* tree leaves combined with fibrolytic enzyme-rich oyster mushrooms (*Pleurotus ostreatus*) in order to improve the nutrition and milk production. Scaling up will involve commercializing a brand of goat yoghurt that will be made using milk from Saanen dairy goats that are fed the pellets.

Output: The novel feed pellets will address the problem of poor nutrition of livestock, especially during the winter and/or dry season, as well as their infestation with parasites that further worsen their body condition and productivity. Also, *M. azedarach* is an invasive plant in many SADC countries; hence its use as an animal feed and medicine will help control its rapid proliferation.

Lessons learnt: There is huge potential in using invasive species for feed formulation. Trials have to be done to understand the effects of products on meat quality and cultural acceptability of any apparent changes e.g. in colour or texture.

PROJECT 7 – NON-DAIRY PROBIOTIC ENRICHED BEVERAGES

Background: Southern Africa suffers from food insecurity due to recurrent drought conditions and poverty. Nutrition outcomes are also very poor because of over-reliance on maize, which has a poor nutrition profile, and is prone to drought and mycotoxin contamination (Miller, 2008). Healthier alternatives such as sorghum, a drought-resistant and protein-dense crop, are largely neglected (Emmambux & Taylor, 2009).

Intervention: This project aimed to produce functional food products which combine sorghum with probiotic cultures into an instant beverage. Probiotics have been shown to alleviate a range of gut-linked health challenges, including diarrhoea and oral thrush (Borchers et al., 2009). For better colonization of the gut, probiotic cultures require prebiotics (e.g. fructooligosaccharides) which selectively stimulate probiotic growth in the gut.

Outcome: Based on IP developed at Tshwane University of Technology, two products were developed for two distinct markets (Synmba cereal beverage for the Botswana market and Niselo sorghum drink for the South Africa market). The IP was protected by trade secret and licencing out agreements signed by the main parties. In Botswana, NFTEC/NAPRO are co-owners of IP, while in South Africa, the IP has been licenced out to Nutritica SA, a university spin-out company.

Lessons learnt: Full commercial placement of the products in two countries has been achieved in just under 15 months. The researchers were able to use multiple marketing channels and sell different attributes of the product to different targets (multiple targeting approach). This approach runs contrary to the market segmentation theories of business schools, but the strategy seems to have worked. In pharmacies, the emphasis is on the health attributes (gluten-free, lactose-free, high fibre, etc.) of the product (premium price), while in fruit and veg shops and at commuter taxi ranks, the convenience (meal on the go) and food (meal replacement) aspects are emphasized.

OVERALL LESSONS LEARNT

Although most SADC countries have active research projects, there is generally a gap in the management of IP from idea screening to commercialization, despite efforts to build capacity in this area. Most R&D is funded by public funds or donor agencies, with negligible research happening in the private and non-governmental sectors. Unfortunately, the lack of common understanding of intellectual property rights (IPRs) at the research and innovation nexus means most potential innovations do not see the light of day. The SADC Industrialisation Strategy and Roadmap has prioritized IPR as it relates to the creation of industries. This action is partially due to the fact that very few institutions in the region have the necessary experience and capacity to effectively handle and valorize IPRs through technology transfer and effective commercialization practices, and need help.

The NEPAD SANBio, in collaboration with the Southern Africa Research Innovation Management Association (SARIMA) and CSIR Licencing and Ventures, provides human capacity development support on IP management and commercialization to the region based on the South African practical experience. We focus on how IP can contribute to creating sustainable technology-based jobs in the region and can promote economic development and the competitiveness in the region's biosciences industry. Another finding over the past 10 years of supporting projects was that most projects led by public institutions (public research institutions or universities) had no protocols governing the transfer of the technology to the commercial partner. They also had problems defining or negotiating their role in the projects and had challenges in management of the confidential business information between the partners, in contract negotiations, in commercialization (licencing) agreements, and in IP strategy to enable the university to share in the benefits of research. Moreover, the projects informed by traditional knowledge seemed to lack an understanding of the global frameworks on ABS (i.e. Convention on Biological Diversity and the Nagoya Protocol on Access and Benefit Sharing) and how they affect technology and product commercialization. The projects also lacked knowledge of what to do to comply with these global frameworks and with countries' ABS-related laws, which affect technology and product commercialization by requiring prior informed consent, material transfer, and benefit sharing agreements.

SANBio and partners have designed a capacity building programme to train project teams on technology and on market and regulatory readiness of the projects, including managing IP. The programme includes training in both contract negotiations between the researchers and the commercial partners and management of confidential information between the partners and commercialization agreements. This training will ensure that an IP strategy and associated guidelines on research and technology commercialization and the generic contract are developed based on experiences (through the planned IP and contract workshop) and shared with SADC within the SADC Regional Framework and Guidelines on Intellectual Property, which is currently under development.

Previous work in SADC (e.g. the review of the STI component of the Regional Indicative Strategic Development Plan [RISDP]) has pointed to the weak linkages between industry and innovators (universities and citizen scientists). This deficit is something that we have also picked up on. We have tried to initiate innovation and business pitching contests and to convene events where industry and private funders are invited.

CONCLUSION

There is much potential in developing products from traditional knowledge to solve problems of the modern era. However, there are potentially huge barriers for both the knowledge holders and those that would be able to invest and take up the technology for commercialization. Conventional research funders are also not attracted to this area. SANBio has been able to find this niche and to operate effectively in it with various partners with some measure of success. In doing so, we continue to learn many lessons which have universal resonance.

ACKNOWLEDGEMENTS

We are grateful for funding from the Department of Science and Technology (DST) through the National Research Foundation (NRF) NEPAD – SANBio Flagship Programme Grant (UID: 108837), the Finnish-Southern African Partnership Programme (BioFISA 2) and SANBio member countries' contributions. We also want to thank the SANBio Secretariat, BioFISA 2 Programme management Unit (BPU) and Project Coordinators for the their contributions in the projects.

REFERENCES

Ahmed, S. A., Cruz, M., Go, D. S., Maliszewska, M., & Osorio-Rodarte, I. (2016). How significant is Sub-Saharan Africa's demographic dividend for its future growth and poverty reduction? *Review of Development Economics*. https://doi.org/10.1111/rode.12227.

AU Secretariat. (2005). Consolidated Plan of Action for Africa's Science and Technology. http://austrc.org/docs/stisa/Africa's%20Science%20and%20Technology%20Consolidated-Plan-of-Action-CPA.pdf (accessed 30 March 2019).

Bhebhe, M., Chipurura, B., & Muchuweti, M. (2015). Determination and comparison of phenolic compound content and antioxidant activity of selected local Zimbabwean herbal teas with exotic *Aspalathus linearis*. *South African Journal of Botany*. https://doi.org/10.1016/j.sajb.2015.06.006.

Borchers, A. T., Selmi, C., Meyers, F. J., Keen, C. L., & Gershwin, M. E. (2009). Probiotics and immunity. *Journal of Gastroenterology*. https://doi.org/10.1007/s00535-008-2296-0.

Brown, O., Hammill, A., & McLeman, R. (2007). Climate change as the "new" security threat: Implications for Africa. *International Affairs*. https://doi.org/10.1111/j.1468-2346.2007.00678.x.

Buck, M., & Hamilton, C. (2011). The nagoya protocol on access to genetic resources and the fair and equitable sharing of benefits arising from their utilization to the convention on biological diversity. *Review of European Community and International Environmental Law*. https://doi.org/10.1111/j.1467-9388.2011.00703.x.

Campbell, L. M., & Vainio-Mattila, A. (2003). Participatory development and community-based conservation: Opportunities missed for lessons learned? *Human Ecology*. https://doi.org/10.1023/A:1025071822388.

de Jonge, B. (2011). What is fair and equitable benefit-sharing? *Journal of Agricultural and Environmental Ethics*. https://doi.org/10.1007/s10806-010-9249-3.

Ebert, A. W. (2014). Potential of underutilized traditional vegetables and legume crops to contribute to food and nutritional security, income and more sustainable production systems. *Sustainability*. https://doi.org/10.3390/su6010319.

El Mokhtar, M. A., Anli, M., Ben Laouane, R., Boutasknit, A., Boutaj, H., Draoui, A., … Fakhech, A. (2019). Food Security and Climate Change. In K. Kahime, M. El Hidan, O. El Hiba, D. Sereno, & L. Bounoua (Eds.), *Handbook of Research on Global Environmental Changes and Human Health*. Hershey: IGI Global (pp. 53–73). https://doi.org/10.4018/978-1-5225-7775-1.ch004.

Emmambux, M. N., & Taylor, J. R. N. (2009). Properties of heat-treated sorghum and maize meal and their prolamin proteins. *Journal of Agricultural and Food Chemistry*. https://doi.org/10.1021/jf802672e

Engelhardt, C., Petereit, F., Lechtenberg, M., Liefländer-Wulf, U., & Hensel, A. (20163). Qualitative and quantitative phytochemical characterization of *Myrothamnus flabellifolia* Welw. *Fitoterapia*. https://doi.org/10.1016/j.fitote.2016.08.013.

Eyal, N., Cancedda, C., Kyamanywa, P., & Hurst, S. A. (2016). Non-physician clinicians in Sub-Saharan Africa and the evolving role of physicians. *International Journal of Health Policy and Management*. https://doi.org/10.15171/ijhpm.2015.215.

Heher, A. D. (2006). Return on investment in innovation: Implications for institutions and national agencies. *Journal of Technology Transfer*. https://doi.org/10.1007/s10961-006-0002-z.

Komane, B., Vermaak, I., Summers, B., & Viljoen, A. (2015). Safety and efficacy of *Sclerocarya birrea* (A.Rich.) Hochst (Marula) oil: A clinical perspective. *Journal of Ethnopharmacology*. https://doi.org/10.1016/j.jep.2015.10.037.

Le Houerou, H. N. (2000). Utilization of fodder trees and shrubs in the arid and semiarid zones of West Asia and North Africa. *Arid Soil Research and Rehabilitation*. https://doi.org/10.1080/089030600263058.

Mathers, C. (2017). Global Burden of Disease. In S. R. Quah (Ed.), *International Encyclopedia of Public Health*. Amsterdam: Elsevier (pp. 256–267). https://doi.org/10.1016/B978-0-12-803678-5.00175-2.

McAfee, K. (1999). Selling nature to save it? Biodiversity and green developmentalism. *Environment and Planning D: Society and Space*. https://doi.org/10.1068/d170133.

Miller, J. D. (2008). Mycotoxins in small grains and maize: Old problems, new challenges. *Food Additives and Contaminants – Part A Chemistry, Analysis, Control, Exposure and Risk Assessment.* https://doi.org/10.1080/02652030701744520.

Murray, C. J. L., Vos, T., Lozano, R., Naghavi, M., Flaxman, A. D., Michaud, C., ... Lopez, A. D. (2012). Disability-adjusted life years (DALYs) for 291 diseases and injuries in 21 regions, 1990–2010: A systematic analysis for the Global Burden of Disease Study 2010. *The Lancet.* https://doi.org/10.1016/S0140-6736(12)61689-4.

Odei, M. (2007). Africa's development: The imperatives of indigenous knowledge and values. PhD Thesis. Department of Philosophy, University of South Africa.

Rakotoarisoa, M. A., Iafrate, M., & Paschali, M. (2012). Why has Africa become a net food importer ? In *Trade and Market Division*, FAO.

UNECA. (2016). Agenda 2063 for Africa: First 10 Year Implementation Plan https://www.uneca.org/sites/default/files/uploaded-documents/CoM/com2016/agenda_2063_final_revised_first_ten_year_implementation_plan_12.10.15.pdf (accessed 31 March 2019).

United Nations Department of Economic and Social Affairs Population Division. (2017). E02 World Population Prospects The 2017 Revision: Key Findings and Advance Tables. *World Population Prospects The 2017.* https://doi.org/10.1017/CBO9781107415324.004.

Vermaak, I., Kamatou, G. P. P., Komane-Mofokeng, B., Viljoen, A. M., & Beckett, K. (2011). African seed oils of commercial importance – Cosmetic applications. *South African Journal of Botany.* https://doi.org/10.1016/j.sajb.2011.07.003.

CHAPTER 3

Contemporizing Tribal and Indigenous Medical Knowledge: An Indian Perspective

Darshan Shankar

CONTENTS

ABSTRACT

Background In contemporary times, India's tribal medicine is perceived as a weak and declining tradition compared to the economically powerful Western modern medicine that was imposed a few centuries ago. Brainwashed policymakers and academics may tend to argue for abandonment and burial of a once-rich tribal tradition. However, from the perspective of an Indian cultural understanding of social time, in which it is believed that all cultures follow repetitive cyclic patterns of birth, growth, and decay, followed by rebirth, new growth, and again decline – patterns that repeat over time – the tradition of tribal medicine is merely in a late-declining phase of its cultural cycle in which it is awaiting a natural revitalization. The phenomenon of any tradition being in a state of decline, at a particular stage of its evolution, is a natural part of a societal cyclic process, and the fact of decline does not imply anything fatal or suggest an inability to revive or discount future potential. The Indian cultural view is that the tradition of tribal medicine can, in fact, be revitalized today because it has already fallen to its lowest cyclic depth, at which stage any tradition may actually be poised to turn around. The challenge is to discover an appropriate cultural strategy for revival. The choices appear to be, firstly, to appeal to Western sciences like phytochemistry and biomedicine, or alternatively in the Indian context, to plan its rescue with the aid of Ayurveda, a sophisticated, codified indigenous health tradition that has its roots in tribal medicine but evolved a systemic theoretical framework beyond the empiricism of its cultural parent. This chapter argues that Ayurveda is culturally the appropriate way to effect the rescue of tribal medicine in this cyclic moment of its rebound from a steep decline. Such an agenda of indigenous revitalization requires a deep understanding of cultural processes beyond the politics of knowledge and the pressures of a currently monocultural globalization process.

Relevance This chapter serves to alert policymakers that it is possible to involve indigenous scientists from non-Western cultures in helping revitalize traditional knowledge. South–South cooperation can be more relevant than North–South collaboration with respect to several domains of indigenous knowledge such as medicine, agriculture, animal husbandry, fine and performing arts, architecture, metallurgy, sociology, and political sciences.

SOCIAL CONTEXT – THE COLONIAL FRAUD, TRADITION, AND MODERNITY

The economic and political forces unleashed with the advent of colonialism and its aftermath have created a monocultural world order derived from Western cultural and intellectual traditions. This order has resulted in a terrible cultural insensitivity and growing loss of non-Western cultural, intellectual, and spiritual traditions in dozens of societies (Alvares 1979), whose civilizational evolution was abruptly interrupted by colonial rulers and subsequently by colonized minds within their own societies (Nandy 1982, 1983).

A resultant fallacy perpetuated by the colonization process is the positioning of Tradition and Modernity as opposites. In fact, they lie on a continuum (Shankar and Unnikrishnan 2004). This dichotomy is a mischievous colonial creation. It is a sociological and political red herring. It is this fallacious division that is largely responsible for the skewed pattern of Western ethno-culture inspired national development in many nations, including India. The colonial fraud that was committed on India, as also on other societies in Asia, Africa, and South America, was a powerful dissemination of the lie that modernity had to be imported from an "advanced West" and that indigenous knowledge systems, in spite of being alive, evolving, and functional, were, alas, merely traditional. Thus, Indian modernity, as is the case in other postcolonial societies even in the 21st century, is largely based on the transplanted modernization of European traditions (Dharampal 1971).

Ironically, the modern history of Europe, in fact, bears testimony to the appropriate and natural process of modernization. Its history reveals how a dogmatic Europe in the 14th century drew core inspiration and light essentially from its own classical Greek tradition (Highet 1949), learnt from several other cultures (including Arabic), and moved ahead. European modernity thus has its roots in and is centrally derived from its own traditions. Modernity rooted in traditions is the natural sociological process of modernization. Just as the past, present, and future lie on a continuum, and there can be no present without a past and no future without the present, so also there can be no modernity without tradition. An intelligent and realistic definition of modernity is "Evolving Tradition" (Foundation for Revitalisation of Local Health Traditions Trust [FRLHT] 2013).

India and so many other nations in the Global South today need to rediscover their own roots and broaden and strengthen the skewed Western ethnocentric modernization process in their native lands. Whilst it is very open-minded of the Indian intellectual and political leadership to learn, adapt, and assimilate knowledge from foreign cultures, it is suicidal to give up the endeavour to contemporize one's own endogenous knowledge systems (Gandhi 2010). India can strengthen her modernization processes by substantial investments in demonstrating the contemporary relevance of our indigenous knowledge systems in various fields, including the fine and performing arts, healthcare, agriculture, architecture, music, mathematics, and philosophy.

Thus, global development will flower when we put in place, in every country, multicultural processes of modernization. Global unity in this rich diversity can be discovered by innovating new platforms for collaboration on the intellectual, political, economic, and cultural planes in creative ways. Multicultural globalization is one of the most important and urgent social and political challenges of the 21st century.

A PERSPECTIVE ON TRIBAL MEDICINE – THE INDIAN CONTEXT

There are two key Sanskrit words that describe the nature of the Indian knowledge society. These are *prakriti* and *samskriti*. They correspond to the English words "nature" and "culture". The word *prakrit* (in the context of knowledge) refers to ecosystem- and ethnic community-specific knowledge traditions derived from *prakriti* or nature. Etymologically, *prakriti* refers to natural phenomena (*kriti*) that are primordial (*pra*). *Prakrit* knowledge, therefore, refers to empirical traditions and local knowledge of communities who live close to nature (e.g., rural and tribal people). Such knowledge includes language, arts, music, weaving and other crafts, agriculture, architecture, and of course healthcare.

The etymology of the word *sams-kriti* refers to codified indigenous knowledge (*kriti*) that has been refined or modified (*samskar*) from the prakrit state. It corresponds to the generic English term "culture". This word thus refers in Indian tradition to the various codified indigenous knowledge expressions (*shastras*) such as grammar (*vyakaran*), mathematics (*ganit*), the fine and performing arts (*shilpa, sangeet, nritya*, etc.), agriculture (*krishishastra*), architecture (*vastu shastra*), and healthcare (Ayurveda) (FRLHT 2012).

In Indian thought and praxis, the *samskrit* (codified) traditions enjoy a dialectic, symbiotic relationship with the *prakrit* because *samskriti* (culture: in the broad sense of the term, including all arts and sciences) is derived from *prakriti* (nature) and in turn also modifies it. Thus, Ayurveda (the *samskrit* health tradition) and local community health traditions (*lok swasthya parampara* or the *prakrit* tradition) also enjoy a dialectic relationship.

This relationship is referred to in the earliest medical texts (1500 BC). The Charaka Samhita states that: "Oushadihi naama roopabhyaam jananthe hyajapaa vane, avtpaashchiva gopaashcha ye aha anye vanaasinaha" (Sutrasthaana, Chapter 1, Shloka 120-21). "The goatherds, shepherds, cowherds, and other forest dwellers know the drugs by name and form." Similarly, Sushruta states that: "Gopaalaasthaapasaa vyaadha ye chaanye vana charinaha. Moola jaathihi cha tebhyo bheshaja vyakthi ishyathe" (Sutrasthaana, Chapter 36, shloka 10). "One can know about the drugs from the cowherds, tapasvis, hunters, those who live in the forest and those who live by eating roots and tubers."

Both the folk and codified indigenous health traditions of the country have contemporary relevance. The *samskrit* tradition holds the promise of making original contributions to the world of medicine, whereas the *prakrit* tradition can provide "health security" via its stupendous knowledge of ecosystem-specific flora and fauna to meet the healthcare needs of millions of rural and urban households.

To casual observers, it seems at first ridiculous, then ironic, and perhaps at last tragic that one should plan for primary healthcare in tribal areas, ignoring and bypassing local, living traditions of healthcare based on herbs and concepts of health and disease that have evolved in tribal society for thousands of years. The tragedy is that neglect and destruction has already occurred in hundreds of tribal and other rural areas in the country and is being planned in the case of others even today. Even amongst those who are somewhat sympathetic to tribal culture there remains a lurking suspicion, when they witness the present weaknesses in traditional practices, that the local tradition may be inherently a weak and incoherent system. Whilst they may indulge in local tradition superficially and try to salvage some practical and working elements from it, via schools/centers and research into traditional medicine and ethnobotany, in the back of their minds they think that what is probably needed is a heavy reinforcement from the apparently well-established modern medical sciences. They would be appalled, for instance, if one suggested that Western medicine was inappropriate for some health problems, though it could still be useful for specific conditions. The modern strategy of anaesthesia is desirable. It is, in fact, derived from thc use of cannabis by indigenous communities. The bioactive molecule morphine isolated from the plant is a powerful modern anaesthetic agent and a boon to surgical practice. This is an example of the limited value of modern biomedicine for management of an acute condition. However, for malarial fevers, the strategy of isolating anti-parasitic molecules has completely failed because of the resistance of the parasites. The global research strategy for malaria management might be better positioned if it used traditional herbal formulations which could perhaps be synergized in creative ways and because of their combinatorial nature would never result in resistance.

There appear to be two sets of key factors responsible for the current erosion in tribal health cultures. One group of key factors is intrinsic to tribal medicine and the other is extrinsic to it (FRLHT 2003). Extrinsic factors relate to colonial and postcolonial economic and political forces that favour the Western model of development. External cultural domination in any society, however, can gain entry only in the midst of internal infirmities, and it would be unfair to attribute all internal weaknesses to external influences.

What is the intrinsic factor responsible for the decline of tribal medical knowledge? The emphasis of this chapter is on this intrinsic factor because much has already been written by contemporary political economists on the external factors (see the writings of Claude Alvares [e.g., Alvares 1979] and of scholars based at the Multiversity Malaysia Decolonial International Network). The intrinsic factor pertains to the "entropic" effect on tribal consciousness due to the cycle of *time*. Due to this temporal effect, today the tribal tradition is perhaps in its weakest state of knowledge and practice. It is here, precisely at this point, that one should note that even if today we can actually "see" the tribal tradition in a weakened state, it is cultural insensitivity and ignorance of the processes of societal change over time that prompts even well-wishers to believe that "the tribal medical heritage" should be replaced because of its current weaknesses. It needs an enlightened understanding of temporality in social processes to realize that tribal medical knowledge was not born weak, nor has it remained static over time. Its current weakness is a "fall" from a previous strength, a fall that was inevitable due to the cyclic effects of time (Aurobindo 1991) and therefore a weakness that can be overcome in a culturally sensitive way – not by replacement but by a revitalization process.

AN INDIAN PERSPECTIVE ON THE CYCLES OF TIME

What is this cyclic effect in the movement of time? The general rule (Auribindo 2011), is that each of nature's infinite manifestations, be it an apple or a flower or a rock or a bird or a micro-organism or an atom, or be it the human form, all these natural manifestations undergo a cyclical pattern of change. Change of form is inevitable over time, but it follows a cyclical pattern. What is the pattern? The pattern has three general phases. The first phase is one of birth, of inner freshness, of innocence, and of purity. The second phase is one of steady growth but wherein the inner innocence and purity is generally lost, and the third phase is one of outer decay, leading towards a death of the form and a subsequent transformation. The struggle against death, which is a natural inner instinct, is transformed into the desire to live on and to be reborn, as it were, and this is what happens at the end of the third phase. Thus, a new birth occurs again and the cyclical process goes on.

The Movement of Social Time

In terms of human consciousness and evolution of knowledge systems, the cyclical movement of social time is divided in Indian traditions into four evolutionary periods. The first period is one of birth. It is an age of pure intuition where humans feel the "oneness" of life and nature within themselves. In this age, the human consciousness is "naturally" in tune with the rest of nature's consciousness. The mind is "one" with nature. There is no separation between subject and object. Such an age arrives in its pure state, naturally, just as the innocence of a child at birth and the freshness in a fruit just plucked are natural. The knowledge and truths discovered in this age are the highest and most realistic because they are based on an "essential" and "holistic" understanding of nature.

In the second phase of social time, there is a fall from this natural state of pure consciousness, pure intuition and essential knowledge as humans desire to create symbols to express themselves. Whereas the symbols may be highly creative and meaningful because they express essential truths arrived at in the pure consciousness of an earlier age, they nevertheless distract the human experience from the pure essence of nature in order to reflect essence in symbolic forms. The knowledge of this age is also based on an essential understanding of nature, but the second age, because of its outward expressions, may be called a symbolic age. Whilst intuition is alive, it is no longer pure intuition.

The third age of social time that follows may be referred to as a rational age. It is reached when a tremendous growth and diversity in symbolic expression has taken place. These symbols over time cease to be understood in terms of the essence that they convey and become meaningful as "forms". To make sense of myriad forms requires their classification and analysis; herein the human mind enters into another level of creativity based on experiment, analysis, and inference. Knowledge can no longer be directly experienced through intuition or symbolic expression. It has to be indirectly perceived.

The fourth stage of the social time cycle is an age of decay and dogmatism where methods, procedures, and rules derived earlier by a mode of reason, analysis, and logic to interpret the diversity of forms, now become absolutely authoritative and sacrosanct. This period turns into a ritualistic and finally into a superstitious and dogmatic state where man blindly follows rules that have been laid down. This stage of the cycle is marked by decadence and death.

The Many Pasts

In the terms of such an understanding of the cyclical movement of social time, the past is not one uniform period. It can be broken up into at least four pasts, and one state devolves into the next with an inevitable and inbuilt entropy. It is important, particularly in the midst of a decadent age (the fourth stage of the cycle of time), for those with discrimination to distinguish between these various phases of the past so that the past is not seen as one uniform period. The short-sightedness of modern social scientists is that they gather evidence of the history of the most recent past, which may carry within it many examples of decadent thought, traditions, and institutions, and then make decisions regarding the values of social form and thought without complete understanding of their genesis.

Such time cycles exist not only with respect to social time but in all aspects of the functioning of the universe, ranging from geological time – the evolution of flora and fauna, the solar systems, and galaxies – to the most recent challenge of climate change.

The Position of Our Times in the Social Cycle of Time

According to traditional Indian seers, world civilization entered the fourth phase of the cycle about 2,500 years ago. This was the beginning of a ritualistic and dogmatic age. The ritualistic age is marked by the oppression of the human spirit. The society is in its most decadent state. It is during this age that human consciousness begins its first struggles and strivings to live and rediscover the truth that has been lost to the age. At the start of this struggle, it is forced to look back into its past memories for inspiration and for a take-off point.

The search for truth inevitably involves a process of going back to the roots, a retracing of the past because the present is full of darkness and decay. The memories of the preceding rational age are those which are most easily accessible, because in most societies memories and traditions of the still-earlier symbolic and intuitive ages are altogether lost. It is thus the knowledge, concepts, and spirit of some earlier age that provide a first inspiration and a progressive direction upwards. This seems to be what happened in a modern European civilization that in the 13th and 14th centuries AD (the conclusion of the medieval period, in which Europe began to emerge from its dark age) sought inspiration from classical Greek rationalistic traditions of the second century BC, upon which it has creatively built up the pillars and substance of the modern sciences. It has taken Europe 500 years of hard work to rediscover the spirit of an earlier age and even surpass its level of achievements.

The Privileged East

In some fortunate societies, however, knowledge traditions from the first and second phases of the "time" cycle, the intuitive and symbolic age, may yet be alive, although certainly these traditions would never be found in the mainstream. This is believed to be the advantage of Eastern societies. In societies that have preserved some knowledge and understanding from previous phases, the choice is available to individuals, groups, and small communities to draw deeper inspiration from a past period closer to their roots. The best example of such access to Indians is the earliest treatises of Ayurveda, such as the Sushruta Samhita,* written around 3000 BC. Despite its antiquity, the text continues to illumine clinical practice even today.

* https://archive.org/details/sushrutasamhita, Library of the University of California, digitized internet archives

The Decadent Age of Tribal Medicine and the Way Out

When we see tribal medicine in a state of general decay we assume it is now in the fourth phase, but it is an arrogance and ignorance that captivates us when we say "let us scrap it". We must instead look back and trace its past to see the best times of its cyclical evolution. It is the only way to learn of its full import and potential. Nevertheless, to think in this fashion we must have some idea about the cycle of social time because only then can one be convinced that there was a time when the knowledge was pure and inspired and that its present decay is a fall that has occurred over time. This research of going back to the "roots" is in fact the only "natural" recourse open to any society presently in decay if it wants to revitalize itself.

THE ORIGINAL ROOTS OF TRIBAL MEDICINE

What original roots or sources can tribal medicine in India consult in order to revitalize itself? Within the tribal setting itself, the knowledge of medicine and health has been an oral tradition passed on by word of mouth from generation to generation. An oral tradition is the safest way to maintain the essence or spirit of any knowledge system, because the human mind has undoubtedly far greater capacity to convey essence than the best of printed books. However, once the spirit weakens and memory and practice fail, then the distortion that can creep into oral traditions is perhaps greater than what might be caused by a misunderstanding of written texts. In the case of texts, there is always the hope of proper interpretation and application, but a fully distorted oral tradition is lost forever. It is perhaps for this very reason that traditional knowledge systems in many societies have had both an oral and a parallel written tradition. If the first begins to fail, one could have recourse to the other. However, if a society's oral traditions are entirely lost and it only depends on written traditions, one can infer that its spiritual qualities are very weak.

It can be easily established based on commonality in basic medical concepts such as "Kaph", "Vat", and "Pith" that there is a relationship between the tribal tradition of health care and the mainstream and codified tradition of Ayurveda. This relationship is the inherent one between nature and culture (or *prakriti* and *samskriti*). As anthropologists would understand it, Ayurveda is the synthesis of tribal and other folk traditions. They have a symbiotic relationship. Ayurveda has long interacted with, learnt from, and contributed to local folk traditions (Agnivesa-Agniveśatantra 2015).

CULTURAL CROSS-CURRENTS: THE INAPPROPRIATENESS OF MODERN MEDICINE IN THE TRIBAL CONTEXT

What better way can there be to strengthen tribal medicine than by consulting its roots? Doing this would imply the start of a meaningful dialogue between tribal medical practitioners and serious practitioners of the written and oral Ayurvedic tradition. The tribals have to find those who are really committed to the spirit of Ayurveda, because there are those who are arrogant with their codified knowledge and those who are greedy and will find the tribal state of affairs unappealing or worthy only of exploitation. What place can modern Western medicine, which drew its first lessons from Arab and Greek tradition and therefore belongs to an entirely different ethos and culture and has its roots in a different traditional age (time), have in such a dialogue, except to generally confuse it? What possibilities can one expect from modern medicine to strengthen tribal tradition except to destroy the last of its weakened remains? Why, then, such insistence for collaboration – from where do such suggestions arise?

The claim of modern medicine to be superior and provide better outcomes arises from the experience of its recent medical and technological successes during the last century. The argument put forward is logical enough. Tribal medicine is in a very weak state, and therefore it is not contributing as it should to the primary healthcare of the community. In these circumstances, we should urgently find a competent and expedient technical means to contribute to people's health, and for this, modern medicine appears to be a solution. If the doubt is raised that such an expedient strategy will inevitably destroy thousands of years of tradition and stop whatever chance a tribal society has of rediscovering its roots and thus achieving self-reliance in health, one is confronted with a profound cultural ignorance that fails to comprehend the dilemma.

The controversy (amongst well-meaning people) about the role of Western medicine in the tribal context arises from the improper appreciation of the movement of time and therefore a limited perspective of cultural and civilizational history. Unless the situation is viewed in its proper culture-time context, which is a holistic way of viewing it, the case for strengthening tribal medicine from its own roots can be lost. The implication of this cultural roots argument is not to raise the controversial question as to whether modern or traditional medicine is better. It is not to pick holes in the weaknesses of modern medicine (of which there may be many) or to deny its efficacies and strengths. It is not necessary in a cultural perspective to critically assess modern medicine for its efficacy; one only needs to point to its cultural irrelevance in the tribal context. Herein it is submitted that if we admit modern medicine because of expediency into the tribal context on supposedly humanitarian grounds of relieving health distress of the tribals, we will simultaneously be carrying out a disguised program of cultural imperialism and either destroying the tribal tradition directly or permitting it to commit suicide. It will also mean a gross underestimation of the potential for Ayurveda (the synthesized expression of tribal and folk medicine) to rescue tribal traditions.

In fact, the tribal healthcare situation without modern help is not at all as helpless as it is made out to be. The tribal societies in India still have access to over 6,500 species of ecosystem-specific medicinal plants with enormous health potential (FRLHT 2017). Mainstream Ayurveda can and must be challenged to rescue the tribal peoples' traditions and in the process strengthen its own base. It certainly possesses the capacity. One is not proposing the substitution of tribal medicine with Ayurveda. One is suggesting a revitalization of the tribal health culture with the help of Ayurveda. "Revitalization" means confirming all that is sound in the local practice, adding to that which is incomplete, and discouraging that which has become distorted.

It requires understanding and courage to take a balanced view of the problem. The results may be slower in coming, because creative learning and not blind application or superimposition is involved. The tribal community will thus grow self-reliant over the years. A great tribal tradition will thus be creatively revived. The wheel of time will very slowly change.

CONCLUSION

Existence is nature. Nature is hugely diverse and constantly changing. It can be studied from multiple perspectives. Any domain of knowledge is about some facet of nature. Tribals who live in intimacy with primordial nature understand nature in deep ways that scientists who may work with natural products in labs that attempt to mimic nature, in abstract and reductionist ways, can perhaps never experience. In medicine, they have solutions to problems based on the use of ecosystem-specific resources. The logic underlying the solutions may be derived from their own cultural way of knowing. The validation of their practices lies in the achievement of successful outcomes. There is little doubt that tribal medicine and the medicine of indigenous communities all over the world have had a long global history of success. The story of quinine from the cinchona bark used by Peruvian communities, which gave rise to the first global drug for malaria, is well known. So how did the Peruvians find out the properties of the bark for fever? They did not know any chemistry, yet through trial and error they were able to find something in their environment which could be used to good effect. There are many other instances which show the overwhelming role which indigenous health practices have played in modern medicine.

It is, however, a fact that India's tribal medicine is today in a state of decline. But entropy is a natural phenomenon.* It also applies to culture and knowledge systems, institutions, and to societal development in general. What is less understood is that the phenomena of entropy has a cyclic path similar to biological and physical systems. The phases of the cycle include genesis, growth, steady state, ageing, and decline. In a rich, multicultural world, it is sheer arrogance and patently false to believe that there is only one way of knowing nature, even if that way is evidently illuminating. Western cultural and intellectual traditions dominate the world of knowledge globally, not because they are superior forms of knowing in comparison with other cultural traditions but because of the ugly, relatively recent, political history of colonialism. Contemporizing a cultural tradition through a Western knowledge system that is foreign to the local culture and world view is possible in principle, but it is a complex task that needs epistemological sensitivity of an order that has thus far not been demonstrated through cross-cultural disciplines that have emerged in the last century under the rubric of ethno-sciences. These are still very Eurocentric in their interpretation of ethnic knowledge. The more appropriate cultural strategy in India's

* Lack of order or predictability; gradual decline into disorder. In cultural context, see *https://dl.acm.org/citation.cfm?id=1803696* by Volkenstein (2009).

context is to use its own sophisticated indigenous knowledge of Ayurveda to rescue and revive tribal medicine.

REFERENCES

Agnivesa-Agniveśatantra (2015) *Caraka samhita of Agnivesa* Part I. Revised by Caraka & Drdhabala; introduction by Satya Narayana Sastri; with elaborated Vidyotini Hindi commentary by Kasinatha Sastri & Gorakhanath Chaturvedi. Reprint. Chaukhambha Bharati Academy, Varanasi, India.

Alvares CA (1979) *Homo Faber: Technology and Culture in India, China and the West from 1500 to the Present Day.* Allied Publishers, Bombay, India.

Aurobindo Sri (1991) *The Human Cycle: The Ideal of Human Unity*, new edn. Sri Aurobindo Ashram Publications Department, Pondicherry, India.

Aurobindo Sri (2011) *The Human Cycle, The Ideal of Human Unity, War and Self-Determination.* http://www.arcliive.org/details/englislitranslati03susr, Library of the University of California, digitized internet archives.

Dharampal S (1971) *Indian Science and Technology in the Eighteenth Century: Some Contemporary European Accounts.* Impex India, Delhi; reprinted 1983, Academy of Gandhian Studies, Hyderabad, India.

Foundation for Revitalisation of Local Health Traditions (FRLHT) Trust (2003) Heritage AMRUTH magazine, FRLHT Bangalore, India.

Foundation for Revitalisation of Local Health Traditions (FRLHT) Trust (2012) An appreciation of India's Medical Heritage with special reference to Ayurveda, FRLHT, Bangalore, India.

Foundation for Revitalisation of Local Health Traditions (FRLHT) Trust (2013) Towards an Integration of Traditional Knowledge and Science, Sociological, Political and Epistemological reflections, an Indian perspective, FRLHT, Bangalore, India.

Foundation for Revitalisation of Local Health Traditions (FRLHT) (2017) Trust (FRLHT) database, Trans-Disciplinary University, Bangalore.

Gandhi MK (2010) *Hind Swaraj: A Centenary Edition.* Rajpal & Sons, Delhi, India.

Highet G (1949) *The Classical Tradition: Greek and Roman Influences on Western Literature.* Oxford University Press, Oxford, UK.

Nandy A (1982) The psychology of colonialism: sex, age, and ideology in British India. *Psychiatry* 45: 197–218.

Nandy A (1983) *The Intimate Enemy: Loss and Recovery of Self under Colonialism.* Oxford University Press, Delhi, India.

Shankar D, Unnikrishnan PM (2004) *Challenging the Indian Medical Heritage.* Centre for Environment Education, Ahmedabad, India.

Volkenstein MV (2009) *Entropy and Information.* Birkhäuser, Basel, Switzerland.

CHAPTER 4

Tribal and Indigenous Knowledge in West Africa: The Use of Food Plants in the Management of Diabetes

Oluwaseyi M. Aboyade and David R. Katerere

CONTENTS

ABSTRACT

Background Food plays a major role in the prevention and management of diabetes. Some traditional foods eaten in West Africa are also used medicinally in the treatment and management of diabetes. This chapter aims to document such food plants. Information about medicinal plants traditionally used in West Africa for the management of diabetes was obtained from published literature. At least 157 West African food plant species from 141 genera and 58 plant families have been identified as plants with antidiabetic potential. Of these, 126 species from 50 families have been studied scientifically for their antidiabetic activity. The families contributing the most species were Leguminosae, Malvaceae, Euphorbiaceae, Poaceae, Rutaceae, Rubiaceae, and Moraceae. Common leafy vegetables used regularly in traditional meals across West Africa, including *Vernonia amygdalina* (Bitter leaf), *Ocimum gratissimum* (African basil), *Manihot esculenta* (Cassava), *Telfaria occidentalis* (Fluted pumpkin), and *Corchorus olitorius* (Jute mallow), have been validated for their antidiabetic potential. Seeds of *Telfaria occidentalis*, *Cucumeropsis mannii*, *Parkia biglobosa*, and *Irvingia gabonensis* have also been used.

Relevance These food plants present an important resource to tap into for the discovery of new drugs and functional foods.

INTRODUCTION

Diabetes is a chronic non-communicable disease with life-altering complications. It is characterized by increased blood glucose levels because of either insulin deficiency or impaired response to insulin (Setacci et al. 2009). According to the International Diabetes Federation (IDF), there are about 425 million adults living with diabetes currently worldwide (IDF 2017). Moreover, globally, half of people who have diabetes are undiagnosed (IDF 2017). About 90% of all diabetics have type 2 diabetes (T2D), which is commonly caused by excessive carbohydrate consumption and/or obesity leading to insulin resistance in people whose bodies are still able to produce insulin, and people living in low- and middle-income countries account for two-thirds of these cases (IDF 2017). In sub-Saharan Africa, there were 15.5 million diabetics in 2017, with almost 300,000 recorded deaths (IDF 2017). This is worrisome for the region, as the projections from the World Health Organization (WHO) indicate that non-communicable diseases such as T2D will be the leading cause of mortality in the continent by 2030 (WHO 2011). IDF estimates the prevalence of diabetes in West Africa as less than 5%, which is probably a gross underestimate since data is unreliable. However, based on those estimates, about 3.56 million diabetic patients live in the 17 countries in the region (IDF 2017). The most populous country in the region (Nigeria) is estimated to have about 1.7 million diabetics, followed by Ghana, with approximately 520,000 diabetics (IDF 2017). The increase in the incidence of obesity and related chronic diseases such as diabetes is a result of changes in food consumption patterns which can be attributed to the Westernization of the society.

With a population of over 362 million across 17 countries in the West African region in 2016, several diverse ethnic groups, e.g. Akan, Ashanti, Fulani, Hausa, Igbo, Ijaw, Mandinka, and Yoruba, exist (Nettle 1996). However, similarities exist in their food, culture, and clothing. Food crops such as yam, cassava, millet, sorghum, sweet potatoes, rice, beans, and okra are prepared in similar ways. Diseases prevalent in the region include malaria, yellow fever, Lassa fever, cholera, and typhoid fever. However, in recent times, due to modernization, the incidence and prevalence of chronic communicable and non-communicable diseases, including diabetes, is on the rise.

USE OF FOOD TO MANAGE DIABETES

Pharmacotherapeutic management for T2D is aimed at reducing the blood glucose level by using oral antidiabetics such as metformin to lower the risk of the development of complications (Kavishankar et al. 2011). Alternative treatment and management utilize herbal remedies, supplements, and functional foods (Perera and Li 2012). Consistent with the advice of Hippocrates, "[L]et your food be your medicine and your medicine be your food" (Smith 2004), food plays an important role in the etiology, prevention, management, and treatment of many diseases, including diabetes. Dietary components, primarily processed carbohydrates but also fats and oils (Steyn et al. 2004), have been implicated in the etiology of T2D. However, other classes of food and nutrients have been reported to be beneficial in the prevention and, possibly, the treatment of diabetes (Steyn et al. 2004). Diabetes can be managed with foods that either prevent weight gain or promote weight loss, lower blood glucose levels, or reduce glucose production (Evert et al. 2013). These classes of food that have therapeutic beneficial effects on human health in addition to nutritional benefits are known as functional foods (López Varela et al. 2002).

Functional foods provide additional physiological advantages by preventing, managing, and treating diseases (López Varela et al. 2002, Rudkowska 2009). Research has shown that whole grains, nuts, herbal teas, legumes, spices, fruits, and vegetables can serve as dietary functional foods in diabetes management (Mirmiran et al. 2014). The mechanisms of action of functional foods depend on their macronutrients, micronutrients, and phytochemical components (Pandian 2013). The presence of high amounts of fiber in food results in a feeling of fullness and satiety, thus reducing food intake and postprandial glucose production (Rudkowska 2009). Also, bioactive compounds such as polyphenols and flavonoids found in plants exhibit antioxidant activity that mitigates diabetes-associated inflammation and oxidative stress (Pandian 2013).

In the last few decades, there has been increasing interest in the health-promoting potential of culturally salient food plants for the prevention of chronic diseases. According to Muandu et al. (1999), traditional food plants can either be naturalized or indigenous plants used in a community over a sufficient length of time. Traditional food plants in African cultural communities are either cultivated or gathered in the wild or semi-wild conditions (Udenta et al. 2014). These dietary food plants are either prepared as traditional vegetable dishes with staple food or as stews or teas or added as condiments to food (Udenta et al. 2014). In West Africa, to manage T2D traditionally, most of the literature available has focused on the use of medicinal plants. This chapter will focus on documenting food plants eaten in West Africa that have the potential not only for nutrition but also for diabetes management and treatment.

METHODS

Information about medicinal plants traditionally used in West Africa for the management of diabetes was obtained from published articles, including ethnobotanical surveys, reviews, and pharmacological studies. Relevant literature was accessed from scientific electronic databases, including Pubmed, Medline, Science Direct, and Google Scholar. Keyword combinations for the search were "antidiabetic", "hypoglycaemic", "West Africa", "medicinal plants", "food plants", and "edible plants". The use of these medicinal plants for dietary purposes was ascertained using databases such as PROTA4U and Useful Tropical Plants, Google searches, and Guinand and Lemessa (2001). Edible plants with published *in vitro* and *in vivo* studies from 2000 to 2017 were included. Botanical names of plants were verified using Kew Botanical Gardens' database: www.theplantlist.org. The resultant plant list was subjected to further search for articles documenting traditional uses for diabetes, including the country where the plants are used as medicine and the plant parts used for food and medicine, and the level of research to date. The plants were then categorized and presented based on families and plant parts used for antidiabetic effect.

RESULTS AND DISCUSSION

This chapter reports on dietary food plants used traditionally in the management of diabetes in West Africa. These plants were obtained from previously conducted ethnobotanical studies of medicinal plants with antidiabetic properties across the region. Table 4.1 summarizes the accepted Latin name and family of these plants, common name(s), plant part(s) used as food, plant part(s) with antidiabetic effect, and country or countries of use. One hundred and fifty-seven (157) dietary plants from 141 genera and 58 plant families were identified as dietary plants with antidiabetic potential in West Africa. The most frequently encountered families were Leguminosae, Malvaceae, Euphorbiaceae, Poaceae, Rutaceae, Rubiaceae, and Moraceae.

Table 4.2 is a compilation of the dietary plants that have at least one form of scientific validation for their traditional claim to the management and treatment of diabetes published in the literature. Of the 157 dietary plants listed in Table 4.1, 126 species from 50 families have been studied scientifically for their antidiabetic activity (Table 4.2). Studies were either pre-clinical (*in vitro* and *in vivo* studies) and/or clinical. All the plants with *in vivo* evidence of efficacy have been studied using animal models such as rats, mice, dogs, and rabbits. The majority (98%) of these studies conducted used chemically induced diabetic models, especially streptozotocin and alloxan. At least one (Kuate et al. 2015) used diet-induced animal models to study the potential of plant foods in preventing hyperglycemia and reducing blood glucose.

African scientists and traditional health practitioners have recognized the need to clinically document the claims of plants with medicinal potential on the continent

Table 4.1 List of Dietary Food Plants Traditionally Used in the Treatment and Management of Diabetes in West Africa

Latin name	Common name	Family name	Plant part as food	Country of use	Plant part with antidiabetic potential
Abelmoschus esculentus (L.) Moench	Lady's finger	Malvaceae	Fruit	Nigeria	Fruit, Seed
Abrus precatorius L.	Indian liquorice	Leguminosae	Seed, leaves, roots	Guinea	Leaves, seed
Acalypha wilkesiana Müll. Arg.	Acalypha	Euphorbiaceae	Young shoot	Nigeria	Roots, leaves
Adansonia digitata L.	Baobab	Malvaceae	Fruit, young leaves	Guinea, Nigeria	Fruit pulp, stem bark
Adenia cissampeloides (Planch. ex Hook.) Harms		Passifloraceae	Leaves	Guinea	Leaves
Aframomum alboviolaceum (Ridl.) K. Schum	Grape-seeded amomum	Zingiberaceae	Leaves	Guinea, Nigeria	Stem bark, Leaves
Aframomum melegueta (Roscoe) K. Schum.	Grains of paradise	Zingiberaceae	Seed, fruit	Guinea	Fruit
Afzelia africana Pers.	African mahogany	Leguminosae	Fruit	Nigeria	Root bark, stem bark
Ageratum conyzoides (L.) L.	Goat weed	Compositae	Leaves	Nigeria	Leaves, shoot, whole plant
Albizia zygia (DC.) J.F. Macbr.	West African albizia	Leguminosae	Young leaves	Guinea	Stem bark
Alchornea cordifolia (Schumach. & Thonn.) Müll. Arg.	Christmas bush	Euphorbiaceae	Leaves, fruits	Nigeria, Togo	Leaves
Allium cepa L.	Shallot	Amaryllidaceae	Bulb, leaves, flowers	Guinea, Nigeria, Ghana	Bulb
Allium sativum L.	Garlic	Amaryllidaceae	Bulb, leaves, sprouted seed	Guinea, Nigeria	Bulb

(Continued)

Table 4.1 (Continued) List of Dietary Food Plants Traditionally Used in the Treatment and Management of Diabetes in West Africa

Latin name	Common name	Family name	Plant part as food	Country of use	Plant part with antidiabetic potential
Anacardium occidentale L.	Cashew	Anacardiaceae	Fruit, seed, leaves	Guinea, Nigeria, Ghana	Leaves, inner bark
Annona muricata L.	Soursop	Annonaceae	Fruit, young shoot, leaves	Togo	Leaves, whole plant
Annona senegalensis Pers.	Wild custard apple	Annonaceae	Fruit, leaves, flowers	Nigeria, Togo, Guinea	Roots, stem bark, leaves
Anogeissus leiocarpa (DC.) Guill. & Perr.	African birch	Combretaceae	Fruit, tender young leaves, gum	Guinea, Togo, Benin	Root
Arachis hypogaea L.	Peanut	Leguminosae	Seed	Nigeria	Seed
Avicennia germinans (L.) L.	Black mangrove	Acanthaceae	Fruit, leaves, root	Guinea	Root
Azadirachta indica A. Juss.	Neem	Meliaceae	Tender shoot, flower	Ghana, Nigeria	Leaves
Bambusa vulgaris Schrad.	Common bamboo	Poaceae	Young shoot	Guinea	Leaves
Bauhinia rufescens Lam.	Silver butterfly tree	Leguminosae	Fruit (seedpod)	Nigeria	Leaves
Bauhinia thonningii Schumach.	Camel's foot	Leguminosae	Leaves, fruit	Guinea	Stem
Bridelia ferruginea Benth.	Mitzeerie	Phyllanthaceae	Root bark	Nigeria, Togo, Ghana	Leaves, bark
Bridelia micrantha (Hochst.) Baill.	Wild coffee	Euphorbiaceae	Fruit, stem bark	Senegal, Nigeria	Leaves
Canna indica L.	Indian shot	Cannaceae	Root, young shoot	Guinea	Root
Carica papaya L.	Papaya	Caricaceae	Fruit, seed, leaves	Guinea, Nigeria	Unripe pulp, seed, leaves
Carum carvi L.	Caraway	Apiaceae	Fruit	Nigeria	Fruits, seeds

(*Continued*)

Table 4.1 (Continued) **List of Dietary Food Plants Traditionally Used in the Treatment and Management of Diabetes in West Africa**

Latin name	Common name	Family name	Plant part as food	Country of use	Plant part with antidiabetic potential
Ceiba pentandra (L.) Gaertn	Kapok tree	Malvaceae	Fruit, seed, flower, leaves	Nigeria	Stem bark
Centella asiatica (L.) Urb.	Gotu kola	Apiaceae	Leaves	Guinea	Whole plant, leaves
Chrysophyllum cainito L.	Star apple	Sapotaceae	Fruit, seeds	Cote d' Ivoire	Fruit, leaves
Cissus aralioides (Welw. ex Baker) Planch.		Vitaceae	Fruit	Guinea	Root, leaves
Citrus aurantium L.	Bitter orange	Rutaceae	Fruit	Nigeria	Fruit
Citrus medica L.	Citron	Rutaceae	Fruit	Guinea	Peel, seed
Citrus paradisi Macfad.	Grapefruit	Rutaceae	Fruit	Nigeria	Fruit juice
Clausena lansium (Lour.) Skeels	Wampee	Rutaceae	Fruit	Nigeria	Stem bark
Cnidoscolus aconitifolius (Mill.) I.M. Johnst.	Tree spinach	Euphorbiaceae	Young shoot, leaves	Nigeria	Leaves
Cola nitida Vent.	Cola nut	Malvaceae	Seed	Guinea	Seed
Combretum micranthum G. Don	Geza	Combretaceae	Leaves	Guinea, Nigeria	Leaves
Commelina africana L.	Wandering Jew	Commelinaceae	Leaves	Nigeria	Leaves
Corchorus olitorius L.	Jute mallow	Malvaceae	Leaves, fruit	Guinea	Seed
Cordia myxa L.	Sapistan plum	Boraginaceae	Leaves, fruit, flower, seed	Guinea	Fruit pulp
Costus afer Ker Gawl.	Spiral ginger	Costaceae	Leaves	Guinea	Stem
Coula edulis Baill.	African walnut	Olacaceae	Seed	Guinea	Leaves
Craterispermum laurinum (Poir.) Benth.		Rubiaceae	Bark	Guinea	Stem bark
Crescentia cujete L.	Calabash tree	Bignoniaceae	Leaves, seed, fruit	Guinea	Leaves

(*Continued*)

Table 4.1 (Continued) List of Dietary Food Plants Traditionally Used in the Treatment and Management of Diabetes in West Africa

Latin name	Common name	Family name	Plant part as food	Country of use	Plant part with antidiabetic potential
Cucumeropsis mannii Naudin	Egusi	Cucurbitaceae	Fruit, seeds	Nigeria	Seed
Curcuma longa L.	Turmeric	Zingiberaceae	Rhizome	Nigeria	Root
Cussonia arborea Hochst. ex A. Rich.		Araliaceae	Fruit	Guinea	Stem bark, root bark
Cymbopogon citratus (DC.) Stapf	Lemongrass	Poaceae	Young shoots, leafy shoots, leaves,oil	Nigeria	Leaves, essential oil
Daniella oliveri Hutch. & Dalziel	African copaiba, balsam tree	Bignoniaceae	Young tender leaves	Nigeria	Roots
Detarium microcarpum Guill. & Perr	Sweet dattock	Leguminosae	Fruit, seed, leaves	Guinea, Nigeria	Roots
Dioscorea abyssinica Hochst. ex Kunth		Dioscoreaceae	Tuber	Guinea, Nigeria	Tuber
Dysphania ambrosioides (L.) Mosyakin & Clemants	Mexican tea	Amaranthaceae	Leaves, seed	Nigeria	Leaves
Elaeis guineensis Jacq.	African oil palm	Arecaceae	Seed oil, nuts	Guinea	Leaves, fruit
Entada africana Guill. & Perr.		Leguminosae	Tender young leaves	Guinea	Stem bark
Euphorbia hirta L.	Asthma weed	Euphorbiaceae	Leaves, shoots	Guinea	Leaves, flower and stem
Faidherbia albida (Delile) A. Chev.	Apple ring acacia	Leguminosae	Seed	Nigeria	Stem bark
Ficus asperifolia Miq.	Fig tree	Moraceae	Leaves	Nigeria	Stem
Ficus glumosa Delile	Mountain fig	Moraceae	Fruit, young leaves	Guinea	Leaves
Ficus sur Forssk.	Cape fig, Broom cluster fig	Moraceae	Figs	Guinea	Leaves
Garcinia kola Herkel	Bitter kola	Clusiaceae	Seed	Guinea, Nigeria	Seed

(*Continued*)

Table 4.1 (Continued) List of Dietary Food Plants Traditionally Used in the Treatment and Management of Diabetes in West Africa

Latin name	Common name	Family name	Plant part as food	Country of use	Plant part with antidiabetic potential
Harrisonia abyssinica Oliv.		Rutaceae	Fruit	Guinea	Stem bark
Harungana madagascariensis Lam.ex Poir.	Haronga	Hypericaceae	Fruit	Guinea	Stem bark
Heinsia crinita (Afzel.) G. Taylor	Bush apple	Rubiaceae	Fruit, leaves	Nigeria	Leaf
Hibiscus asper Hook.f.	Bush roselle	Malvaceae	Leaves, fruits	Guinea	Leaves
Hibiscus sabdariffa L.	Roselle	Malvaceae	Leaves, flowers, seeds, root	Nigeria	Fruit juice, flower
Hoslundia opposita Vahl.	Bird gooseberry	Lamiaceae	Tender young leaves, fruit	Guinea	Leaves
Hymenocardia acida Tul.	Heart fruit	Phyllanthaceae	Fruit, young leafy shoot	Guinea	Leaves
Impatiens irvingii Hook.f		Balsaminaceae	Leaves	Guinea	Flowers
Imperata cylindrica (L.) Raeusch.	Cogon grass	Poaceae	Shoots, roots	Guinea	Leaves
Indigofera arrecta A. Rich.	Bengal indigo	Leguminosae	Young leaves	Ghana	Leaves,
Ipomoea batatas (L.) Lam.	Sweet potato	Convolvulaceae	Root, young shoot	Ghana	Leaves, root tuber
Irvingia gabonensis (Aubrey-Lecomte ex O'Rorke) Baill.	Dika nut	Irvingiaceae	Seed, fruit	Guinea, Nigeria	Seed powder
Jatropha curcas L.	Physic nut	Euphorbiaceae	Shoots, nuts	Guinea	Leaves
Khaya senegalensis (Desv.) A. Juss.	Dry zone mahogany	Meliaceae	Seed	Guinea, Nigeria	Stem bark, root
Landolphia dulcis Sabine ex G. Don		Apocynaceae	Fruit	Guinea	Leaves
Landolphia heudelotii A.DC.		Apocynaceae	Fruit pulp	Guinea	Leaves

(*Continued*)

Table 4.1 (Continued) List of Dietary Food Plants Traditionally Used in the Treatment and Management of Diabetes in West Africa

Latin name	Common name	Family name	Plant part as food	Country of use	Plant part with antidiabetic potential
Landolphia hirsuta (Hua) Pichon	Ibo tree of Lagos	Apocynaceae	Fruit	Guinea	Stem bark
Lannea acida A. Rich		Anacardiaceae	Fruit, leaves, gum	Guinea, Benin	Leaves
Lecaniodiscus cupanioides Planch. ex Benth.		Sapindaceae	Seed, fruit	Guinea	Stem bark
Leptadenia lancifolia (Schumach. & Thonn.) Decne.		Apocynaceae	Young leaves, shoots, flowers	Nigeria	Leaf
Lippia chevalieri Moldenke		Verbenaceae	Leaves	Guinea	Root, stem bark
Lophira lanceolata Tiegh. ex Keay	Dwarf red ironwood	Ochnaceae	Seed	Guinea	Leaves
Mammea africana Sabine	African mammea apple	Calophyllaceae	Fruit, seed	Nigeria	Stem bark
Mangifera indica L.	Mango	Anacardiaceae	Fruit, seeds, flowers, leaves	Guinea, Nigeria	Leaves
Manihot esculenta Crantz	Cassava	Euphorbiaceae	Leaves, root	Nigeria	Leaves, tubers
Manniophyton fulvum Mull. Arg.	Gasso nut	Euphorbiaceae	Seed	Guinea, Nigeria	Bud, leaves
Maytenus senegalensis Lam.	Spike thorn	Celastraceae	Leaves, flowers, fruit	Nigeria	Root bark
Melanthera scandens (Schumach. & Thonn.) Roberty		Compositae	Leaves	Nigeria	Leaves
Microdesmis puberula Hook.f. ex Planch.		Pandaceae	Seed, leaves, fruit	Guinea	Bud
Milicia excelsa (Welw.) C.C. Berg	African teak	Moraceae	Leaves, fruits	Guinea	Stem bark

(*Continued*)

Table 4.1 (Continued) List of Dietary Food Plants Traditionally Used in the Treatment and Management of Diabetes in West Africa

Latin name	Common name	Family name	Plant part as food	Country of use	Plant part with antidiabetic potential
Momordica charantia L.	Bitter melon	Cucurbitaceae	Fruit, young shoots, leaves	Nigeria, Ghana	Leaves, fruit juice, seed
Morinda citrifolia L.	Great morinda	Rubiaceae	Fruit, leaves	Nigeria	Fruits, leaves
Morinda lucida Benth.	Brimstone tree	Rubiaceae	Roots	Guinea, Nigeria	Stem bark
Morus alba L.	Mulberry	Moraceae	Fruit, young leaves, shoot, inner bark	Ghana	Leaves, root bark
Mucuna pruriens (L.) DC.	Cowitch	Leguminosae	Seed	Guinea, Nigeria	Seeds
Murraya koenigii (L.) Spreng.	Curry leaves	Rutaceae	Leaves, fruit	Guinea, Nigeria	Leaves
Musa × *paradisiaca* L.	Banana	Musaceae	Fruit, male inflorescence, inner shoot	Nigeria	Fruit, fruit peel, leaves, root, stem
Musanga cecropioides R. Br. ex Tedlie	Corkwood	Urticaceae	Aerial root, fruit	Nigeria	Stem bark, leaves
Neocarya macrophylla (Sabine) Prance ex F. White	Gingerbread plum	Chrysobalanaceae	Fruit, kernel	Nigeria	
Ocimum gratissimum L.	Clove basil	Lamiaceae	Leaves	Guinea, Nigeria	Leaves
Opilia amentacea Roxb.	Opilia	Opiliaceae	Fruit, leaves	Burkina Faso, Benin, Togo	Leaves
Oxytenanthera abyssinica (A. Rich.) Munro	Savanna bamboo	Poaceae	Young shoot, seed	Benin	Leaves
Parinari curatellifolia Planch. ex Benth.	Mobola plum	Chrysobalanaceae	Fruit, seed	Guinea	Seed

(*Continued*)

Table 4.1 (Continued) List of Dietary Food Plants Traditionally Used in the Treatment and Management of Diabetes in West Africa

Latin name	Common name	Family name	Plant part as food	Country of use	Plant part with antidiabetic potential
Parinari excelsa Sabine	Guinea plum	Chrysobalanaceae	Fruit, seed	Nigeria, Senegal	Bark
Parkia biglobosa (Jacq.) G. Don	African locust bean	Leguminosae	Seed, seed pods	Guinea, Nigeria	Seeds
Pennisetum purpureum Schumach.	Elephant grass	Poaceae	Leaves, shoots	Guinea	Root
Persea americana Mill.	Avocado	Lauraceae	Fruit, leaves	Guinea, Nigeria	Leaves, seed
Philenoptera cyanescens (Schumach. & Thonn.) Roberty	African indigo	Leguminosae	Leaves	Nigeria	Leaves, stem oils
Phyllanthus emblica L.	Emblica	Phyllanthaceae	Fruit	Nigeria	Fruit
Piliostigma thonningii (Schumach.) Milne-Redh.	Camel's foot	Fabaceae	Fruit, leaves	Nigeria	Stem bark
Platostoma africanum P. Beauv.		Lamiaceae	Leaves	Guinea, Nigeria	Fruit
Plukenetia conophora Mull. Arg.	Owusa nut	Euphorbiaceae	Seed, fruit, leaves, young shoots	Guinea	Seed
Prosopis africana (Guili. & Perr.) Taub.	African mesquite	Leguminosae	Seed	Guinea, Nigeria	Leaves
Pseudocedrela kotschyi (Schweinf.) Harms	Wooden banana	Meliaceae	Leaves, flowers, fruit	Nigeria	Roots
Psidium guajava L.	Guava	Myrtaceae	Fruit	Guinea	Leaves, fruit
Pteridium aquilinum (L.) Kuhn	Bracken	Dennstaedtiaceae	Root, young leaves	Guinea	Bud
Pterocarpus erinaceus Poir.	African kino	Leguminosae	Leaves, seeds	Benin	Stem bark

(Continued)

Table 4.1 (Continued) List of Dietary Food Plants Traditionally Used in the Treatment and Management of Diabetes in West Africa

Latin name	Common name	Family name	Plant part as food	Country of use	Plant part with antidiabetic potential
Pterocarpus mildbraedii Harms		Leguminosae	Leaves	Guinea	Leaves
Raphia hookeri G. Mann & H. Wendl.	Ivory Coast raffia palm	Arecaceae	Fruit, sap	Nigeria	Seed
Ravenala madagascariensis Sonn.	Traveler's tree	Strelitziaceae	Fruit, seeds	Guinea	Leaves
Sarcocephalus latifolius (Sm.) Bruce	African peach	Rubiaceae	Fruit	Guinea, Nigeria, Benin	Roots, leaves
Scoparia dulcis L.	Sweet broom	Plantaginaceae	Seed infusion	Guinea	Whole plant
Securinega virosa (Roxb.) Baill.	Snowberry tree	Euphorbiaceae	Fruit	Nigeria	Leaves
Senna occidentalis (L.) Link	Wild coffee	Leguminosae	Seed	Togo	Whole plant, root
Sida rhombifolia L.	Arrowleaf sida	Malvaceae	Leaves	Guinea	Aerial parts, stem bark
Solanum macrocarpon L.	African eggplant	Solanaceae	Fruit	Guinea	Leaves
Spathodea campanulata P. Beauv.	African tulip tree	Bignoniaceae	Seed	Guinea	Stem bark
Sphenocentrum jollyanum Pierre	Akerejupon	Menispermaceae	Fruit	Nigeria	Roots, leaves
Sphenostylis stenocarpa (A. Rich.) Harms	African yam bean	Leguminosae	Immature seed pods, seed, root, leaves	Guinea	Seed
Spondias mombin L.	Yellow mombin	Anacardiaceae	Young leaves, fruit, seed	Nigeria	Leaf
Stachytarpheta indica (L.) Vahl	Blue porterweed	Verbenaceae	Leaves	Nigeria	Whole plant

(*Continued*)

Table 4.1 (Continued) List of Dietary Food Plants Traditionally Used in the Treatment and Management of Diabetes in West Africa

Latin name	Common name	Family name	Plant part as food	Country of use	Plant part with antidiabetic potential
Sterculia tragacantha Lindl.	Karaya gum	Malvaceae	Young shoots, leaves, seeds, gum	Guinea	Stem bark, leaves
Stevia rebaudiana (Bertoni) Bertoni	Stevia	Compositae	Leaves	Ghana	Leaves
Syzygium guineense (Willd.) DC.	Water berry	Myrtaceae	Fruit	Guinea	Leaves
Tamarindus indica L.	Tamarind	Leguminosae	Seed pods, mature seeds, young leaves	Guinea	Bark, fruit
Telfairia occidentalis Hook.f.	Fluted gourd, fluted pumpkin	Cucurbitaceae	Fruit pulp, leaves, seeds	Nigeria	Leaves, seed
Terminalia catappa L.	Indian almond	Combretaceae	Fruit, seed	Guinea, Nigeria, Cote d' Ivoire	Leaves, fruit
Tetrapleura tetraptera (Schumach. & Thonn.) Taub.	Aidan tree	Leguminosae	Fruit	Nigeria	Leaves, fruit pulp
Theobroma cacao L.	Cocoa tree	Malvaceae	Seeds, pulp	Nigeria	Cocoa pods
Treculia africana Decne.	African breadfruit	Moraceae	Seed	Guinea, Nigeria	Root
Trema orientalis (L.) Blume	Charcoal tree	Cannabaceae	Leaves, fruit	Guinea	Stem bark
Trichilia emetica Vahl	Natal mahogany, Banket mahogany	Meliaceae	Seeds	Burkina Faso	Leaves
Triplochiton scleroxylon K. Schum.	African obeche tree	Malvaceae	Leaves	Nigeria	Bark

(*Continued*)

Table 4.1 (Continued) List of Dietary Food Plants Traditionally Used in the Treatment and Management of Diabetes in West Africa

Latin name	Common name	Family name	Plant part as food	Country of use	Plant part with antidiabetic potential
Uvaria chamae P. Beauv.	Bush banana or finger root	Annonaceae	Fruit, root bark	Guinea	Root
Vernonia amygdalina Delile	Bitter leaf	Compositae	Leaves	Nigeria	Leaves
Vigna subterranea (L.) Verdc.	Bambara Groundnut	Leguminosae	Seed, leaves	Guinea	Seed
Vigna unguiculata (L.) Walp.	Cowpea	Fabaceae	Seed, root	Guinea	Seed
Vitellaria paradoxa C.F. Gaertn.	Shea butter tree	Sapotaceae	Flowers, seed kernel	Guinea	Stem bark
Vitex doniana Sweet	Black plum	Lamiaceae	Young twigs, leaves, fruit	Nigeria	Leaves, stem bark
Xeroderris stuhlmannii (Taub.) Mendonça & E.P. Sousa	Wing pod, wing bean	Leguminosae	Seeds	Guinea	Stem bark
Ximenia americana L.	Tallow wood	Olacaceae	Fruit, flower pstals, young leaves, roasted seeds	Guinea	Leaves
Xylopia aethiopica (Dunal) A. Rich.	Ethiopian pepper	Annonaceae	Fruit, seed, leaves	Guinea	Leaves, fruits
Zea mays L.	Maize	Poaceae	Seed, uripe cobs, stem pith, young leaves	Guinea	Corn silk
Zingiber officinale Roscoe	Ginger	Zingiberaceae	Root	Nigeria, Guinea	Root
Ziziphus spina-christi (L.) Desf.	Christ's thorn	Rhamnaceae	Fruit	Guinea	Leaves

Sources: Baldé et al. (2006), Abo et al. (2008), Olabanji et al. (2008), Gbolade (2009), N'Guessan et al. (2009a), Etuk et al. (2010), Diallo et al. (2012), Dieye et al. (2008), Lawin et al. (2015), Karou et al. (2011), Mohammed et al. (2014), Mentreddy (2007), Tchacondo et al. (2012), Udenta et al. (2014), Ezuruike and Prieto (2014).

Table 4.2 Plant List with Scientific Validation

Latin name	Plant part with antidiabetic potential	In vitro	In vivo animal	In vivo human	References
Abelmoschus esculentus (L.) Moench	Fruit, seed	–	+		Erfani et al. (2018), Khatun et al. (2011)
Abrus precatorius L.	Leaves, seed	–	+	–	Monago and Alumanah (2005), Nwanjo (2008)
Acalypha wilkesiana Müll. Arg.	Roots, leaves	–	+	–	Al-Attar (2010), Odoh et al. (2013), El-Khateeb et al. (2014), Olukunle et al. (2016)
Adansonia digitata L.	Fruit pulp, stem bark	–	+	–	Tanko et al. (2008), Bhargav et al. (2009), Gwarzo and Bako (2013)
Aframomum melegueta K. Schum.	Fruit	+	+	–	Adefegha and Oboh (2012), Adesokan et al. (2010)
Afzelia africana Pers.	Root bark, stem bark	–	+	–	Oyedemi et al. (2011), Odo et al. (2012)
Ageratum conyzoides (L.) L.	Leaves, shoot, whole plant	–	+	–	Nyunaï et al. (2006), Nyunaï et al. (2009), Egunyomi et al. (2011), Rahman et al. (2013)
Alchornea cordifolia (Schumach. & Thonn.) Müll. Arg.	Leaves	–	+	–	Mohammed et al. (2013), Thomford et al. (2015)
Allium cepa L.	Bulb	–	+	+	El-Demerdash et al. (2005), Campos et al. (2003), Taj Eldin et al. (2010)
Allium sativum L.	Bulb	–	+	+	El-Demerdash et al. (2005), Eidi et al. (2006)
Anacardium occidentale L.	Leaves, inner bark	–	+	–	Olatunji et al. (2005), Okpashi et al. (2014)
Annona muricata L.	Leaves, whole plant	–	+	+	Adeyemi et al. (2009), Florence et al. (2014), Arroyo et al. (2009)
Arachis hypogaea L.	Seed	–	+	–	Bilbis et al. (2002), Emekli-Alturfan et al. (2008)
Azadirachta indica A. Juss.	Leaves	–	+	–	Khosla et al. (2000)
Bambusa vulgaris Schrad.	Leaves	–	+	–	Haque et al. (2015), Senthilkumar et al. (2011)
Bauhinia rufescens Lam.	Leaves	–	+	–	Aguh et al. (2013)
Bauhinia thonningii Schumach.	Stem	–	+	–	Asuzu & Nwaehujor (2013)

(*Continued*)

Table 4.2 (Continued) Plant List with Scientific Validation

Latin name	Plant part with antidiabetic potential	In vitro	In vivo animal	In vivo human	References
Bridelia ferruginea Benth.	Leaves, bark	–	+	–	Kolawole et al. (2006), Njamen et al. (2012), Adewale and Oloyede (2012), Aja et al. (2013)
Bridelia micrantha (Hochst.) Baill.	Leaves	–	+	–	Adika et al. (2012), Omeh et al. (2014)
Canna indica L.	Root	–	+	–	Purintrapiban et al. (2006), Chen et al. (2013)
Carica papaya L.	Unripe pulp, seed, leaves	–	+	–	Osadolor et al. (2011), Sasidharan et al. (2011), Juárez-Rojop et al. (2012), Maniyar and Bhixavatimath (2012)
Carum carvi L.	Fruits, seeds	–	+	–	Eddouks et al. (2002)
Ceiba pentandra (L.) Gaertn.	Stem bark	–	+	–	Ladeji et al. (2003)
Centella asiatica (L.) Urb.	Whole plant, Leaves	–	+	–	Kabir et al. (2014)
Chrysophyllum cainito L.	Fruit, leaves	–	+	–	N'Guessan et al. (2009b)
Citrus aurantium L.	Fruit	–	+	–	Opajobi et al. (2011)
Citrus medica L.	Fruit peel, seed	–	+	–	Conforti et al. (2007), Sah et al. (2011)
Citrus paradisi Macfad.	Fruit juice	–	+	–	Adeneye (2008)
Clausena lansium (Lour.) Skeels	Stem bark	–	+	–	Adebajo et al. (2009)
Cnidoscolus aconitifolius (Mill.) I.M. Johnst.	Leaves	–	+	–	Oyagbemi et al. (2010), Iwuji et al. (2014)
Cola nitida Vent.	Seed	+	+	–	Iliemene et al. (2014), Oboh et al. (2014)
Combretum micranthum G. Don	Leaves	–	+	–	Chika and Bello (2010)
Commelina africana L.	Leaves	–	+	–	Agunbiade et al. (2012)
Corchorus olitorius L.	Seed	–	+	–	Egua et al. (2014)
Cordia myxa L.	Fruit pulp	–	+	–	Mishra and Garg (2011)
Costus afer Ker Gawl.	Stem	+	+	–	Tchamgoue et al. (2015), Monago et al. (2016)
Crescentia cujete L.	Leaves	–	+	–	N'Guessan et al. (2008)
Cucumeropsis mannii Naudin	Seed	–	+	–	Teugwa et al. (2013)

(*Continued*)

Table 4.2 (Continued) Plant List with Scientific Validation

Latin name	Plant part with antidiabetic potential	In vitro	In vivo animal	In vivo human	References
Curcuma longa L.	Root	–	+	+	Kuroda et al. (2005), Chuengsamarn et al. (2012)
Cussonia arborea Hochst. ex A. Rich.	Stem bark, root bark	–	+	–	Aba et al. (2014)
Cymbopogon citratus (DC.) Stapf	Leaves, essential oil	–	+	–	Adeneye and Agbaje (2007), Bharti et al. (2013)
Daniella oliveri Hutch. & Dalziel	Roots	–	+	–	Iwueke and Nwodo (2008)
Detarium microcarpum Guill. & Perr.	Roots	–	+	–	Okolo et al. (2012)
Dysphania ambrosioides (L.) Mosyakin & Clemants	Leaves	–	+	–	Song et al. (2011)
Elaeis guineensis Jacq.	Leaves, fruit	–	–	+	Kalman et al. (2013), Sharif et al. (2015)
Euphorbia hirta L.	Leaves, flower and stem	–	+	–	Subramanian et al. (2011)
Faidherbia albida (Delile) A. Chev.	Stem bark	–	+	–	Umar et al. (2014)
Ficus asperifolia Miq.	Stem	–	+	–	Omoniwa et al. (2009)
Ficus glumosa Delile	Leaves	–	+	–	Umar et al. (2013)
Ficus sur Forssk.	Leaves	–	+	–	Akomas et al. (2014)
Garcinia kola Herkel	Seed	–	+	–	Udenze et al. (2012), Duze et al. (2012)
Harungana madagascariensis Lam. ex Poir.	Stem bark	–	+	–	Iwalewa et al. (2008)
Heinsia crinita (Afzel.) G. Taylor	Leaves	–	+	–	Ebong (2014), Okokon et al. (2009)
Hibiscus sabdariffa L.	Fruit juice, flower	–	+	–	Ajiboye et al. (2015), Mozaffari-Khosravi et al. (2009)
Hoslundia opposita Vahl.	Leaves	–	+	–	Akolade et al. (2014)
Hymenocardia acida Tul.	Leaves	–	+	–	Ezeigbo and Asuzu (2012), Ezeigbo et al. (2015)
Imperata cylindrica (L.) Raeusch.	Leaves	–	+	–	Ayu Suraya A et al. (2012)
Indigofera arrecta A. Rich.	Leaves,	–	+	+	Nyarko et al. (1993), Sittie and Nyarko (1998), Addy and Myarko (1988)

(*Continued*)

Table 4.2 (Continued) Plant List with Scientific Validation

Latin name	Plant part with antidiabetic potential	In vitro	In vivo animal	In vivo human	References
Ipomoea batatas (L.) Lam.	Leaves, tuber	–	+	+	Ludvic et al. (2004), Zhao et al. (2007), Ijaola et al. (2014)
Irvingia gabonensis (Aubrey-Lecamte ex O'Rorke) Baill.	Seed powder	–	+	–	Hossain et al. (2012)
Jatropha curcas L.	leaves	–	+	–	Mishra et al. (2010)
Khaya senegalensis (Desv.) A. Juss.	Stem bark, root	+	+	–	Kolawole et al. (2012), Ibrahim et al. (2014)
Lophira lanceolata Tiegh. ex Keay	Leaves	–	+	–	Awede et al. (2015)
Mammea africana Sabine	Stem bark	–	+	–	Okokon et al. (2007), Tchamadeu et al. (2010)
Mangifera indica L.	Leaves	–	+	+	Waheed et al. (2006), Gondi et al. (2015)
Manihot esculenta Crantz	Leaves, tubers	–	+	+	Mvitu-Muaka et al. (2010)
Manniophyton fulvum Müll. Arg.	Bud, leaves	–	+	–	Onyemairo et al. (2015)
Maytenus senegalensis Lam.	Root bark	–	+	–	Mann et al. (2014)
Melanthera scandens (Schumach. & Thonn.) Roberty	Leaves	–	+	–	Akpan et al. (2012)
Momordica charantia L.	Leaves, fruit juice, seed	–	+	–	Ahmed et al. (2004)
Morinda citrifolia L.	Fruits, leaves	–	+	–	Nayak et al. (2011)
Morinda lucida Benth.	Stem bark	–	+	–	Odutuga et al. (2010)
Morus alba L.	Leaves, root bark	–	+	+	Singab et al. (2005), Mudra et al. (2007), Mohammadi and Naik (2008)
Mucuna pruriens (L.) DC.	Seeds	–	+	–	Majekodunmi et al. (2011), Akhtar et al. (1990), Bhaskar et al. (2008)
Murraya koenigii (L.) Spreng.	Leaves	–	+	–	Khan et al. (1995), El-Amin et al. (2013)
Musa × paradisiaca L.	Fruit, fruit peel, leaves, root, stem	–	+	–	Ojewole and Adewunmi (2003), Lakshmi et al. (2014)
Musanga cecropioides R. Br. ex Tedlie	Stem bark, leaves	–	+	–	Adeneye et al. (2007), Ajayi and Igboekwe (2013)

(*Continued*)

Table 4.2 (Continued) Plant List with Scientific Validation

Latin name	Plant part with antidiabetic potential	In vitro	In vivo animal	In vivo human	References
Ocimum gratissimum L.	Leaves	–	+	–	Oguanobi et al. (2012)
Opilia amentacea Roxb.	Leaves	–	+	–	Konaté et al. (2014)
Oxytenanthera abyssinica (A. Rich.) Munro	Leaves	–	+	–	Ezeja et al. (2014)
Parinari curatellifolia Planch. ex Benth.	Seed	–	+	–	Ogunbolude et al. (2009), Ogbonnia et al. (2011)
Parinari excelsa Sabine	Bark	–	+	–	Ndiaye et al. (2008)
Parkia biglobosa (Jacq.) G. Don	Seeds	–	+	–	Fred-Jaiyesimi and Abo (2009)
Persea americana Mill.	Leaves, seed	–	+	–	Lima et al. (2012), Marrero-Faz et al. (2014)
Philenoptera cyanescens (Schumach. & Thonn.) Roberty	Leaves, stem oils	+	+	–	Kazeem and Davies (2016)
Phyllanthus emblica L.	Fruit	–	+	+	Qureshi et al. (2009), Krishnaveni et al. (2010), Usharani et al. (2013)
Piliostigma thonningii (Schumach.) Milne-Redh.	Stem bark	–	+	–	Asuzu and Nwaehujor (2013)
Plukenetia conophora Mull. Arg.	Seed	–	+	–	Famobuwa et al. (2015)
Prosopis africana (Guill. & Perr.) Taub.	Leaves	–	+	–	Shittu (2009)
Pseudocedrela kotschyi (Schweinf.) Harms	Roots	–	+	–	Ojewale et al. (2014)
Psidium guajava L.	Leaves, fruit	–	+	–	Shen et al. (2008), Soman et al. (2013), Okpashi et al. (2014)
Raphia hookeri G. Mann & H. Wendl.	Seed	–	+	–	Mbaka et al. (2012)
Ravenala madagascariensis Sonn.	Leaves	+	+	–	Priyadarshini et al. (2010)
Sarcocephalus latifolius (Sm.) Bruce	Roots, leaves	–	+	–	Iwueke and Nwodo (2008)
Scoparia dulcis L.	Whole plant	+	+	–	Mishra et al. (2013)

(*Continued*)

Table 4.2 (Continued) Plant List with Scientific Validation

Latin name	Plant part with antidiabetic potential	In vitro	In vivo animal	In vivo human	References
Securinega virosa (Roxb.) Baill.	Leaves	–	+	–	Tanko et al. (2008)
Senna occidentalis (L.) Link	Whole plant, root	–	+	–	Verma et al. (2010), Sharma et al. (2014)
Sida rhombifolia L.	Aerial parts, Stem bark	–	+	–	Ghosh et al. (2011), Dhalwal et al. (2010)
Solanum macrocarpon L.	Leaves	–	+	–	Emiloju and Chinedu (2016)
Spathodea campanulata P. Beauv.	Stem bark	–	+	–	Niyonzima et al. (1993)
Sphenocentrum jollyanum Pierre	Roots, leaves	–	+	–	Mbaka et al. (2009, 2010)
Sphenostylis stenocarpa (A. Rich.) Harms	Seed	–	+	–	Onoagbe et al. (2010), Ubaka and Ukwe (2010)
Spondias mombin L.	Leaves	+	–	–	Fred-Jaiyesimi et al. (2009)
Stachytarpheta indica (L.) Vahl.	Whole plant	–	+	–	Silambujanaki et al. (2009)
Stevia rebaudiana (Bertoni) Bertoni	Leaves	–	+	–	Chen et al. (2005), Suanarunsawat et al. (2004)
Syzygium guineense (Willd.) DC.	Leaves	–	+	–	Worku (2009)
Tamarindus indica L.	Bark, fruit	–	+	–	Al-Ahdab (2015)
Telfairia occidentalis Hook.f.	Leaves, seed	–	+	–	Eseyin et al. (2010)
Terminalia catappa L.	Leaves, fruit	–	+	–	Nagappa et al. (2003), N'guessa et al. (2011)
Tetrapleura tetraptera (Schumach. & Thonn.) Taub.	Leaves, fruit pulp	–	+	–	Ojewole and Adewunmi (2004), Atawodi et al. (2014), Kuate et al. (2015)
Theobroma cacao L.	Cocoa pods	–	+	–	Ruzaidi et al. (2005, 2008), Dare et al. (2014)
Treculia africana Decne.	Root	–	+	–	Oyelola et al. (2007)
Trema orientalis (L.) Blume	Stem bark	–	+	–	Dimo et al. (2006)
Trichilia emetica Vahl	Leaves	–	+	–	Konate et al. (2014)
Triplochiton scleroxylon K. Schum.	Bark	–	+	–	Prohp and Onoagbe (2009, 2011, 2013)

(Continued)

Table 4.2 (Continued) Plant List with Scientific Validation

Latin name	Plant part with antidiabetic potential	In vitro	In vivo animal	In vivo human	References
Uvaria chamae P. Beauv.	Root	–	+	–	Emordi et al. (2015)
Vernonia amygdalina Delile	Leaves	–	+	–	Akah and Okafor (1992), Ong et al. (2011), Atangwho et al. (2014)
Vigna subterranea (L.) Verdc.	Seed	–	+	–	Rotimi et al. (2010)
Vigna unguiculata (L.) Walp.	Seed	–	+	–	Ashraduzzaman et al. (2011)
Vitellaria paradoxa C.F. Gaertn.	Stem bark	–	+	–	Coulibaly (2014)
Vitex doniana Sweet	Leaves, stem bark	–	+	–	Obasi et al. (2013), Nwaneri-Chidozie et al. (2014), Oche et al. (2014)
Ximenia americana L.	Leaves	–	+	–	Siddaiah et al. (2011)
Xylopia aethiopica (Dunal) A. Rich.	Leaves, fruits	–	+	–	Mohammed et al. (2016), Okpashi et al. (2014)
Zea mays L.	Corn silk	–	+	+	Suzuki et al. (2005), Gomez (2015)
Zingiber officinale Roscoe	Root	+	+	+	Andallu et al. (2003), Nammi et al. (2009), Eyo et al. (2011), Rani et al. (2012), Shidfar et al. (2015)
Ziziphus spina-christi (L.) Desf.	Leaves	–	+	–	Glombitza et al. (1994)

(Siegfried and Hughes 2012). This is because clinical trials are today considered the gold standard for demonstrating the safety and efficacy of any drug. Our study revealed that only 10% of the food plants had their efficacy in either the management of diabetes, especially T2D, or the prevention of hyperglycemia documented in human clinical studies. The much greater cost of clinical trials relative to animal studies is certainly a limiting factor.

No scientific studies were found for 31 of the food plants listed in Table 4.1. Of these, 90% are used traditionally in Guinea for managing diabetes. These plants are, however, not endemic to Guinea, as they are found in tropical West Africa from Senegal to Nigeria. Reasons for the lack of literature regarding the efficacy of such plants might be the lack of funds to conduct scientific studies, use of inappropriate study design, and lack of positive results from such studies. Anecdotal evidence suggests that although many clinical trials have been conducted in the region, very few have been published. This might be attributed to a lack in the quality of the trials conducted, which might not meet the standard of publishing journals (Willcox et al. 2012).

Plant Parts with Antidiabetic Potential

Leafy vegetables, seed oils, fruits, and spices are used as adjuvants in traditional medicine. Leaves (48.7%) were the plant parts most commonly reported as used for diabetes in this chapter. Plant leaves are mostly prescribed as water infusions for the treatment of diabetes. However, others, such as the leaves of *Vernonia amygdalina,* are eaten as a leafy relish with staple diets. Leaves of plants such as *Mangifera indica, Ficus* species, and *Vernonia amygdalina* have been found to have anti-hyperglycemic effects (Aderibigbe et al. 1999, Erasto et al. 2009, Olaokun et al. 2014). Other plant parts used for diabetes include seeds (14.1%), fruits (14.1%), stem and stem barks (25.6%), and roots and root barks (16.7%). Numbers sum to >100% because multiple plant parts of some species are used as medicine.

Parts Used as Food

The West African diet is dominated by tubers, rhizomes, roots, starchy fruits, cereals, grain legumes, and several traditional bean varieties (Udenta et al. 2014). The food plants with antidiabetic potential recorded in this chapter are both exotic and indigenous to the region. Examples of exotic food plants are *Carica papaya* (papaya), *Zingiber officinale* (ginger), *Allium sativum* (garlic), and *Curcuma longa* (turmeric), while indigenous food plants include *Telfairia occidentalis* (fluted gourd), *Tetrapleura tetraptera* (aidan tree), *Hibiscus sabdariffa* (roselle), and *Cola nitida* (cola nut).

In the reviewed species, the parts most frequently used as food were the fruits and the leaves (each 55.0%), followed by the seeds (40.1%) and shoots (13.4). Other plant parts used as foods include flowers, roots, stems, and bark. Africa is known to have over 1,000 plant species used as vegetables, with leafy vegetables constituting over 80% (Maundu et al. 2009). Leaves are used either cooked fresh or dried (steamed, boiled, or stewed), eaten raw in relishes or dried for preservation (Udenta et al. 2014). They provide vitamins and minerals such as iron, beta carotenoids, and other phytochemicals that are essential for good health and disease prevention (Heber 2004). Leafy vegetables are also good sources of dietary fiber. They are either cultivated (*Brassica oleracea, Apium graveolens, Manihot esculenta,* etc.) or collected from the wild (*Urtica dioica, Heinsia crinita, Commelina africana, Euphorbia hirta,* etc.). Wild vegetables grow in unattended places during specific seasons or throughout the year depending on the climate, and they have great nutritional as well as medicinal value (Bvenura and Sivakumar 2017). The high prevalence of edible vegetable species reported in this chapter may be attributed to high cultural diversity and high dependence on locally available biological resources in West Africa (Maundu et al. 2009). Food diversity in Southern Africa is poor, particularly among low-income residents and black people (Labadarios et al. 2011).

Plant Parts Used as Food with Antidiabetic Potential

Platel and Srinivasan (1997) and Srinivasan (2005) outlined the importance of some common spices and vegetables in the management of diabetes. This section focuses on edible parts of these plants that have been studied for their antidiabetic potential. From Table 4.2, 59.5% (n = 75) of the edible parts of food plants also used to treat diabetes have been studied scientifically. Of these, 46.7% are leaves, 26.7% fruits, and 22.7% seeds.

Leafy Vegetables

Six plant families mainly represent the leafy vegetables with antidiabetic potential. These families are Amaranthaceae, Solanaceae, Compositae (Asteraceae), Cucurbitaceae, Leguminosae (Fabaceae), and Malvaceae. Leafy vegetables form an important part of the traditional diet (Smith and Eyzaguirre 2007). Meta-analyses on the effect of higher intake of fruits, vegetables, or their fibers have concluded that a higher intake of leafy vegetables is significantly associated with a reduced risk of T2D (Carter et al. 2010, Wang et al. 2016). The intake of vegetables in West Africa (204 g/day) is comparable to the global average (208.8 g/day) (Misha et al. 2015). Regular consumption of leafy vegetables is significantly associated with improved glycemic control in diabetics and reduced triglyceride, stress, and inflammatory marker levels (Takahashi et al. 2012). The therapeutic activity of leafy vegetables is linked to the presence of polyphenols, vitamins (C and E), fibers, magnesium, potassium, and folate (Carter et al. 2010, Wang et al. 2016). Polyphenols, such as flavonoids, and carotenoids are known to prevent oxidative damage, which may interfere with the uptake of glucose by the cells (Wang et al. 2016). The high content of dietary fiber in green leafy vegetables also improves the body's ability to secrete insulin and inhibits postprandial glucose load (Montonen et al. 2005).

Common leafy vegetables such as *Vernonia amygdalina*, *Ocimum gratissimum*, *Manihot esculentus*, *Telfaria occidentalis*, and *Corchorus olitorius*, used regularly in traditional meals across West Africa, have been validated for their antidiabetic potential (Eseyin et al. 2010, Mvitu-Muaka et al. 2010, Oguanobi et al. 2012, Atangwho et al. 2014). The inclusion of these vegetables with antidiabetic potentials in daily diets has been reported to complement the action of oral hypoglycemic drugs and/or insulin injections (Platel and Srinivasan 1997).

Fruits

Of the 68 species of edible fruits in West Africa listed in this chapter, only 19 fruits have been studied scientifically for their antidiabetic properties. Consumption of the recommended minimum daily intake of fruits has been known to prevent non-communicable diseases (Ruel et al. 2005). Although fruit consumption among adults ≥20 years in sub-Saharan Africa is very low, adults ≥20 years in West Africa consume 93 g/day of fruits compared to the global average of 81.3 g/day (Misha et al. 2015). Higher intakes of fruits have been correlated with a lower risk of T2D (Wang et al. 2016). Fruits such as *Ficus sur*, *Adansonia digitata*, *Citrus* species, and *Terminalia catappa*, especially those containing high levels of vitamin C (Afkhami-Ardekani and Shojaoddiny-Ardekani 2007) and polyphenols (Kim et al. 2016), have been shown to have antidiabetic properties.

Seeds and Nuts

Plant seeds, nuts, and oils are commonly used in West Africa either as condiments in soups and stews or as soups alone. West African adults over the age of 20 years consume 16.3 g/day of nuts and seeds compared to the global amount of 8.9 g/day, according to Misha et al. (2015). In that study, only 15 West African edible seeds were found to have been validated scientifically for their use in the management of diabetes. These include seeds of *Telfaria occidentalis*, *Cucumeropsis mannii*, *Parkia biglobosa*, and *Irvingia gabonensis*, which are commonly used as condiments in soups within the region, while others, such as *Cola nitida*, *Garcinia kola*, and *Elaeis guineensis*, are eaten raw. Plant seeds contain chemical constituents such as trigonelline, charantin, and ferrulic acid, which have been shown to play a role in the treatment and management of diabetes (Joseph and Jini 2013). With the exception of *Momordica charantia*, very little information exists in the literature about the potential of chemical constituents present in West African seed and nut food plants as antidiabetic agents.

BIOACTIVE CONSTITUENTS

Several phytochemical compounds isolated from these dietary food plants have also been shown to play a role in the treatment, prevention, and management of T2D. Food plants contain secondary metabolites such as phenolic acids, triterpenoids, polyphenols, and alkaloids, which are responsible for their antidiabetic potential (Daniel 2006, Vinayagam and Xu 2015). Dietary polyphenols, especially flavonoids, are known to improve insulin and glucose homeostasis (Vinayagam and Xu 2015). This homeostasis is achieved either through increasing glucose uptake in insulin-sensitive tissues, inhibiting glucose absorption in the intestine, modulating glucose release from the liver, stimulating insulin secretion in β-cells, inhibiting α-glucosidase and α-amylase, activating insulin receptors, or modulating intracellular signalling pathways (Kim et al. 2016). *In vivo* and *in vitro* studies have shown that subclasses of flavonoids, including anthocyanins, flavonols, and flavan-3-ols, improve glucose metabolism, reduce insulin resistance, and reduce β-cell dysfunction by regulating GLUT4 (Vinayagam and Xu 2015). However, the effectiveness of polyphenols as antidiabetic agents is inconsistent (Kim et al. 2016), possibly due to methodological flaws of clinical studies (small sample sizes, poor statistical analysis, and inconsistent data collection).

Curcuma longa (turmeric) is a well-known spice in West Africa, even though it originates from India. Curcumin, a phenolic compound isolated from the rhizome of this plant, has a long history of use in the treatment of T2D and prevention of its onset in prediabetic human patients (Chuengsamarn et al. 2012). It has the ability to delay onset of T2D by reducing insulin resistance, improving pancreatic β-cell function, and preventing cell death (Zhang et al. 2013). Another culinary spice of Indian origin used in West Africa, *Zingiber officinalis* (ginger), contains 6-gingerol, which has been shown to reduce insulin and blood glucose levels and improve glucose tolerance in type 2 diabetic animal models (Shivashankara et al. 2013).

Several phytochemicals isolated from *Momordica charantia* have been implicated in the mitigation of T2D. Momordicine I and II have been shown to stimulate insulin secretion in MIN6 β-cells (Keller et al. 2011). Oleanolic acid glycosides improve glucose tolerance in type 2 DM by preventing glucose from being absorbed in the intestines (Hui et al. 2009). Moreover, 5-β,19-epoxy-3-β,25-dihydroxycucurbita-6,23(E)-dienes, and 3-β,7-β,25-trihydroxycucurbita-5,23(E)-dien- 19-al have been shown to improve glucose tolerance and hypoglycemic effects in various mice models (Harinantenaina et al. 2006). Charantin and polypeptide-p (or p-insulin), phytochemicals also found in *Momordica charantia,* possess an insulin-like effect and also the ability to activate insulin receptors (Joseph and Jini 2013). Traditionally, young shoots and leaves of *Momordica charantia* are harvested from the wild and eaten as a vegetable.

From *Mangifera indica*, 1,2,3,4,6 penta-O-galloyl-β-d-glucose was isolated and was found to improve diabetes caused by intake of high fat diet in C57BL/6 mice (Mohan et al. 2013). β-sitosterol, a steroid found in *Azadirachta indica, Solanum macrocarpon, Ficus* species, and *Ziziphus spina-christi*, has also been shown to increase fasting insulin levels and regulate glucose uptake and lipolysis in fat cells (Chai et al. 2011). Garlic and onions (*Allium* spp) contain sulfur compounds such as allicin, allylpropyl disulphide, and S-allyl cysteine sulfoxide, with glucose lowering properties (Nasim and Dhir 2013). However, they are highly labile when exposed to cooling. From *Ocimum gratissimum*, chicoric acid has been isolated. This compound has been shown to possess hypoglycemic activity in various animal models (Casanova et al. 2014) by stimulating the AMP-activated kinase (AMPK) pathway (Schlernitzauer et al. 2013).

For food plants originating from West Africa with anti-diabetic potentials, very little chemistry and bioassay screening has been conducted to determine the chemical constituents responsible for their activity and their mechanisms of action. There is an urgent need to conduct phytochemical investigations of plants such as *Ficus* species, *Cola nitida*, *Garcinia kola*, *Albizia zygia* and *Afzelia africana* that are endemic to the region.

Micronutrients

Micronutrients such as calcium, chromium, magnesium, and potassium found in food plants have been shown to play a role in modulating glucose (Barbagallo and Dominguez 2015). Enzymes involved in glucose metabolism utilize magnesium as an essential cofactor (Suárez et al. 1995). These enzymes are involved in the maintenance of glucose, metabolism of insulin, and the development of T2D (Barbagallo et al. 2003). In animal studies, diets low in magnesium result in impaired insulin secretion (Suárez et al. 1995). This observation has been corroborated in clinical studies which highlight the importance of magnesium supplementation for glucose metabolism, insulin action, and sensitivity (Paolisso et al. 1994, Rodriguez-Moran and Guerrero-Romero 2003). Reduced intake of dietary magnesium has been linked to an increase in the incidence of T2D (Kao et al. 1999), and a meta-analysis of data from 18 controlled clinical trials confirms that bioavailable magnesium supplements can improve glucose levels in people with T2D or at risk of it (Veronese et al. 2016).

Popular foods in the West African diet found to be rich in magnesium include *Vigna unguiculata* beans, *Adansonia digitata* leaves, *Parkia biglobosa* seed and fruit pulp, *Arachis hypogaea* seed, *Landolphia heudelotti* fruit pulp, *Cucumeropsis mannii* seed, and *Anarcadium occidentale* seed (Stadlmayr et al. 2010). The mean dietary magnesium intake in the region (649 mg/day) was three times higher than the WHO recommendation (217 mg/day) (Joy et al. 2013). As high dietary magnesium is associated with improved insulin sensitivity (Cahill et al. 2013), one potential contributor to the increasing prevalence of diabetes in this region might be the lower intake, in Westernized diets, of these traditional foods rich in magnesium.

Another important micronutrient is calcium. It is essential in insulin-mediated processes in the muscle and adipose tissues. It has been suggested that a deficiency in calcium results in an increased risk of T2D (Villagas et al. 2009). *Vigna unguiculata* beans, *Adansonia digitata* leaves, *Parkia biglobosa* seed and fruit pulp, *Irvingia gabonensis* seed, *Ficus* species fruits, *Vernonia amygdalina* leaves, *Hibiscus asper* leaves, *Hibiscus sabdariffa* leaves, *Corchorus olitorius* leaves, *Abelmoschus esculentus* leaves and fruit, *Solanum macrocarpum* leaves, and *Tamarindus indica* seed are some of the calcium-rich food plants in West Africa (Boukari et al. 2001, Stadlmayr et al. 2010). Most of these food plants are leaves which are included in soups made in the region. However, the study of Joy et al. (2014) revealed that West Africans get their calcium supply mostly from roots and tubers. Compared to other regions in Africa, West Africa has the second lowest risk (36%) of calcium deficiency (Joy et al. 2014).

Chromium is linked to the regulation of glucose homeostasis in the body. Experimental evidence has shown that chromium deficiency results in hyperglycemia and glucose intolerance (Vincent 2000). However, because the symptoms of chromium deficiency are similar to metabolic syndrome, this has led to conflicting reports on chromium's role in glucose control (Vincent 2000). More recently, multiple clinical trials have demonstrated that chromium supplementation, especially with chromium picolinate, improves glycemic control and hemoglobin A1c (HbA1c) levels in people with T2D (Suksomboon et al. 2014). Hence, the protective role of dietary chromium now appears to be well supported.

Another essential nutrient that has been implicated in diabetes is potassium. Potassium modulates the control of blood glucose because potassium-induced cell depolarization results in the secretion of insulin from the pancreatic β-cells (Ekmekcioglu et al. 2016). An increase in the insulin level in the plasma is associated with a corresponding increase in the absorption of potassium into the cells (DeFronzo et al. 1980). Foods such as Landolphia, Bambara nuts, locust beans, cowpea, and baobab fruit are rich in potassium, ranging from 1182 to 2010 mg per 100 g edible portion (Stadlmayr et al. 2010).

EFFECT OF CULTURAL AND ETHNIC DIVERSITY ON FOOD DIVERSITY

According to Smith (2013), sub-Saharan Africa has a rich, diverse, agricultural biodiversity that is part of local traditional food systems, and this biodiversity should contribute to food and livelihood security. A wide range of locally available food resources are included in the daily diets of different communities. The wide variety of indigenous food crops in West Africa can be attributed to the region's biological and agricultural diversity which allows for diverse diets and better nutritional security (Smith 2013). In some parts of West Africa, some of the plants listed in Table 4.1 are considered wild food plants. Wild food plants are defined as "all plant resources outside of agricultural areas that are harvested or collected for the purpose of human consumption in forests, savannah and other bush land areas" (Guinand and Lemessa 2001). Harris and Mohammed (2003) reported that wild foods are important sources of diversity, vitamins, and minerals in the diet in rural Africa. Available literature reports about 216 different species of wild foods in the Igbo indigenous community in Southern Nigeria (Bharucha and Pretty 2010). Seven edible wild leaves from the Republic of the Niger were reported to be rich in minerals, amino acids, and fatty acids (Freiberger et al. 1998).

CONSERVATION OF INDIGENOUS CROPS

Many indigenous food plant species with antidiabetic properties are going extinct due to neglect, deforestation, and adoption of Western lifestyles. Since leaves usually regrow after they have been harvested in an appropriate manner, there is a sustainable availability of species from which the leaves are used. However, wild species for which roots, tubers, bark, or whole plants are harvested are at much more risk. Means of preserving them include encouraging sustainable harvest methods and bringing them into cultivation. In addition, there are possible methods of regenerating plants from their leaves that serve as a conservation technique (Li et al. 2013).

The work being done by the Global Environment Facility, a non-governmental organization being funded by donor governments and private funders, is helping West Africa, especially countries like Benin, Ghana, Guinea Bissau, Burkina Faso, and Liberia. This organization is helping communities by training them on best practices for conservation of plant natural resources, especially those with medicinal value, and reintroduction of drought-resistant indigenous crops with more nutritional value than the readily available commercial crops (Global Environmental Facility 2010). The ability of West African communities to revive good practices (and attach some value to them) in the traditional management of indigenous resources with the help of the various governments within these countries will help in increasing the knowledge of the usefulness of these food plant species.

FUTURE DIRECTIONS

This chapter has revealed that an abundant number of dietary food plants eaten in West Africa have the potential to be used in the prevention and treatment of diabetes. More awareness of the benefits of these foods needs to be generated within the region to reduce the prevalence and mortality rates of diabetes and to prevent its incidence as part of an integrated approach to the management of diabetes. Awareness of the potential for these foods is also relevant for drug discovery research. West African cultural and food diversity present an important resource to tap into for the discovery of new drugs and functional foods. Research should prioritize the development of innovative evidence-based products which appeal to the modern palate.

REFERENCES

Aba PE, Asuzu IU, Odo RI (2014) Antihyperglycaemic and antioxidant potentials of *Cussonia arborea* in alloxan-induced diabetic rats. *Comp Clin Path* 23(2):451–458.

Abo KA, Fred-Jaiyesimi AA, Jaiyesimi AEA (2008) Ethnobotanical studies of medicinal plants used in the management of diabetes mellitus in South Western Nigeria. *J Ethnopharmacol* 115:67–71.

Addy ME, Nyarko AK (1988) Diabetic patients' response to oral administration of aqueous extract of *Indigofera arrecta*. *Phytother Res* 2:192–195.

Adebajo AC, Iwalewa EO, Obuotor EM et al. (2009) Pharmacological properties of the extract and some isolated compounds of *Clausena lansium* stem bark: Anti-trichomonal, antidiabetic, anti-inflammatory, hepatoprotective and antioxidant effects. *J Ethnopharmacol* 122:10–19.

Adefegha SA, Oboh G (2012) Inhibition of key enzymes linked to type 2 diabetes and sodium nitroprusside-induced lipid peroxidation in rat pancreas by water extractable phytochemicals from some tropical spices. *Pharm Biol* 50:857–865.

Adeneye AA (2008) Hypoglycemic and hypolipidemic effects of methanol seed extract of *Citrus paradisi* Macfad (Rutaceae) in alloxan-induced diabetic Wistar rats. *Nig Q J Hosp Med* 18:211–215.

Adeneye AA, Agbaje E (2007) Hypoglycemic and hypolipidemic effects of fresh leaf aqueous extract of *Cymbopogon citratus* Stapf. in rats. *J Ethnopharmacol* 112:440–444.

Adeneye AA, Ajagbonna OP Ayodele OW (2007) Hypoglycemic and antidiabetic activities on the stem bark aqueous and ethanol extracts of *Musanga cecropioides* in normal and alloxan-induced diabetic rats. *Fitoterapia* 78:502–505.

Aderibigbe AO, Emudianughe TS, Lawal B (1999) Antihyperglycaemic effect of *Mangifera indica* in rat. *Phytother Res* 13:504–507.

Adesokan A, Akanji MA, Adewara GS (2010) Evaluation of hypoglycaemic efficacy of aqueous seed extract of *Aframomum melegueta* in alloxan-induced diabetic rats. *Sierra Leone J Biomed Res* 2:91–94.

Adewale OB, Oloyede OI (2012) Hypoglycemic activity of aqueous extract of the bark of *Bridelia ferruginea* in normal and alloxan-induced diabetic rats. *Prime Res Biotechnol* 2(4):53–56.

Adeyemi DO, Komolafe OA, Adewole OS, Obuotor EM, Adenowo TK (2009) Anti hyperglycemic activities of *Annona muricata* (Linn). *Afr J Tradit Complement Altern Med* 6(1):62–69.

Adika OK, Madubunyi II, Asuzu IU (2012) Antidiabetic and antioxidant effects of the methanol extract of *Bridelia micrantha* (Hochst) Baill. (Euphorbiaceae) leaves on alloxan-induced diabetic albino mice. *Comp Clin Path* 21:945–951.

Afkhami-Ardekani M, Shojaoddiny-Ardekani A (2007) Effect of vitamin C on blood glucose, serum lipids & serum insulin in type 2 diabetes patients. *Indian J Med Res* 126(5):471–474.

Aguh B, Nock I, Ndams I, Agunu A, Ukwubile C (2013) Hypoglycaemic activity and nephroprotectective effect of *Bauhinia rufescens* in alloxan-induced diabetic rats. *Int J Adv Pharm Biol Chem* 2:249–255.

Agunbiade OS, Ojezele OM, Ojezele JO, Ajayi AY (2012) Hypoglycaemic activity of *Commelina africana* and *Ageratum conyzoides* in relation to their mineral composition. *Afr Health Sci* 12:198–203.

Ahmed I, Adeghate E, Cummings E, Sharma AK, Singh J (2004) Beneficial effects and mechanism of action of *Momordica charantia* juice in the treatment of streptozotocin-induced diabetes mellitus in rats. *Mol Cell Biochem* 261:63–70.

Aja PM, Nwafor EJ, Ibiam AU, Orji OU, Ezeani N, Nwali BU (2013) Evaluation of antidiabetic and liver enzymes activity of aqueous extracts of *Moringa oleifera* and *Bridelia ferruginea* leaves in alloxan induced diabetic albino rats. *Int J Biochem Res Rev* 3(3):248–258.

Ajayi GO, Igboekwe NA (2013) Evaluation of anti-diabetic potential of the leaves of Musanga cecropioides R. Brown. *Planta Med* 79:PE7. https://doi.org/10.1055/s-0033-1352026.

Ajiboye TO, Raji HO, Adeleye AO et al. (2015) *Hibiscus sabdariffa* calyx palliates insulin resistance, hyperglycemia, dyslipidemia and oxidative rout in fructose-induced metabolic syndrome rats. *J Sci Food Agric* 96(5):1522–1531. https://doi.org/10.1002/jsfa.7254.

Akah PA, Okafor CL (1992) Blood sugar lowering effect of *Vernonia amygdalina* Del, in an experimental rabbit model. *Phytother Res* 6:171–173.

Akhtar MS, Qureshi AQ, Iqbal J (1990) Antidiabetic evaluation of *Mucuna pruriens*, Linn seeds. *J Pakistan Med Assoc* 40:147–150.

Akolade JO, Usman LA, Okereke OE, Muhammad NO (2014) Antidiabetic potentials of essential oil extracted from the leaves of *Hoslundia opposita* Vahl. *J Med Food* 17:1122–1128.

Akomas SC, Okafor AI, Ijioma SN (2014) Glucose level, haematological parameters and lipid profile in *Ficus sur* treated diabetic rats. *Compr J Agric Biol Sci* 2(1):5–11.

Akpan EJ, Okokon JE, Offong W (2012) Antidiabetic and hypolipidemic activities of ethanolic leaf extract and fractions of *Melanthera scandens*. *Asian Pac J Trop Biomed* 2:523–527.

Al-Ahdab MA (2015) Anti-hyperglycemic effect of *Tamarindus indica* extract in streptozotocin-induced diabetes in male rats. *World Appl Sci J* 33(12):1940–1948.

Al-Attar AM (2010) Physiological study on the effect of *Acalypha wilkesiana* leaves extract on streptozotocin-induced experimental diabetes in male mice. *Am Med J* 1(1):51–58.

Andallu B, Radhika B, Suryakantham V (2003) Effect of aswagandha, ginger and mulberry on hyperglycemia and hyperlipidemia. *Plant Food Hum Nutr* 58(3):1–7.

Arroyo J, Martínez J, Ronceros G et al. (2009) Efecto hipoglicemiante coadyuvante del extracto etanólico de hojas de *Annona muricata* L (guanábana), en pacientes con diabetes tipo 2 bajo tratamiento de glibenclamida. *An Fac Med* 70:163–167.

Ashraduzzaman M, Alam MA, Khatun S, Banu S, Absar N (2011) *Vigna unguiculata* (L.) Walp. seed oil exhibiting anti-diabetic effects in alloxan-induced diabetic rats. *Malaysian J Pharm Sci* 9: 13–23.

Asuzu IU, Nwaehujor CO (2013) The anti-diabetic, hypolipidemic and anti-oxidant activities of D-3-O-methylchiroinositol in alloxan-induced diabetic rats. *Hygeia J Drug Med* 5(2):27–33.

Atangwho IJ, Yin KB, Umar MI, Ahmad M, Asmawi MZ (2014) *Vernonia amygdalina* simultaneously suppresses gluconeogenesis and potentiates glucose oxidation via the pentose phosphate pathway in streptozotocin-induced diabetic rats. *BMC Complement Altern Med* 14:426. https://doi.org/10.1186/1472-6882-14-426.

Atawodi SEO, Yakubu OE, Liman ML, Iliemene DU (2014) Effect of methanolic extract of *Tetrapleura tetraptera* (Schum and Thonn) Taub leaves on hyperglycemia and indices of diabetic complications in alloxan-induced diabetic rats. *Asian Pac J Trop Biomed* 4:272–278.

Awede B, Houetchegnon P, Djego JG, Djrolo F, Gbenou J, Laleye A (2015) Effects of *Lophira Lanceolata* and of three species of *Gardenia* leaves aqueous extracts on blood glucose and lipids in Wistar rat. *J Phys Pharm Adv* 5(10):757–765.

Ayu SA, Samsul BS, Fattepur S, Halijah H (2012) Antidiabetic activity of ethanolic extract of *Imperata cylindrical* [sic] (lalang) leaves in alloxan induced diabetic rats. *Arch Pharm Pract* 3(1):46.

Baldé NM, Youla A, Baldé MD et al. (2006) Herbal medicine and treatment of diabetes in Africa: An example from Guinea. *Diabetes Metab* 32:171–175.

Barbagallo M, Dominguez LJ (2015) Magnesium and type 2 diabetes. *World J Diabetes* 6(10):1152–1157. https://doi.org/10.4239/wjd.v6.i10.1152.

Barbagallo M, Dominguez LJ, Galioto A et al. (2003) Role of magnesium in insulin action, diabetes and cardio-metabolic syndrome X. *Mol Aspects Med* 24(1–3):39–52.

Bhargav B, Rupal AV, Reddy AS, Narasimhacharya AVRL (2009) Antihyperglycemic and hypolipidemic effects of *Adansonia digitata* L. on alloxan induced diabetic rats. *J Cell Tissue Res* 9:1879–1882.

Bharti SK, Kumar A, Prakash O, Krishnan S, Gupta AK (2013) Essential oil of *Cymbopogon citratus* against diabetes: Validation by in vivo experiments and computational studies. *J Bioanal Biomed* 5:194–203.

Bharucha Z, Pretty J (2010) The roles and values of wild foods in agricultural systems. *Philos Trans R Soc Lond B Biol Sci* 365(1554):2913–2926. https://doi.org/10.1098/rstb.2010.0123.

Bhaskar A, Vidhya VG, Ramya M (2008) Hypoglycemic effect of *Mucuna pruriens* seed extract on normal and streptozotocin-diabetic rats. *Fitoterapia* 79:539–543.

Bilbis LS, Shehu RA, Abubakar MG (2002) Hypoglycemic and hypolipidemic effects of aqueous extract of *Arachis hypogaea* in normal and Alloxan-induced diabetic rats. *Phytomedicine* 9:553–555.

Boukari I, Shier NW, Fernandez RXE et al. (2001) Calcium analysis of selected Western African foods. *J Food Compos Anal* 14(1):37–42.

Bvenura C, Sivakumar D (2017) The role of wild fruits and vegetables in delivering a balanced and healthy diet. *Food Res Int* 99(1):15–30.

Cahill F, Shahidi M, Shea J, et al. (2013) High dietary magnesium intake is associated with low insulin resistance in the Newfoundland population. *PLoS ONE* 8(3):e58278. https://doi.org/10.1371/journal.pone.0058278.

Campos KE, Diniz YS, Cataneo AC, Faine LA, Alves MJQF, Novelli ELB (2003) Hypoglycaemic and antioxidant effects of onion, *Allium cepa*: Dietary onion addition, antioxidant activity and hypoglycaemic effects on diabetic rats. *Int J Food Sci Nutr* 54:241–246.

Carter P, Gray LJ, Troughton J, Khunti K, Davies MJ (2010) Fruit and vegetable intake and incidence of type 2 diabetes mellitus: Systematic review and meta-analysis. *BMJ* 341:c4229. https://doi.org/10.1136/bmj.c4229.

Casanova LM, Da Silva D, Sola-Penna M et al. (2014) Identification of chicoric acid as a hypoglycemic agent from *Ocimum gratissimum* leaf extract in a biomonitoring in vivo study. *Fitoterapia* 93:132–141.

Chai JW, Lim SL, Kanthimathi MS, Kuppusamy UR (2011) Gene regulation in β-sitosterol-mediated stimulation of adipogenesis, glucose uptake, and lipid mobilization in rat primary adipocytes. *Genes Nutr* 6(2):181–188.

Chen HJ, Chen CN, Sung ML, Wu YC, Ko PL, Tso TK (2013) *Canna indica* L. attenuates high-glucose- and lipopolysaccharide-induced inflammatory mediators in monocyte/macrophage. *J Ethnopharmacol* 148:317–321.

Chen TH, Chen SC, Chan P, Chu YL, Yang HY, Cheng JT (2005) Mechanism of the hypoglycemic effect of stevioside, a glycoside of *Stevia rebaudiana*. *Planta Med* 71:108–113.

Chika A, Bello SO (2010) Antihyperglycaemic activity of aqueous leaf extract of *Combretum micranthum* (Combretaceae) in normal and alloxan-induced diabetic rats. *J Ethnopharmacol* 129:34–37.

Chuengsamarn S, Rattanamongkolgul S, Luechapudiporn R, Phisalaphong C, Jirawatnotai S (2012) Curcumin extract for prevention of type 2 diabetes. *Diabetes Care* 35:2121–2127.

Conforti F, Statti GA, Tundis R, Loizzo MR, Menichini F (2007) In vitro activities of *Citrus medica* L. cv. Diamante (Diamante citron) relevant to treatment of diabetes and Alzheimer's disease. *Phytother Res* 21:427–433.

Coulibaly FA (2014) Evaluation of the antidiabetic activity of the extracts of Vitellaria paradoxa in Oryctolaguscuniculus rabbit (Lagomorph). *Experiment* 24(3):1673–1682.

Daniel M (2006) *Medicinal Plants: Chemistry and Properties*. Science Publishers, Enfield, NH.

Dare CA, Onwumelu RN, Oyedapo OO (2014) Studies on the effect of polyphenol of cocoa (*Theobroma cacao* L.) seeds on specific carbohydrate-degrading enzymes in streptozotocin-induced diabetic rats. *Greener J Biochem Biotech* 1(1):23–34.

DeFronzo RA, Felig P, Ferrannini E, Wahren J (1980) Effect of graded doses of insulin on splanchnic and peripheral potassium metabolism in man. *Am J Physiol* 238:E421–E427.

Dhalwal K, Shinde VM, Singh B, Mahadik KR (2010) Hypoglycemic and hypolipidemic effect of *Sida rhombifolia* ssp. *retusa* in diabetic induced animals. *Int J Phytomed* 2:160–165.

Diallo A, Traore MS, Keita SM et al. (2012) Management of diabetes in Guinean traditional medicine: An ethnobotanical investigation in the coastal lowlands. *J Ethnopharmacol* 144:353–361.

Dieye AM, Sarr A, Diop SN et al. (2008) Medicinal plants and the treatment of diabetes in Senegal: Survey with patients. *Fundam Clin Pharmacol* 22:211–216.

Dimo T, Ngueguim FT, Kamtchouing P, Dongo E, Tan PV (2006) Glucose lowering efficacy of the aqueous stem bark extract of *Trema orientalis* (Linn) Blume in normal and streptozotocin diabetic rats. *Pharmazie* 61:233–236.

Duze BN, Sewani-Rusike CR, Nkeh-Chungag BN (2012) Effects of an ethanolic extract of *Garcinia kola* on glucose and lipid levels in streptozotocin induced diabetic rats. *Afr J Biotechnol* 11(33):8309–8315.

Ebong PE, Igile GO, Mgbeje BIA et al. (2014) Hypoglycemic, hepatoprotective and nephroprotective effects of methanolic leaf extract of *Heinsia crinita* (Rubiaceae) in alloxan-induced diabetic albino Wistar rats. *IOSR J Pharm* 4(1):37–43.

Eddouks M, Maghrani M, Lemhadri A, Ouahidi ML, Jouad H (2002) Ethnopharmacological survey of medicinal plants used for the treatment of diabetes mellitus, hypertension and cardiac diseases in the south-east region of Morocco (Tafilalet). *J Ethnopharmacol* 82:97–103.

Egua MO, Etuk EO, Bello SO, Hassan SW (2014) Antidiabetic potential of liquid-liquid partition fractions of ethanolic seed extract of *Corchorus olitorius*. *J Pharmacognosy Phytother* 6(1):4–9.

Egunyomi A, Gbadamosi IT, Animashahun MO (2011) Hypoglycaemic activity of the ethanol extract of *Ageratum conyzoides* Linn. shoots on alloxan-induced diabetic rats. *J Med Plants Res* 5(22):5347–5350.

Eidi A, Eidi M, Esmaeili E (2006) Antidiabetic effect of garlic (*Allium sativum* L.) in normal and streptozotocin-induced diabetic rats. *Phytomedicine* 13:624–629.

Ekmekcioglu C, Elmadfa I, Meyer AL, Moeslinger T (2016) The role of dietary potassium in hypertension and diabetes. *J Physiol Biochem* 72(1):93–106.

El-Amin M, Virk P, Elobeid MA et al. (2013) Anti-diabetic effect of *Murraya koenigii* (L) and *Olea europaea* (L) leaf extracts on streptozotocin induced diabetic rats. *Pak J Pharm Sci* 26:359–365.

El-Demerdash FM, Yousef MI, El-Naga NIA (2005) Biochemical study on the hypoglycemic effects of onion and garlic in alloxan-induced diabetic rats. *Food Chem Toxicol* 43:57–63.

El-Khateeb AY, Azzaz NAKE, Mahmoud HI (2014) Phytochemical constituents, hypoglycemic and haematological effects of methanolic *Acalypha wilkesiana* leaves extract on streptozotocin-induced diabetic rats. *Eur J Chem* 5(3):430–438.

Emekli-Alturfan E, Kasikci E, Yarat A (2008) Peanut (*Arachis hypogaea*) consumption improves glutathione and HDL-cholesterol levels in experimental diabetes. *Phytother Res* 22:180–184.

Emiloju OC, Chinedu SN (2016) Effect of *Solanum aethiopicum* and *Solanum macrocarpon* fruits on weight gain, blood glucose and liver glycogen of Wistar rats. *World J Nutr Health* 4:1–4.

Emordi JE, Agbaje EO, Oreagba IA, Iribhogbe OI (2015) Preliminary phytochemical screening and evaluation of hypoglycemic properties of the root extract of *Uveria chamae*. *Bangladesh J Pharmacol* 10:326–331.

Erasto P, Van De Venter M, Roux S, Grierson DS, Afolayan AJ (2009) Effect of leaf extracts of *Vernonia amygdalina* on glucose utilization in chang-liver, C2C12 muscle and 3T3-L1 cells. *Pharm Biol* 47:175–181.

Erfani MN, Tabandeh MR, Shahriari A, Soleimani Z (2018) Okra (*Abelmoscus esculentus*) improved islets structure, and down-regulated PPARs gene expression in pancreas of high-fat diet and streptozotocin-induced diabetic rats. *Cell J* 20(1):31–40. http://doi.org/10.22074/cellj.2018.4819.

Eseyin O, Ebong P, Eyong E, Awofisayo O, Agboke A (2010) Effect of *Telfairia occidentalis* on oral glucose tolerance in rats. *Afr J Pharm Pharmacol* 4(6):368–372.

Etuk EU, Bello SO, Isezuo SA, Mohammed BJ (2010) Ethnobotanical survey of medicinal plants used for the treatment of Diabetes mellitus in the north western region of Nigeria. *Asian J Exp Biol Sci* 1:55–59.

Evert AB, Boucher JL, Cypress M et al. (2013) Nutrition therapy recommendations for the management of adults with diabetes. *Diabetes Care* 37(S1):S120–S143.

Eyo JE, Ozougwu JC, Echi PC (2011) Hypoglycaemic effects of *Allium cepa*, *Allium sativum* and *Zingiber officinale* aqueous extracts on alloxan-induced diabetic *Rattus novergicus* [sic]. *Med J Islamic World Acad Sci* 19:121–126.

Ezeigbo II, Asuzu IU (2012) Anti-diabetic activities of the methanol leaf extracts of *Hymenocardia acida* (Tul.) in alloxan-induced diabetic tats. *Afr J Tradit Complement Altern Med* 9(2):204–209.

Ezeigbo II, Gibbons S, Cleasby ME (2015) Extracts of *Hymenocardia acida* ameliorate insulin resistance in skeletal muscle cells. *Endocrine Abstracts* 38:P261. https://doi.org/10.1530/endoabs.38.P261.

Ezeja MI, Omeh YS, Mbagwu C (2014) Antidiabetic potentials of the methanol leaf extract of *Oxytenanthera abyssinica*. *Int J Diabetes Dev Ctries* 34:116–120.

Ezuruike UF, Prieto JM (2014) The use of plants in the traditional management of diabetes in Nigeria: Pharmacological and toxicological considerations. *J Ethnopharmacol* 155:857–924.

Famobuwa OE, Akinlami OO, Agbowuro AA (2015) Oral glucose tolerance test in normal control and glucose induced hyperglycemic rats with *Tetracarpidium conophorum* (Nigerian walnuts). *J Nat Prod Biomed Res* 1(2):48–50.

Florence NT, Benoit MZ, Jonas K et al. (2014) Antidiabetic and antioxidant effects of *Annona muricata* (Annonaceae), aqueous extract on streptozotocin-induced diabetic rats. *J Ethnopharmacol* 151:784–790.

Fred-Jaiyesimi AA, Abo KA (2009) Hypoglycaemic effects of *Parkia biglobosa* (Jacq) Benth seed extract in glucose-loaded and NIDDM rats. *Int J Biol Chem Sci* 3(3): 545–550.

Fred-Jaiyesimi AA, Wilkins MR, Abo KA (2009) Hypoglycaemic and amylase inhibitory activities of leaves of *Spondias mombin* Linn. *Afr J Med Med Sci* 38:343–349.

Freiberger, CE, Vanderjagt DJ, Pastuszyn A et al. (1998) Nutrient content of the edible leaves of seven wild plants from Niger. *Plant Foods Hum. Nutr.*, 53: 57–69.

Gbolade AA (2009) Inventory of antidiabetic plants in selected districts of Lagos State, Nigeria. *J Ethnopharmacol* 121:135–139.

Ghosh G, Subudhi BB, Mishra SK (2011) Antihyperglycemic activity of root bark of *Polyalthia longifolia* var. *pendula* and aerial parts of *Sida rhombifolia* Linn. and its relationship with antioxidant property. *Asian J Chem* 23(1):141–144.

Global Environmental Facility (2010) GEF's Programmatic approach to biodiversity conservation in West and Central Africa. https://www.thegef.org/sites/default/files/publications/west-africa-BIO_0.pdf.

Glombitza KW, Mahran GH, Mirhom YW, Michel KG, Motawi TK (1994) Hypoglycemic and antihyperglycemic effects of *Zizyphus spina-christi* in rats. *Planta Med* 60:244–247.

Gomez EC (2015) 113: The effects of corn (*Zea mays*) in the dietary management of patients with Type 2 diabetes mellitus. *BMJ Open* 5(S1):113.

Gondi M, Basha SA, Bhaskar JJ, Salimath PC, Rao UJ (2015) Anti-diabetic effect of dietary mango (*Mangifera indica* L.) peel in streptozotocin-induced diabetic rats. *J Sci Food Agric* 95:991–999.

Guinand Y, Lemessa D (2001) Wild-food plants in Ethiopia: Reflections on the role of "wild foods" and "famine foods" at a time of drought. In: Kenyatta C, Henderson A (eds) *The Potential of Indigenous Wild Foods*. USAID/OFDA, Southern Sudan, p. 32–46.

Gwarzo MY, Bako HY (2013) Hypoglycemic activity of methanolic fruit pulp extract of *Adansonia digitata* on blood glucose levels of alloxan induced diabetic rats. *Int J Anim Vet Adv* 5(3):108–113.

Haque AKMM, Das AK, Bashar SS, Al-Mahamud R, Rahmatullah M (2015) Analgesic and antihyperglycemic activity evaluation of *Bambusa vulgaris* aerial parts. *J Appl Pharm Sci* 5(9):127–130.

Harinantenaina L, Tanaka M, Takaoka S et al. (2006) *Momordica charantia* constituents and antidiabetic screening of the isolated major compounds. *Chem Pharm Bull* 54:1017–1021.

Harris FMA, Mohammed S 2003. Relying on nature: wild foods in northern Nigeria. *AMBIO: A Journal of the Human Environment* 32(1):24–29.

Heber D (2004) Vegetables, fruits and phytoestrogens in the prevention of diseases. *J Postgrad Med* 50:145–149.

Hossain MS, Sokeng S, Shoeb M et al. (2012) Hypoglycemic effect of Irvingia gabonensis (Aubry-Lacomate Ex. Ororke), Baill in type 2 diabetic Long-Evans rats. *Dhaka Univ J Pharm Sci* 11(1):19–24.

Hui H, Tang G, Go VLW (2009) Hypoglycemic herbs and their action mechanisms. *Chin Med* 4:11.

Ibrahim MA, Koorbanally NA, Islam MS (2014) Antioxidative activity and inhibition of key enzymes linked to type-2 diabetes (α-glucosidase and α-amylase) by *Khaya senegalensis*. *Acta Pharm* 64(3):311–324.

Ijaola TO, Osunkiyesi AA, Taiwo AA et al. (2014) Antidiabetic effect of *Ipomoea batatas* in normal and alloxan induced diabetic rats. *IOSR J Appl Chem* 7(5):16–25.

Iliemene UD, Shekins OO, Eze ED, Liman LM (2014) Phytochemical constituents and antidiabetic property of *Cola nitida* seeds on alloxan induced diabetes mellitus in rats. *Br J Pharm Res* 4(23):2631–2641.

International Diabetes Federation (2017) *Diabetes Atlas*, 8th edition. International Diabetes Federation, Brussels, Belgium. http://www.diabetesatlas.org.

Iwalewa EO, Adewale IA, Taiwo BJ et al. (2008) Effects of *Harungana madagascariensis* stem bark extract on the antioxidant markers in alloxan induced diabetic and carrageenan induced inflammatory disorders in rats. *J Complement Integr Med* 5(1):2. https://doi.org/10.2202/1553-3840.1088.

Iwueke AV, Nwodo OFC (2008) Antihyperglycaemic effect of aqueous extract of *Daniella oliveri* and *Sarcocephalus latifolius* roots on key carbohydrate metabolic enzymes and glycogen in experimental diabetes. *Biokemistri* 20(2):63–70.

Iwuji S, Nwafor A, Egwurugwu J, Chikezie H (2014) Antihyperglycaemic efficacy of *Cnidoscolus aconitifolius* compared with glibenclamide in alloxan-induced diabetic Wistar rats. *Int Res J Med Sci* 2(3):1–4.

Joseph B, Jini D (2013) Antidiabetic effects of *Momordica charantia* (bitter melon) and its medicinal potency. *Asian Pac J Trop Dis* 3(2):93–102. https://doi.org/10.1016/S2222-1808(13)60052-3.

Joy EJM, Ander EL, Young SD et al. (2014) Dietary mineral supplies in Africa. *Physiol Plant* 151(3):208–229. https://doi.org/10.1111/ppl.12144.

Joy EJM, Young SD, Black CR, Ander EL, Watts MJ, Broadley MR (2013) Risk of dietary magnesium deficiency is low in most African countries based on food supply data. *Plant Soil* 368:129–137.

Juárez-Rojop IE, Díaz-Zagoya JC, Ble-Castillo JL et al. (2012) Hypoglycemic effect of *Carica papaya* leaves in streptozotocin-induced diabetic rats. *BMC Complement Alternat Med* 12:1–11.

Kabir AU, Samad MB, D'Costa NM, Akhter F, Ahmed A, Hannan JMA (2014) Antihyperglycemic activity of *Centella asiatica* is partly mediated by carbohydrase inhibition and glucose-fiber binding. *BMC Complement Alternat Med* 14:31.

Kalman DS, Schwartz HI, Feldman S, Krieger DR (2013) Efficacy and safety of *Elaeis guineensis* and *Ficus deltoidea* leaf extracts in adults with pre-diabetes. *Nutr J* 12:1–7.

Kao W, Folsom AR, Nieto F, Mo J, Watson RL, Brancati FL (1999) Serum and dietary magnesium and the risk for type 2 diabetes mellitus: The atherosclerosis risk in communities study. *Arch Intern Med* 159:2151–2159.

Karou SD, Tchacondo T, Djikpo Tchibozo MA et al. (2011) Ethnobotanical study of medicinal plants used in the management of diabetes mellitus and hypertension in the Central Region of Togo. *Pharm Biol* 49:1286–1297.

Kavishankar GB, Lakshmidevi N, Murthy SM, Prakash HS, Niranjana SR (2011) Diabetes and medicinal plants – A review. *Int J Pharm Biomed Sci* 2(3):65–80.

Kazeem MI, Davies TC (2016) Hypoglycaemic potential of leaf extracts of *Lonchocarpus cyanescens* (Schum. and Thonn) Benth. *Trans R Soc S Afr* 71:1–6.

Keller AJ, Ma J, Kavalier A, He K, Brillantes AMB, Kennelly EJ (2011) Saponins from the traditional medicinal plant *Momordica charantia* stimulate insulin secretion in vitro. *Phytomedicine* 19:32–37.

Khan BA, Abraham A, Leelamma S (1995) Hypoglycemic action of *Murraya koenigii* (curry leaf) and *Brassica juncea* (mustard): Mechanism of action. *Indian J Biochem Biophys* 32:106–108.

Khatun H, Rahman A, Biswas M, Islam AU (2011) Water-soluble fraction of *Abelmoschus esculentus* L interacts with glucose and metformin hydrochloride and alters their absorption kinetics after coadministration in rats. *ISRN Pharm* 2011:260537. https://doi.org/10.5402/2011/260537.

Khosla P, Bhanwra S, Singh J, Seth S, Srivastava RK (2000) A study of hypoglycaemic effects of *Azadirachta indica* (neem) in normal and alloxan diabetic rabbits. *Indian J Physiol Pharmacol* 44:69–74.

Kim Y, Keogh JB, Clifton PM (2016) Polyphenols and glycemic control. *Nutrients* 8(1):17. https://doi.org/10.3390/nu8010017.

Kolawole OM, Oladoyinbo SO, Agbede OO, Adu FD (2006) The effect of *Bridelia ferruginea* and *Senna alata* on plasma glucose concentration in normoglycemic and glucose induced hyperglycemic rats. *Ethnobot Leaflets* 10:209–218.

Kolawole OT, Kolawole SO, Ayankunle AA, Olaniran OI (2012) Anti-hyperglycemic effect of *Khaya senegalensis* stem bark aqueous extract in wistar rats. *Eur J Med Pl* 2:66–73.

Konate K, Yomalan K, Sytar O, Zerbo P, Brestic M (2014) Free radicals scavenging capacity, antidiabetic and antihypertensive activities of flavonoid-rich fractions from leaves of *Trichilia emetica* and *Opilia amentacea* in an animal model of type 2 diabetes mellitus. *Evid Based Complement Alternat Med* 2014:867075.

Krishnaveni M, Mirunalini S, Karthishwaran K, Dhamodharan G (2010) Antidiabetic and antihyperlipidemic properties of *Phyllanthus emblica* Linn. (Euphorbiaceae) on streptozotocin induced diabetic rats. *Pak J Nutr* 9(1):43–51.

Kuate D, Kengne AP, Biapa CP, Azantsa BG, Wan Muda WAMB (2015) *Tetrapleura tetraptera* spice attenuates high-carbohydrate, high-fat diet-induced obese and type 2 diabetic rats with metabolic syndrome features. *Lipids Health Dis* 14:50. http://doi.org/10.1186/s12944-015-0051-0.

Kuroda M, Mimaki Y, Nishiyama T et al. (2005) Hypoglycemic effects of turmeric (*Curcuma longa* L. rhizomes) on genetically diabetic KK-Ay mice. *Biol Pharm Bull* 28:937–939.

Labadarios, D, Steyn NP, Nel J (2011). How diverse is the diet of adult South Africans? *Nutrition Journal* 10(33).

Ladeji O, Omekarah I, Solomon M (2003) Hypoglycemic properties of aqueous bark extract of *Ceiba pentandra* in streptozotocin-induced diabetic rats. *J Ethnopharmacol* 84:139–142.

Lakshmi V, Agarwal SK, Ansari JA, Mahdi AA, Srivastava AK (2014) Antidiabetic potential of *Musa paradisiaca* in streptozotocin-induced diabetic rats. *J Phytopharmacol* 3(2):77–81.

Lawin IF, Lalèyè FOA, Agbani OP et al. (2015) Ethnobotanical assessment of the plant species used in the treatment of diabetes in the Sudano-Guinean zone of Benin. *J Anim Plant Sci* 26(3):4108–4123.

Li Q, Deng M, Zhang J et al. (2013) Shoot organogenesis and plant regeneration from leaf explants of *Lysionotus serratus* D. Don. *Sci World J* 2013:280384. https://doi.org/10.1155/2013/280384.

Lima CR, Vasconcelos CF, Costa-Silva JH et al. (2012) Anti-diabetic activity of extract from *Persea americana* Mill. leaf via the activation of protein kinase B (PKB/Akt) in streptozotocin-induced diabetic rats. *J Ethnopharmacol* 141:517–525.

López-Varela S, González-Gross M, Marcos A (2002) Functional foods and the immune system: A review. *Eur J Clin Nutr* 56:S29–S33.

Ludvik B, Neuffer B, Pacini G (2004) Efficacy of *Ipomoea batatas* (Caiapo) on diabetes control in type 2 diabetic subjects treated with diet. *Diabetes Care* 27:436–440.

Majekodunmi SO, Oyagbemi AA, Umukoro S, Odeku OA (2011) Evaluation of the anti-diabetic properties of *Mucuna pruriens* seed extract. *Asian Pac J Trop Med* 4:632–636.

Maniyar Y, Bhixavatimath P (2012) Antihyperglycemic and hypolipidemic activities of aqueous extract of *Carica papaya* Linn. leaves in alloxan-induced diabetic rats. *J Ayurveda Integr Med* 3:70–74.

Mann A, Roheem FO, Saidu AN, Yisa J, Fadipe LA, Ogbadoyi EO (2014) Assessment of phytoconstituents and antidiabetic activity of the crude extract and partitioned fractions of *Maytenus senegalensis* (Lam.) Exell (Celastraceae) root bark. *Int Res J Pure Appl Chem* 4(6): 46–761.

Marrero-Faz E, Sanchez-Calero J, Young L, Harvey A (2014) Inhibitory effect of *Persea americana* Mill leaf aqueous extract and its fractions on PTP1B as therapeutic target for type 2 diabetes. *Bol Latinoam Caribe Plant Med Aromat* 13(2):144–151.

Maundu P, Achigan-Dako E, Morimoto Y (2009) Biodiversity of African vegetables. In: Shackleton CM, Pasquini MW, Drescher AW (eds) *African Indigenous Vegetables in Urban Agriculture*. Earthscan, London, UK, p. 65–104.

Maundu P, Kabuye CHS, Ngugi GW (1999) *Traditional Food Plants of Kenya*. KENRIK, National Museum of Kenya, Nairobi, Kenya.

Mbaka GO, Adeyemi OO, Adesina SA (2010) Anti-diabetic activity of the seed extract of *Sphenocentrum jollyanum* and morphological changes on pancreatic beta cells in alloxan-induced diabetic rabbits. *J Med Med Sci* 1(11):550–556.

Mbaka GO, Adeyemi OO, Ogbonnia SO, Noronha CC, Okanlawon OA (2009) The protective role of *Sphenocentrum jollyanum* (Pierre) root ethanol extract on alloxan -induced diabetic rabbits. *Planta Med* 75:PH47.

Mbaka GO, Ogbonnia SO, Banjo AE (2012) Activity of *Raphia hookeri* root extract on blood glucose, lipid profile and glycosylated haemoglobin on alloxan induced diabetic rats. *J Morphol Sci* 29(3):214–222.

Mentreddy SR (2007) Medicinal plant species with potential antidiabetic properties. *J Sci Food Agric* 87:743–750.

Micha R, Khatibzadeh S, Shi P on behalf of the Global Burden of Diseases Nutrition and Chronic Diseases Expert Group (NutriCoDE), et al. (2015) Global, regional and national consumption of major food groups in 1990 and 2010: A systematic analysis including 266 country-specific nutrition surveys worldwide. *BMJ Open* 5:e008705. http://doi.org/10.1136/bmjopen-2015-008705.

Mirmiran P, Bahadoran Z, Azizi F (2014) Functional foods-based diets a novel dietary approach for management of type 2 diabetes and its complications: A review. *World J Diabetes* 5(3):267–281.

Mishra A, Garg GP (2011) Antidiabetic activity of fruit pulp of *Cordia dichotoma* in alloxan induced diabetic rats. *Int J Pharm Sci Res* 2(9):2314–2319.

Mishra MR, Mishra A, Pradhan DK, Panda AK, Behera RK, Jha S (2013) Antidiabetic and antioxidant activity of *Scoparia dulcis* Linn. *Indian J Pharm Sci* 75:610–614.

Mishra SB, Vijayakumar M, Ojha SK, Verma A (2010) Antidiabetic effect of *Jatropha curcas* L. leaves extract in normal and alloxan induced diabetic rats. *Int J Pharm Sci* 2(1):482–487.

Mohammadi J, Naik PR (2008) Evaluation of hypoglycemic effect of *Morus alba* in an animal model. *Indian J Pharmacol* 40:15–18.

Mohammed A, Ibrahim MA, Islam MS (2014) African medicinal plants with antidiabetic potentials: A review. *Planta Med* 80:354–377.

Mohammed A, Koorbanally NA, Islam MS (2016) Anti-diabetic effect of *Xylopia aethiopica* (Dunal) A. Rich. (Annonaceae) fruit acetone fraction in a type 2 diabetes model of rats. *J Ethnopharmacol* 180:131–139.

Mohammed RK, Ibrahim S, Atawodi SE et al. (2013) Anti-diabetic and haematological effects of n-butanol fraction of *Alchornea cordifolia* leaf extract in streptozotocin-induced diabetic wistar rats. *Sci J Biol Sci* 2(3):45–53.

Mohan CG, Viswanatha GL, Savinay G, Rajendra CE, Halemani PD (2013) 1,2,3,4,6 Penta-O-galloyl-β-d-glucose, a bioactivity guided isolated compound from *Mangifera indica* inhibits 11β-HSD-1 and ameliorates high fat diet-induced diabetes in C57BL/6 mice. *Phytomedicine* 20:417–426.

Monago CC, Alumanah EO (2005) Antidiabetic effect of chloroform-methanol extract of *Abrus precatorius* Linn seed in alloxan diabetic rabbit. *J Appl Sci Environ Manage* 9(1):85–88.

Monago C, Anacletus FC, Nwauche TK (2016) Hypoglycaemic activity of the aqueous extract of *Costus afer* stems alone and in combination with metformin. *FASEB J* 30:1101.3.

Montonen J, Knekt P, Harkanen T et al. (2005) Dietary patterns and the incidence of type 2 diabetes. *Am J Epidemiol* 161:219–27.

Mozaffari-Khosravi H, Jalali-Khanabadi BA, Afkhami-Ardekani M, Fatehi F (2009) Effects of sour tea (*Hibiscus sabdariffa*) on lipid profile and lipoproteins in patients with type II diabetes. *J Alternat Complement Med* 15(8):899–903.

Mudra M, Ercan-Fang N, Zhong L, Furne J, Levitt M (2007) Influence of mulberry leaf extract on the blood glucose and breath hydrogen response to ingestion of 75 g sucrose by type 2 diabetic and control subjects. *Diabetes Care* 30:1272–1274.

Mvitu Muaka M, Longo-Mbenza B, Tulomba Mona D, Nge Okwe A (2010) Reduced risk of metabolic syndrome due to regular intake of vegetables rich in antioxidants among African type 2 diabetics. *Diabetes and Metab Syndr Clin Res Rev* 4(3):132–136.

Nagappa AN, Thakurdesai PA, Venkat Rao N, Singh J (2003) Antidiabetic activity of *Terminalia catappa* Linn fruits. *J Ethnopharmacol* 88:45–50.

Nammi S, Sreemantula S, Roufogalis BD (2009) Protective effects of ethanolic extract of *Zingiber officinale* rhizome on the development of metabolic syndrome in high-fat diet-fed rats. *Basic Clin Pharmacol Toxicol* 104:366–373.

Nasim SA, Dhir B (2013) Reduction in serum glucose with garlic extracts. In: Watson RR, Preedy VR (eds) *Bioactive Food as Dietary Interventions for Diabetes*. Academic Press, San Diego, CA, p. 97–109.

Nayak BS, Marshall JR, Isitor G, Adogwa A (2011) Hypoglycemic and hepatoprotective activity of fermented fruit juice of *Morinda citrifolia* (noni) in diabetic rats. *Evid Based Complement Alternat Med* 2011:875293. https://doi.org/10.1155/2011/875293.

Ndiaye M, Diatta W, Sy AN, Dieye AM, Faye B, Bassene E (2008) Antidiabetic properties of aqueous barks extract of *Parinari excelsa* in alloxan-induced diabetic rats. *Fitoterapia* 79:267–270.

Nettle D (1996) Language diversity in West Africa: an ecological approach. *Journal of Anthropological Archaeology* 15(4): 403–438.

N'Guessan K, Aké-Assi E, Adou YC, Traoré D (2008) Effet de l'extrait aqueux des feuilles de *Crescentia cujete* sur la glycémie de lapins diabétiques. *Revue Méd Pharm Afr* 21:89–96.

N'Guessan K, Fofie NGBY, Zirihi GN (2011) Effect of aqueous extract of *Terminalia catappa* leaves on the glycaemia of rabbits. *J Appl Pharm Sci* 1(8):59–64.

N'Guessan K, Kouassi K, Kouadio K (2009a) Ethnobotanical study of plants used to treat diabetes in traditional medicine, by Abbey and Krobou people of Agboville, Cote-d'Ivoire. *Am J Sci Res* 4:45–58.

N'Guessan K, Amoikon KE, Tiebre MS, Kadja B, Zirihi GN (2009b) Effect of aqueous extract of *Chrysophyllum cainito* leaves on the glycaemia of diabetic rabbits. *Afr J Pharm Pharmacol* 3(10):501–506.

Niyonzima G, Scharpé S, Van Beeck L, Vlietinck AJ, Laekeman GM, Mets T (1993) Hypoglycaemic activity of *Spathodea campanulata* stem bark decoction in mice. *Phytotherapy Res* 7:64–67.

Njamen D, Nkeh-Chungag BN, Tsala E, Fomum ZT, Mbanya JC, Ngufor GF (2012) Effect of *Bridelia ferruginea* (Euphorbiaceae) leaf extract on sucrose-induced glucose intolerance in rats. *Trop J Pharm Res* 11(5):759–765.

Nwaneri-Chidozie VO, Yakubu OE, Jatto OS, Paul P, Lele KC (2014) Lipid profile status of streptozotocin-induced diabetic rats treated with ethanol, n-hexane and aqueous extracts of *Vitex doniana* leaves. *Res J Pharm Biol Chem Sci* 5(2):40–49.

Nwanjo HU (2008) Hypoglycemic and hypolipidemic effects of aqueous extracts of *Abrus precatorius* Linn seeds in streptozotocin-induced diabetic Wistar rats. *J Herbs Spices Med Plants* 14:68–76.

Nyarko AK, Sittie AA, Addy ME (1993) The basis for the antihyperglycemic activity of *Indigofera arrecta* in the rat. *Phytotherapy Res* 7:1–4.

Nyunaï N, Manguelle-Dicoum A, Njifutié N, Abdennebi E, Gérard C (2009) Antihyperglycaemic effect of *Ageratum conyzoides* L. fractions in normoglycemic and diabetic male Wistar rats. *Int J Biomed Pharm Sci* 4(1):38–42.

Nyunaï N, Njikam N, Mounier C, Pastoureau P (2006) Blood glucose lowering effect of aqueous leaf extracts of *Ageratum conyzoides* in rats. *Afr J Tradit Complement Alternat Med* 3(3):76–79.

Obasi NA, Kalu KM, Okorie U, Otuchristian G (2013) Hypoglycemic effects of aqueous and methanolic leaf extracts of *Vitex doniana* on alloxan induced diabetic albino rats. *J Med Sci* 13:700–707.

Oboh G, Nwokocha KE, Akinyemi AJ, Ademiluyi AO (2014) Inhibitory effect of polyphenolic-rich extract from *Cola nitida* (Kolanut) seed on key enzyme linked to type 2 diabetes and Fe(2+) induced lipid peroxidation in rat pancreas in vitro. *Asian Pac J Trop Biomed* 4:S405–S412.

Oche O, Sani I, Chilaka NG, Samuel NU, Samuel A (2014) Pancreatic islet regeneration and some liver biochemical parameters of leaf extracts of *Vitex doniana* in normal and streptozotocin-induced diabetic albino rats. *Asian Pac J Trop Biomed* 4:124–130.

Odo RI, Asuzu IU, Aba PE (2012) The antidiabetic activities of the methanolic root bark extract of *Afzelia africana* in alloxan-induced diabetic mice. *J Complement Integrat Med* 9(1):1553–3840. https://doi.org/10.1515/1553-3840.1649.

Odoh UE, Ndubuokwu RI, Inyagha SI, Ezejiofor M (2013) Antidiabetic activity and chemical characterization of the *Acalypha wilkesiana* (Euphorbiaceae) Mull Arg. roots in alloxan-induced diabetic rats. *Planta Med* 79:SL10.

Odutuga AA, Dairo JO, Minari JB, Bamisaye FA (2010) Anti-diabetic effect of *Morinda lucida* stem bark extracts on alloxan-induced diabetic rats. *Res J Pharmacol* 4(3): 78–82.

Ogbonnia SO, Mbaka GO, Anyika EN, Ladiju O, Igbokwe HN (2011) Evaluation of anti-diabetics and cardiovascular effects of *Parinari curatellifolia* seed extract and *Anthoclista vogelli* root extract individually and combined on postprandial and alloxan-induced diabetic albino rats. *Br J Med Med Res* 1(3):146–162.

Oguanobi NI, Chijioke CP, Ghasi SI (2012) Effects of aqueous leaf extract of *Ocimum gratissimum* on oral glucose tolerance test in type-2 model diabetic rats. *Afr J Pharm Pharmacol* 6(9):630–635.

Ogunbolude Y, Ajayi MA, Ajagbawa TM, Igbakin AP, Rocha JBT, Kade IJ (2009) Ethanolic extracts of seeds of *Parinari curatellifolia* exhibit potent antioxidant properties: A possible mechanism of its antidiabetic action. *J Pharmacognosy Phytother* 1(6):67–75.

Ojewale AO, Olaniyan OT, Faduyile FA, Odukanmi OA, Oguntola JA, Dare BJ (2014) Testiculo protective effects of ethanolic roots extract of *Pseudocedrela kotschyi* on alloxan induced testicular damage in diabetic rats. *Br J Med Med Res* 4(1):548.

Ojewole JA, Adewunmi CO (2003) Hypoglycemic effect of methanolic extract of *Musa paradisiaca* (Musaceae) green fruits in normal and diabetic mice. *Methods Find Exp Clin Pharmacol* 25:453–456.

Ojewole JA, Adewunmi CO (2004) Anti-inflammatory and hypoglycaemic effects of *Tetrapleura tetraptera* (Taub) [Fabaceae] fruit aqueous extract in rats. *J Ethnopharmacol* 95(2–3):177–182.

Okokon JE, Antia BS, Osuji LC, Udia PM (2007) Antidiabetic and hypolipidemic effects of *Mammea africana* (Guttiferae) in streptozotocin induced diabetic rats. *J Pharmacol Toxicol* 2:278–283.

Okokon JE, Umoh EE, Etim EI, Jackson CL (2009) Antiplasmodial and antidiabetic activities of ethanolic leaf extract of *Heinsia crinata. J Med Food* 12:131–136.

Okolo CE, Akah PA, Uzodinma SU (2012) Antidiabetic activity of the root extract of *Detarium microcarpum* (Fabacaee) Guill and Perr. *Phytopharmacology* 3(1):12–18.

Okpashi VE, Bayim BPR, Obi-Abang M (2014) Comparative effects of some medicinal plants: *Anacardium occidentale*, *Eucalyptus globulus*, *Psidium guajava*, and *Xylopia aethiopica* extracts in alloxan-induced diabetic male Wistar albino rats. *Biochem Res Int* 2014:203051. https://doi.org/10.1155/2014/203051.

Olabanji SO, Omobuwajo OR, Ceccato D, Adebajo AC, Buoso MC, Moschini G (2008) Accelerator-based analytical technique in the study of some anti-diabetic medicinal plants of Nigeria. *Nucl Instrum Methods Phys Res B* 266:2387–2390.

Olaokun OO, McGaw LJ, Eloff JN, Naidoo V (2013) Evaluation of the inhibition of carbohydrate hydrolysing enzymes, antioxidant activity and polyphenolic content of extracts of ten African *Ficus* species (Moraceae) used traditionally to treat diabetes. *BMC Complement Alternat Med* 13:94.

Olatunji LA, Okwusidi JI, Soladoye AO (2005) Antidiabetic effect of *Anacardium occidentale*. Stem-bark in fructose-diabetic rats. *Pharm Biol* 43:589–593.

Olukunle JO, Jacobs EB, Oyewusi JA (2016) Hypoglycemic and hypolipidemic effects of *Acalypha wilkesiana* leaves in alloxan induced diabetic rats. *FASEB J* 30:1269.3.

Omeh YN, Onoja SO, Ezeja MI, Okwor PO (2014) Subacute antidiabetic and in vivo antioxidant effects of methanolic extract of *Bridelia micrantha* (Hochst Baill) leaf on alloxan-induced hyperglycaemic rats. *J Complement Integrat Med* 11(2):99–105. https://doi.org/10.1515/jcim-2013-0067.

Omoniwa BP, Luka CD (2009) Antidiabetic and toxicity evaluation of aqueous stem extract of *Ficus asperifolia* in normal and alloxan-induced diabetic albino rats. *Asian J Exp Biol Sci* 3(4):726–732.

Ong KW, Hsu A, Song L, Huang D, Tan BK (2011) Polyphenols-rich *Vernonia amygdalina* shows anti-diabetic effects in streptozotocin-induced diabetic rats. *J Ethnopharmacol* 133:598–607.

Onoagbe IO, Negbenebor EO, Ogbeide VO, Omonkhua AA (2010) A study of the anti-diabetic effects of *Urena lobata* and *Sphenostylis stenocarpa* in streptozotocin-induced diabetic rats. *Eur J Sci Res* 43(1):6–14.

Onyemairo NJ, Patrick-Iwuanyanwu KC, Monago CC (2015) Evaluation of the anti-diabetic properties of *Manniophyton fulvum*. *J Biol Genet Res* 1(8):23–30.

Opajobi AO, Esume CO, Campbell P, Onyesom I, Osasuyi A (2011) Effects of aqueous extracts of *Rauvolfia vomitoria* and *Citrus aurantium* on liver enzymes of streptozotocin-induced diabetic and normal rabbits. *Cont J Med Res* 5:1–5.

Osadolor HB, Ariyo II, Emokpae MA, Anukam KC (2011) Hypoglycemic effects of unripe pawpaw on streptozotocin induced diabetic albino rats. *Res J Med Plants* 5:90–94.

Oyagbemi AA, Odetola AA, Azeez OI (2010) Antidiabetic properties of ethanolic extract of *Cnidoscolus aconitifolius* on alloxan induced diabetes mellitus in rats. *Afr J Med Med Sci* 39(Suppl):171–178.

Oyedemi S, Adewusi E, Aiyegoro O, Akinpelu D (2011) Antidiabetic and haematological effect of aqueous extract of stem bark of *Afzelia africana* (Smith) on streptozotocin–induced diabetic Wistar rats. *Asian Pac J Trop Biomed* 1:353–358.

Oyelola OO, Moody JO, Odeniyi MA, Fakeye TO (2007) Hypoglycemic effect of *Treculia africana* Decne root bark in normal and alloxan-induced diabetic rats. *Afr J Tradit Complement Alternat Med* 4:387–391.

Pandian RS (2013) Functional foods in managing diabetes. *Int J Pharm Bio Sci* 4(2): B572–B579.

Paolisso G, Scheen A, Cozzolino D et al. (1994) Changes in glucose turnover parameters and improvement of glucose oxidation after 4-week magnesium administration in elderly noninsulin-dependent (type II) diabetic patients. *J Clin Endocrinol Metab* 78:1510–1514.

Perera PK, Li Y (2012) Functional herbal food ingredients used in type 2 diabetes mellitus. *Pharmacognosy Rev* 6(11):37–45.

Platel K, Srinivasan K (1997) Plant foods in the management of diabetes mellitus: Vegetables as potential hypoglycaemic agents. *Nahrung* 41:68–74.

Priyadarsini SS, Vadivu R, Jayshree N (2010) In vitro and in vivo antidiabetic activity of the leaves of *Ravenala madagascariensis* Sonn. in alloxan induced diabetic rats. *J Pharm Sci Tech* 2(9):312–317.

Prohp TP, Onoagbe IO (2009) Anti-diabetic properties and toxicological studies of *Triplochiton scleroxylon* on the heart enzymes in normal and streptozotocin-induced diabetic rabbits. *Pak J Nutr* 8(7):1025–1029.

Prohp TP, Onoagbe IO (2011) Anti-diabetic studies of aqueous extract of *Triplochiton scleroxylon* on platelets and associated parameters in alloxan-induced diabetic rabbits. *Afr J Plant Sci* 5(12):697–701.

Prohp TP, Onoagbe IO (2012) Effects of aqueous extract of *Triplochiton scleroxylon* on some haematological parameters and blood glucose concentrations in non-diabetic rabbits. *Afr J Biotechnol* 11(1):198–202.

Purintrapiba J, Suttajit M, Forsberg NE (2006) Differential activation of glucose transport in cultured muscle cells by polyphenolic compounds from *Canna indica* L. root. *Biol Pharm Bull* 29:1995–1998.

Qureshi SA, Asad W, Sultana V (2009). The effect of *Phyllantus emblica* Linn. on Type-II diabetes, triglycerides and liver-specific enzyme. *Pak J Nutr* 8(2):125–128.

Rahman MA, Sultana R, Rasheda Akter R, Islam MS (2013) Antidiarrheal and antidiabetic effect of ethanol extract of whole *Ageratum conyzoides* L. in albino rat model. *Afr J Pharm Pharmacol* 7(23):1537–1545.

Rani MP, Krishna MS, Padmakumari KP, Raghu KG, Sundaresan A (2012) *Zingiber officinale* extract exhibits antidiabetic potential via modulating glucose uptake, protein glycation and inhibiting adipocyte differentiation: An in vitro study. *J Sci Food Agric* 92:1948–1955.

Rodriguez-Moran M, Guerrero-Romero F (2003) Oral magnesium supplementation improves insulin sensitivity and metabolic control in type 2 diabetic subjects: A randomized double-blind controlled trial. *Diabetes Care* 26:1147–1152.

Rotimi SO, Olayiwola I, Ademuyiwa O, Adamson I (2010) Inability of legumes to reverse diabetic-induced nephropathy in rats despite improvement in blood glucose and antioxidant status. *J Med Food* 13:163–169.

Rudkowska I (2009) Functional foods for health: Focus on diabetes. *Maturitas* 62:263–269.

Ruel MT, Minot N, Smith L (2005) Patterns and determinants of fruit and vegetable consumption in sub-Saharan Africa: A multi-country comparison. Background paper for the Joint FAO/WHO Workshop on Fruit and Vegetables for Health, 1–3 September 2004, Kobe, Japan. World Health Organization, Geneva, Switzerland.

Ruzaidi A, Amin I, Nawalyah AG, Hamid M, Faizul HA (2005) The effect of Malaysian cocoa extract on glucose levels and lipid profiles in diabetic rats. *J Ethnopharmacol* 98:55–60.

Ruzaidi AMM, Abbe MMJ, Amin I, Nawalyah AG, Muhajir H (2008) Protective effect of polyphenol-rich extract prepared from Malaysian cocoa (*Theobroma cacao*) on glucose levels and lipid profiles in streptozotocin-induced diabetic rats. *J Sci Food Agric* 88:1442–1447.

Sah AN, Joshi A, Juyal V, Kumar T (2011) Antidiabetic and hypolipidemic activity of *Citrus medica* Linn. seed extract in streptozotocin induced diabetic rats. *Pharmacognosy J* 3:80–84.

Sasidharan S, Sumathi V, Jegathambigai NR, Latha LY (2011) Antihyperglycaemic effects of ethanol extracts of *Carica papaya* and *Pandanus amaryfollius* leaf in streptozotocin-induced diabetic mice. *Nat Prod Res* 25:1982–1987.

Schlernitzauer A, Oiry C, Hamad R et al. (2013) Chicoric acid is an antioxidant molecule that stimulates AMP kinase pathway in L6 myotubes and extends lifespan in *Caenorhabditis elegans*. *PLoS ONE* 8(11):e78788. https://doi.org/10.1371/journal.pone.0078788.

Senthilkumar MK, Sivakumar P, Changanakkattil F, Rajesh V, Perumal P (2011) Evaluation of anti-diabetic activity of *Bambusa vulgaris* leaves in streptozotocin induced diabetic rats. *Int J Pharm Sci Drug Res* 3(3):208–210.

Setacci C, De Donato G, Setacci F, Chisci E (2009) Diabetic patients: Epidemiology and global impact. *J Cardiovasc Surg* 50(3):263.

Sharif F, Hamid M, Ismail A, Adam Z (2015) Antihyperglycemic activity of oil palm *Elaeis guineensis* fruit extract on streptozotocin-induced diabetic rats. *Jurnal Sains Kesihatan Malaysia* 13(2):37–43.

Sharma S, Choudhary M, Bhardwaj S, Choudhary N, Rana A (2014) Hypoglycemic potential of alcoholic root extract of *Cassia occidentalis* Linn. in streptozotocin induced diabetes in albino mice. *Bull Fac Pharm Cairo Univ* 52:211–217.

Shidfar F, Rajab A, Rahideh T, Khandouzi N, Hosseini S, Shidfar S (2015) The effect of ginger (*Zingiber officinale*) on glycemic markers in patients with type 2 diabetes. *J Complement Integrat Med* 12:165–170.

Shittu H (2009) Biological and chemical characterization of potential antidiabetic principles from selected Nigerian medicinal plants. PhD Thesis submitted to the Department of Pharmaceutical Sciences, University of Strathclyde, Glasgow.

Shivashankara AR, Haniadka R, Sandhya P, Palatty PL, Baliga MS (2013) Ginger (*Zingiber officinale* Roscoe) in the treatment of diabetes and metabolic syndrome: Preclinical observations. In: Watson RR, Preedy VR (eds) *Bioactive Food as Dietary Interventions for Diabetes*. Academic Press, San Diego, CA, p. 571–582.

Siddaiah M, Jayaveera KN, Souris K, Krishna JPY, Kumar PV (2011) Phytochemical screening and anti-diabetic activity of methanolic extract of leaves of *Ximenia americana* in rats. *Int J Innov Pharm Res* 2:78–83.

Siegfried NL, Hughes G (2012) Herbal medicine, randomised controlled trials and global core competencies. *S Afr Med J* 102:912–913.

Silambujanaki P, Chitra V, Soni D, Raju D, Sankari M (2009) Hypoglycemic activity of *Stachytarpheta indica* on streptozotocin induced wistar strain rats. *Int J PharmTech Res* 1(4):1564–1567.

Singab ANB, El-Beshbishy HA, Yonekawa M, Nomura T, Fukai T (2005). Hypoglycemic effect of Egyptian *Morus alba* root bark extract: Effect on diabetes and lipid peroxidation of streptozotocin-induced diabetic rats. *J Ethnopharmacol* 100:333–338.

Sittie AA, Nyarko AK (1998) *Indigofera arrecta*: Safety evaluation of an antidiabetic plant extract in non-diabetic human volunteers. *Phytotherapy Res* 12:52–54.

Smith FI, Eyzaguirre P (2007) African leafy vegetables: Their role in the World Health Organization's global fruit and vegetables initiative. *Afr J Food Agric Nutr Dev* 7(3–4):1–8.

Smith IF (2013) Sustained and integrated promotion of local, traditional food systems for nutrition security. In: Fanzo J, Hunter D, Borelli T, Mattei F (eds) *Diversifying Food and Diets: Using Agricultural Biodiversity to Improve Nutrition and Health.* Routledge, Abingdon and New York, pp. 122–139.

Smith R (2004) "Let food be thy medicine..." *BMJ* 328(7433):0.

Soman S, Rajamanickam C, Rauf AA, Indira M (2013) Beneficial effects of *Psidium guajava* leaf extract on diabetic myocardium. *Exp Toxicol Pathol* 65:91–95.

Song MJ, Lee SM, Kim DK (2011) Antidiabetic effect of *Chenopodium ambrosioides*. *Phytopharmacology* 1(2):12–15.

Srinivasan K (2005) Plant foods in the management of diabetes mellitus: Spices as beneficial antidiabetic food adjuncts. *Int J Food Sci Nutr* 56(6):399–414.

Stadlmayr B, Charrondière UR, Addy P et al. (eds) (2010) *Composition of Selected Foods from West Africa.* Food and Agriculture Organization of the United Nations, Rome, Italy.

Steyn NP, Mann J, Bennett PH et al. (2004) Diet, nutrition and the prevention of type 2 diabetes. *Public Health Nutr* 7:147–165.

Suanarunsawat T, Klongpanichapak S, Rungseesantivanon S, Chaiyabutr N (2004) Glycemic effect of stevioside and *Stevia rebaudiana* in streptozotocin-induced diabetic rats. *Eastern J Med* 9:51–56.

Suárez A, Pulido N, Casla A, Casanova B, Arrieta FJ, Rovira A (1995) Impaired tyrosine-kinase activity of muscle insulin receptors from hypomagnesaemic rats. *Diabetologia* 38:1262–1270.

Subramanian SP, Bhuvaneshwari S, Prasath GS (2011) Antidiabetic and antioxidant potentials of *Euphorbia hirta* leaves extract studied in streptozotocin-induced experimental diabetes in rats. *Gen Physiol Biophys* 30:278–285.

Suksomboon N, Poolsup N, Yuwanakorn A (2014) Systematic review and meta-analysis of the efficacy and safety of chromium supplementation in diabetes. *J Clin Pharm Ther* 39(3):292–306.

Suzuki R, Okada Y, Okuyama T (2005) The favorable effect of style of *Zea mays* L. on streptozotocin induced diabetic nephropathy. *Biol Pharm Bull* 28:919–920.

Taj Eldin IM, Ahmed EM, Elwahab HMA (2010) Preliminary study of the clinical hypoglycemic effects of *Allium cepa* (red onion) in type 1 and type 2 diabetic patients. *Envir Health Insights* 4:71–77.

Takahashi K, Kamada C, Yoshimura H et al. (2012) Effects of total and green vegetable intakes on glycated hemoglobin A1c and triglycerides in elderly patients with type 2 diabetes mellitus: The Japanese Elderly Intervention Trial. *Geriatr Gerontol Int* 12(Suppl 1):50–58.

Tanko Y, Yerima M, Mahdi MA, Yaro AH, Musa KY, Mohammed A (2008) Hypoglycemic activity of methanolic stem bark of *Adansonnia digitata* extract on blood glucose levels of streptozocin-induced diabetic Wistar rats. *Int J Appl Res Nat Prod* 1:32–36.

Tchacondo T, Karou SD, Agban A et al. (2012) Medicinal plants use in central Togo (Africa) with an emphasis on the timing. *Pharmacognosy Res* 4:92–103.

Tchamadeu MC, Dzeufiet PD, Nouga CC et al. (2010) Hypoglycaemic effects of *Mammea africana* (Guttiferae) in diabetic rats. *J Ethnopharmacol* 127:368–372.

Tchamgoue AD, Tchokouaha LRY, Tarkang PA, Kuiate JR, Agbor GA (2015) *Costus afer* possesses carbohydrate hydrolyzing enzymes inhibitory activity and antioxidant capacity in vitro. *Evid Based Complement Alternat Med* 2015:987984. https://doi.org/10.1155/2015/987984.

Teugwa CM, Boudjeko T, Tchinda BT, Mejiato PC, Zofou D (2013) Anti-hyperglycaemic globulins from selected Cucurbitaceae seeds used as antidiabetic medicinal plants in Africa. *BMC Complement Alternat Med* 13:63. https://doi.org/10.1186/1472-6882-13-63.

Thomford AK, Thomford KP, Ayertey F et al. (2015) The ethanolic leaf extract of *Alchornea cordifolia* (Schum. & Thonn.) Muell. Arg inhibits the development of dyslipidaemia and hyperglycaemia in dexamethasone-induced diabetic rats. *J App Pharm Sci* 5(9):52–55.

Ubaka CM, Ukwe CV (2010) Antidiabetic effect of the methanolic seed extract of *Sphenostylis stenocarpa* (Hoechst ex. A. Rich. Harms) in rats. *J Pharm Res* 3:2192–2194.

Udenta E, Obizoba I, Oguntibeju O (2014) Anti-diabetic effects of Nigerian indigenous plant foods/diets. In: Oguntibejo O (ed.) *Antioxidant-Antidiabetic Agents and Human Health*. London, IntechOpen (pp. 9–93). https://doi.org/10.5772/57240.

Udenze ECC, Braide VB, Okwesilieze CN, Akuodor GC (2012) Pharmacological effects of *Garcinia kola* seed powder on blood sugar, lipid profile and atherogenic index of alloxan induced diabetes in rats. *Pharmacologia* 3(12):693–699.

Umar IA, Mohammed A, Ndidi US, Abdulazeez AB, Olisa WC, Adam M (2014) Anti-hyperglycemic and anti-hyperlipidemic effects of aqueous stem bark extract of *Acacia albida* Delile. in alloxan-induced diabetic rats. *Asian J Biochem* 9(4):170–178.

Umar ZU, Moh'd A, Tanko Y (2013) Effects of ethanol leaf extract of *Ficus glumosa* on fasting blood glucose and serum lipid profile in diabetic rats. *Niger J Physiol Sci* 28:99–104.

Usharani P, Fatima N, Muralidhar N (2013) Effects of *Phyllanthus emblica* extract on endothelial dysfunction and biomarkers of oxidative stress in patients with type 2 diabetes mellitus: A randomized, double-blind, controlled study. *Diabetes Metab Syndr Obes* 6:275–284.

Verma L, Singour PK, Chaurasiya PK, Rajak H, Pawar RS, Patil UK (2010) Effect of ethanolic extract of *Cassia occidentalis* Linn. for the management of alloxan-induced diabetic rats. *Pharmacognosy Res* 2:132–137.

Veronese N, Watutantrige-Fernando S, Luchini C, et al. (2016) Effect of magnesium supplementation on glucose metabolism in people with or at risk of diabetes: A systematic review and meta-analysis of double-blind randomized controlled trials. *Eur J Clin Nutr* 70(12):1354–1359.

Villegas R, Gao YT, Dai Q et al. (2009) Dietary calcium and magnesium intakes and the risk of type 2 diabetes: The Shanghai Women's Health Study. *Am J Clin Nutr* 89(4):1059–1067.

Vinayagam R, Xu B (2015) Antidiabetic properties of dietary flavonoids: A cellular mechanism review. *Nutr Metab* 12:60.

Vincent JB (2000) Elucidating a biological role for chromium at a molecular level. *Acc Chem Res* 33(7):503–510.

Waheed A, Miana GA, Ahmad SI (2006) Clinical investigation of hypoglycemic effect of leaves of *Mangifera inidca* in type-2 (NIDDM) diabetes mellitus. *Pak J Pharmacol* 23(2):13–18.

Wang PY, Fang JC, Gao ZH, Zhang C, Xie SY (2016) Higher intake of fruits, vegetables or their fiber reduces the risk of type 2 diabetes: A meta-analysis. *J Diabetes Investig* 7:56–69.

Willcox M, Siegfried N, Johnson Q (2012) Capacity for clinical research on herbal medicines in Africa. *J Alternat Complement Med* 18:622–628.

Worku T (2009) Investigation of antihyperglycemic and hypoglycemic activity of Ajuga remota and Syzygium guineense on mice. Masters' dissertation submitted to the School of Graduate Studies of Addis Ababa, Addis Ababa University, Addis Ababa.

World Health Organization (2011) *Global Status Report on Non-Communicable Diseases: 2010*. World Health Organization, Geneva, Switzerland.

Zhang DW, Fu M, Gao SH, Liu JL (2013) Curcumin and diabetes: A systematic review. *Evid Based Complement Alternat Med* 2013:636053. https://doi.org/10.1155/2013/636053.

Zhao R, Li Q, Long L, Li J, Yang R, Gao D (2007) Antidiabetic activity of flavone from *Ipomoea batatas* leaf in non-insulin dependent diabetic rats. *Int J Food Sci Tech* 42:80–85.

CHAPTER 5

Traditional and Local Knowledge Systems in the Caribbean: Jamaica as a Case Study

David Picking and Ina Vandebroek

CONTENTS

ABSTRACT

In Jamaica, like elsewhere in the Caribbean, traditional and local knowledge systems represent a blend of knowledge, worldviews, beliefs, skills, and practices derived from Amerindian, European, African, and other cultural groups. Through a process called creolization, new and unique knowledge systems were developed around Old and New World plant species, and this creative process still plays an important role in the island's culture today. The historical movement of people and plants offers a solid basis to better understand Jamaica's present-day applications of traditional knowledge in areas as diverse as agriculture and diet, folk or traditional medicine, development of the national nutraceutical industry and regulation of Intellectual Property Rights (IPR), consumption of bush teas and fermented beverages called root tonics, and plant conservation. Jamaica's rich multicultural knowledge systems consist of intertwined elements of biological, cultural, and linguistic diversity that merit deeper understanding, protection, and sustainable development for the benefit of local communities across the island. In order to better appreciate the multiple dimensions of these biocultural knowledge systems, counteract their loss, and recognize their value, equitable and dynamic collaborations are needed across multiple sectors, including social and natural scientists, policy and decision makers, the private sector, educators, and local communities. Continued documentation and valuation of these systems are urgently needed, especially in the face of challenges such as climate change and loss of biological diversity that disproportionally affect island nations.

INTRODUCTION

Traditional knowledge (TK) and local knowledge (LK) systems have been defined as "dynamic and complex bodies of know-how, practices and skills that are developed and sustained by peoples/communities with shared histories and experiences" (Beckford and Barker 2007). These systems represent the way of life, traditions, and nature of subsistence of communities, and form the basis for their decision-making in relation to healthcare, food security, plant management, and other needs. They are dynamic and holistic; unique to a particular culture and environment, and often associated with rural communities; usually have been developed informally and transmitted verbally; and remain vastly under-documented around the globe (Vandebroek et al. 2011). TK and LK systems can vary in scale from the village level to the nation and beyond. In this chapter, we use the terms TK and LK to represent these knowledge systems across Jamaica.

HISTORICAL CONTEXT

The national motto "Out of Many, One People," aptly represents the blend and influence of Jamaica's multiracial past and the many peoples that have influenced, and continue to influence, all areas of Jamaican culture and national identity, including dance, music, religion, language, folklore, agriculture, and cuisine (JIS 2017a). In order to understand contemporary TK and LK systems in Jamaica, it is important to review the cultural history of the island, representing Amerindian (Taíno), European, and African groups, as well as Asian immigrants. Some aspects of Jamaican TK and LK are common to the Caribbean through a unique process known as creolization, in which Old and New World cultures came together and coexisted through colonization, forced transportation, and slavery (Senior 2003, Hall 2015). The process of creolization makes tracing the exact origins of Jamaican TK and LK difficult, which is further complicated by the fact that most of the earliest records were documented by Europeans and written, invariably, from a colonial perspective (Senior 2003).

Indigenous Knowledge of the Earliest Documented Inhabitants of Jamaica

Table 5.1 summarizes key periods and dates in Jamaican history. The earliest recorded inhabitants of the West Indies, the Lithic and Archaic groups, represented hunter-gatherer communities who relied on marine resources and native flora and fauna. The Saladoid inhabitants, a pre-Columbian indigenous culture, represented larger, more settled agricultural groups whose subsistence appears similar to that of

Table 5.1 Key Dates – Mapping Jamaican History (Hart 1985, CBAJ 2017)

Key Dates	Description
4000–500 BC	Lithic and Archaic stages – hunter-gatherer communities
500 BC–AD 600	Saladoid stage – larger more settled agricultural groups
AD 600–1492	Post Saladoid stage - Taíno
1494–1550	European contact period – significant destruction of native population
1509 onwards	Spanish settlement. Taíno population estimated to be 60,000
1517	Spanish forced introduction of African peoples to Jamaica
1611	Census: 74 Taíno, 523 Spaniards, 558 enslaved Africans, 107 free Africans
1655–1660	Spanish and English fight for possession. 1,500 people of African descent reported
1660	Spanish withdrawal. Reports of self-supporting villages in rural areas populated by free and escaped Africans, and Taíno
1660–1739	1st Maroon war with the English
1670s	Commercial production of sugar and heavy influx of enslaved Africans from the Gold Coast
1731	Census: 7,648 whites (including indentured workers), 74,525 black slaves, 865 free blacks
1739	1st Maroon Treaty signed between English and Leeward and Windward Maroons
1795	2nd Maroon war between English and Trelawny Town Maroons
1796	2nd Maroon Treaty signed, but the Treaty is broken when 500 Trelawny Town Maroons are shipped to Halifax, Canada, and subsequently to Sierra Leone
1834	Emancipation and system of apprenticeship introduced
1838	Full freedom from slavery granted
1845–1917	36,000 Indian nationals arrived as indentured workers
1854–1884	1,000 Chinese nationals arrived as indentured workers
1925	Chinese population grows to 3,366
1962	Jamaica independence from the United Kingdom

South American tropical forest communities, with evidence pointing to their migration from the American mainland (Hart 1985, Wilson 1997).

The later Taíno populations are identified in the post-Saladoid stage as a people who developed economies that utilized cultivated crops and land and ocean resources, and who were well suited to Jamaica and the other islands of the Greater Antilles (Dominican Republic, Haiti, Puerto Rico, Jamaica, and Cuba). Across the Greater Antilles, the Taíno populations shared a number of common cultural practices related to agriculture, social organization, and religion. They constructed thatched buildings in large permanent villages, cultivated cassava, weaved hammocks and baskets, used ritual objects such as stools, constructed canoes, used tobacco and hallucinogenic substances, and established a complex system of government with a hierarchy of chiefs. Through their use of long canoes, they are known to have traded with, and adopted cultural practices from, populations in Central America. Subsequently, the islands of the Greater Antilles became highly populated by the Taíno people, and their culture became the most highly developed in the Caribbean (Atkinson 2006, Wilson 1997).

Table 5.2 identifies a list of domesticated plants, i.e. those requiring cultivation, that are thought to have been introduced by the Taíno, or their ancestors, the Saladoid people. These plants are not native, that is, they are not indigenous to Jamaica, because they were brought to Jamaica by humans, albeit a very long time ago, and are therefore considered exotic plants. Cassava was the most important crop to the Taíno and was planted using a hardwood digging stick called a *coa*, in large earth mounds called *conucos*. Other important crops, both cultivated and uncultivated, included beans, maize, peppers, calabash, and cotton. A wide variety of other native plants were also used as foodstuffs and for other purposes (Rashford 1991, Rouse 1992, Vega 1996, Veloz Maggiolo 1997, Wilson 1997, Keegan 2000, Atkinson 2006).

While Taíno farming was originally thought to have been relatively simple (Parry 1955), evidence exists of sophisticated horticultural techniques, such as the

Table 5.2 Flora Introduced to Jamaica by the Taíno (Rashford 1991, Atkinson 2006)

Common Jamaican name	Scientific name
Cassava	*Manihot esculenta* Crantz
Sweet potato	*Ipomoea batatas* (L.) Lam.
Yampie	*Dioscorea trifida* L.f.
Arrowroot	*Maranta arundinacea* L.
Coco	*Xanthosoma sagittifolium* (L.) Schott
Maize, corn	*Zea mays* L.
Pine, pineapple	*Ananas comosus* (L.) Merr.
Peanut	*Arachis hypogaea* L.
Squashes	*Cucurbita maxima* Duchesne
Beans	*Phaseolus vulgaris* L.
Tobacco	*Nicotiana rustica* L.

use of *conucos* or mound farming, irrigation, terracing, crop rotation, and agricultural astrology (Rouse 1992, Wilson 1997, Keegan 2000, Atkinson 2006). The use of *conucos* in the cultivation of cassava provided more room and fertile soil for the growth of the tuber, maintained humidity for longer periods, protected the crop from seasonal variations in rainfall, prevented soil erosion, and improved drainage (Rouse 1992, Wilson 1997, Atkinson 2006). These more sophisticated horticultural techniques represented a shift away from solely slash-and-burn agriculture practiced by earlier settlers and were strongly influenced by methods used in the Amazon region and more suited to large tracts of land, rather than small islands (Veloz Maggiolo 1997, Atkinson 2006). Through these techniques, the Taíno were able to successfully transform much of Jamaica's forests into what has been termed "settlement vegetation" in their response to wild plants and their cultivation of, largely introduced, domesticated plants (Table 5.2). The Spanish, Africans, and British were to later continue these processes (Rashford 1991). It is important to note that the majority of Jamaica's economically valuable plants were introduced, and many of these have become naturalized (Parry 1955, Senior 2003).

The Taíno wore few clothes, but cotton was an important wild crop that was harvested to weave headbands worn by unmarried women, skirts worn by married women, ornamental bands, and decorative belts, and to make hammocks that they slept in (Wilson 1997). It was reported that Jamaican Taíno traded cotton cloth and hammocks with Cuba and Hispaniola and made sails for some Spanish ships (Atkinson 2006).

Important fruit trees of the Taíno included bija, achiote, or annatto (*Bixa orellana* L.); genipap or jagua (*Genipa americana* L.); jicaco, hicaco, or coco-plum (*Chrysobalanus icaco* L.); caimito or star apple (*Chrysophyllum cainito* L.); West Indian plum varieties (*Spondias* spp.); mamey or mamee tree (*Mammea americana* L.); papaya (*Carica papaya* L.); guanabana or soursop (*Annona muricata* L.); guayaba or guava (*Psidium guajava* L.); uva de playa or seaside grape (*Coccoloba uvifera* (L.) L.); and anon or sweetsop (*Annona squamosa* L.) (Vega 1996, Wilson 1997).

Trees that were important for their use in construction included caoba, thought to be West Indian mahogany (*Swietenia mahagoni* (L.) Jacq.), damahagua (*Hibiscus tiliaceus* L.), manaka, and ceiba or silk-cotton tree (*Ceiba pentandra* (L.) Gaertn.), which was considered sacred by the Taíno, guao (*Metopium brownei* (Jacq.) Urb.), and guaiacum or lignum-vitae (*Guaiacum officinale* L.) (Wilson 1997, Rashford 1985). Large hardwood trees were used to make canoes, some of which were reported to transport up to 150 people. Other trees provided utensils, such as the higüero or calabash tree (*Crescentia cujete* L.), whose fruit was used as a bowl, and the cabuya (*Furcraea tuberosa* (Mill.) Aiton), which was harvested for thread (Wilson 1997).

The ritual use of plants was well established in Taíno culture. A snuff, *cohoba*, derived from the crushed seeds of the piptadenia tree (*Anadenanthera peregrina* (L.) Speg.), was used by Taíno chiefs (*caciques*) and shamans (*bohuti*) to induce hallucinogenic experiences and to communicate with the spirit world. It was snuffed through polished wood or cane tubes, from a round wooden table, one such ceremony being observed by Columbus in Jamaica in 1495 (Rouse 1992, Saunders 2006).

An hallucinogenic snuff derived from seeds of the same tree, but called *yopo*, was used widely in pre-Hispanic times in the plains or grassland areas of the Orinoco basin of Venezuela and Colombia, in forests of southern Guiana, and in the Rio Branco area of the northern Amazonia of Brazil (Schultes and Hofmann 1992).

The roles of men and women were segregated but complementary in Taíno culture. Women had power due to the key role they played in food production and their control of the production of high-status objects such as stools, headdresses, and clothing used by *caciques*. Women were key agents for the transmission and preservation of culture through the rearing of children with songs, stories, and belief systems. This became a critical aspect of cultural transmission and maintenance following the arrival of the Spanish in 1492, when marriage between Taíno women and Spanish men became common (Wilson 1997).

European Contact and Spanish Colonization

Pre-Columbus, the worldwide distribution of wild and domesticated plants did not include significant exchanges between the Old World (Africa, Asia, Europe) and the New World (the Americas). Columbus began a biological and cultural exchange between Old and New that effectively doubled the resources of both (Rashford 1991).

Columbus arrived in Jamaica in 1494 and Spanish settlement began in 1509. Contact with the European colonizers exposed the Taíno to new diseases for which they had no immunity, which, together with a brutal system of slavery, caused the decimation of their population. A census in 1509 reported 60,000 Taíno, while a subsequent census, a century later, in 1611, reported 74 (Hart 1985). In the 151 years of settlement (1509–1660), the Spanish oversaw what has been reported as the eradication of the Taíno people and their subsequent replacement by the forced settlement of African slaves. However, the extinction theory is challenged by a number of academics, with claims that groups of Taíno escaped to the mountains of inland Jamaica and intermarried with escaped Africans to form what came to be known as Maroon communities (Agorsah 1994, Forte 2004).

In the period of Spanish settlement, new plants and farming techniques were introduced, by the Spanish themselves and the African peoples they forcibly transported, further contributing to the development of Jamaica's settlement vegetation which was established by the Taíno (Rashford 1991). The Spanish settlers protected and cultivated both new, introduced crops (Table 5.3) and native crops, as well as useful trees and shrubs, such as pimento (*Pimenta dioica* (L.) Merr.) and naseberry (*Manilkara zapota* (L.) P. Royen). They tried unsuccessfully to establish wheat and barley but instead adopted cassava and corn as staples. They introduced a number of crops characteristic of the Mediterranean, but only the citruses became widely established in Jamaica. The introduction of sugar cane was particularly successful. However, it was this success that would lead to the forced introduction of African peoples and the destruction of Jamaica's lowland forests. Sugar cane would become the most important crop in the economic and social history of Jamaica, followed by banana and citruses (Rashford 1991).

Table 5.3 Flora Introduced to Jamaica by the Spanish (Grigg 1974, Rashford 1991)

Common Jamaican name	Scientific name
Sweet orange	*Citrus sinensis* (L.) Osbeck
Sour orange	*Citrus × aurantium* L.
Lemon	*Citrus limon* (L.) Osbeck
Lime	*Citrus* spp.
Sugar cane	*Saccharum officinarum* L.
Plantain	*Musa × paradisiaca* L.
Banana	*Musa* spp.
Pomegranate	*Punica granatum* L.
Ginger	*Zingiber officinale* Roscoe
Indigo	*Indigofera tinctoria* L.
Cocoa	*Theobroma cacao* L.
Avocado	*Persea americana* Mill.
Cho-cho	*Sechium edule* (Jacq.) Sw.
Coconut	*Cocos nucifera* L.

The Africans

Following the rapid decline of the Taíno population, the Spanish began in 1517 to forcibly introduce African peoples to Jamaica, and their exploitation for agriculture and livestock rearing formed the basis of the Spanish occupation for over a century (Rashford 1991). In 1612 and 1613, the total number of documented African slaves who entered Jamaica, under Spanish rule, was 503 (Deason et al. 2012). However, these numbers quickly escalated when the English seized Jamaica in 1655 and developed a plantation economy to supply the growing demand for sugar, transforming the island into a major slave destination. By 1665, many coastal forests had been converted to plantations, to be followed later in the 18th century by the clearance of large tracts of mountain forest for coffee production, representing a rapid transformation to industrial-scale agriculture (Parry 1955, Mistry et al. 2009).

From the middle of the 15th century to the end of the 19th century, the transatlantic slave trade was responsible for the forced migration of 12 to 15 million African men, women, and children to countries in the Western Hemisphere (Carney and Rosomoff 2009, USI 2017). It is estimated that over 650,000 (and up to 927,000) people were transported to Jamaica between 1655 and 1807 (Franco 1978, Deason et al. 2012). However, by 1834, when chattel slavery was abolished, the population reportedly had dropped to 311,070 (Deason et al. 2012).

Modern genealogical research, based on the mitochondrial genome that is inherited entirely through mothers, has shown that the matriline of Jamaica hails almost entirely from West Africa, and thus shows under-representation of Eurasian or Asian/New World matrilines. This result coincides with historical reports of high importation rates of Africans, a small resident population of female Europeans, and the decimation of indigenous communities. Modern Jamaicans share a close genetic similarity with groups from the present day Gold Coast (Ghana), despite importation of high numbers of Africans from the Bight of Biafra (Gulf of Guinea) and West-Central Africa towards the end of the British slave trade (Deason et al. 2012).

In order to feed the rapidly growing African workforce on European plantations, the transatlantic slave trade also saw the widespread transport of African and other Old World plants, and the establishment of *provision gardens* or *grounds*, areas of land where plantation owners allowed slaves to cultivate plants for subsistence (Carney and Rosomoff 2009). Yam, for example, a preferred food of Africans, proved more accessible and sustainable than European cereals. Yams thus became vital to the *provision grounds*. They were less demanding on the soil than cereal crops, and their long growth and low maintenance made them ideal for Africans who often had to travel miles to reach them (Parry 1955, Carney and Rosomoff 2009, Beckford et al. 2011, DeLoughrey 2011). The *provision grounds* were on the borderline of the estates, often in the foothills, but as the number of Africans increased to meet the increasing demand for sugar, the *provision grounds* were pushed further into the hills (Parry 1955). Thus, the use of pantropical plants of value for food, medicine, cordage, and dyes, and for the practice of syncretic religions, played a significant role in the survival of African-Jamaicans. *Provision grounds* were also a critical economic factor in the process of emancipation with the transition to self-sufficient rural communities (Parry 1955).

Many of these plantation workers, coming from different cultural groups in Africa, were known to be skilled in the use of medicinal plants, and there are written reports of both captive and captor turning to skilled African herbalists for treatment during times of disease (Carney 2003, Payne-Jackson and Alleyne 2004). African traditional medicine stood in sharp contrast to the orthodox European medical treatments of that time, the latter including venesection (blood-letting), cupping, blistering, purging, and leeching. While most of these earlier orthodox European practices have disappeared, herbal medicine continues to be widely practiced and used in Jamaica to this day (Carney 2003, Picking 2011).

Many herbalists were also practitioners of the Akan-inspired system of belief, which came to be known as Obeah or Obi. The practitioners of Obeah were seen as a significant threat to the plantation system and the subjugation of Africans. This came to a head with the largest slave uprising, in 1760, Tacky's rebellion, named after its leader, a Gold Coast African who was closely advised by an Obeah priest. Following the rebellion, the practice of Obeah became a crime punishable by death, with the introduction of a comprehensive law intended to prevent further slave rebellions. The original law was amended in 1898 with the introduction of the Obeah Act, with punishment changed from death to imprisonment and flogging. The practice of flogging was abolished in 2012; however, the Obeah Act remains in force today (Bilby 2012, MOJ 2017).

The 18th and 19th centuries saw an influx of European missionaries. This led to widespread conversion to Christianity amongst the African population and dismissal and discouragement of African traditions as "pagan" and "backward," attitudes that still exist in some sections of Jamaican society to this day (Senior 2003).

Table 5.4 identifies some of the plants that are known to have been introduced to Jamaica by Africans, through forced transportation by the Spanish and subsequently

Table 5.4 Flora Introduced to Jamaica by the Africans (Grigg 1974, Rashford 1991)

Common Jamaican name	Scientific name
Introduced during the period of Spanish slavery 1517–1655	
Melon	*Citrullus lanatus* (Thunb.) Matsum. & Nakai
Cow pea	*Vigna unguiculata* (L.) Walp
Yam	*Dioscorea* spp.
Gungo pea	*Cajanus cajan* (L.) Millsp.
Sorrel	*Hibiscus sabdariffa* L.
Tamarind	*Tamarindus indica* L.
Bissy	*Cola acuminata* (P.Beauv.) Schott & Endl.
Castor oil	*Ricinus communis* L.
John Crow bead	*Abrus precatorius* L.
Okra	*Abelmoschus esculentus* (L.) Moench
Introduced from 1655 – after the Spanish left Jamaica	
Ackee	*Blighia sapida* K.D. Koenig
Mango	*Mangifera indica* L.
Bamboo	*Bambusa vulgaris* Schrad.
Dasheen	*Colocasia esculenta* (L.) Schott

the British (Rashford 1991). Different types of yam, gungo peas, melon, and sorrel were just some of the earliest plant introductions, which have gone on to become mainstays of Jamaican cuisine. Bissy, known as an antidote for poisoning, can be found in many Jamaican households today, castor oil is commonly used for skin and hair care, and tamarind continues to be used as a source of food and medicine. John Crow bead, popularly used in jewelry, has a history of use in African religious practices, as a source of poison and as a source of medicine. Ackee has gone on to become the national fruit of Jamaica, and mango the most loved fruit of all Jamaicans (Rashford 1991, Warner 2007).

Other foods that were introduced to feed the rapidly expanding African population included breadfruit (*Artocarpus altilis* (Parkinson) Fosberg) and Otaheite apple (*Syzygium malaccense* (L.) Merr. & L. M. Perry). Both were introduced from Haiti in 1793 by British naval officer Captain Bligh following the infamous naval mutiny that prevented completion of his voyage to Jamaica in 1789 (Parry 1955, Senior 2003, Higman 2008).

Historically, Jamaica has been closely associated with the plant *Cannabis sativa* L. Known locally as ganja, the plant is thought to have been introduced to Jamaica by Africans, its presence as hemp being documented in the New World with the arrival of the Spanish (Rubin and Comitas 1975). However, the timing of its use as a psychoactive plant is not clear, with some authors stating that such use only started after 1845, following emancipation and the introduction of indentured workers from India (Rubin and Comitas 1975, Senior 2003, Higman 2008, Abel 2011). Certainly, the species' use in India is well documented, with ancient Hindu texts detailing its religious and medicinal properties dating back to 2000–1400 BC (Senior 2003,

Beckford and Barker 2007). Other authors point to evidence of its psychoactive use in West Africa, an area from which many Africans were forcibly transported to Jamaica. Here, the species was referred to as "Congo tobacco" and used as a narcotic called *dyamba* in the Kingdom of Congo (West-Central Africa, now northern Angola, Cabinda, Republic of Congo, the western portion of the Democratic Republic of Congo and the southernmost part of Gabon). Additionally, its use as a narcotic, *diamba*, or *maconha*, was widely reported amongst African slaves in Brazil, and it seems likely that this knowledge would have found its way into other New World communities (Senior 2003).

The growing, selling, and consuming of ganja was criminalized in Jamaica in 1913. However, the establishment and rise of the Rastafari movement, an Afro-Jamaican syncretic "roots" movement that claims Africa as its spiritual homeland, from the 1930s in urban Jamaica saw the use of the plant as an integral part of religious practice. Closely associated with reggae, ganja became increasingly popular on the international music scene. Jamaica went on to become a major cultivator and exporter of ganja, to meet the growing international illicit demand, particularly from the United States. Unofficially, ganja became Jamaica's most valuable agricultural export (Senior 2003, Abel 2011).

The Maroons

Maroon identity is based less on cultural or linguistic distinctiveness and more on history, land, and social values (Brandon 2004). The Maroons are best known for successfully fighting for and maintaining their freedom for over 150 years while Jamaica was under the control of British colonial forces (Agorsah 1994). The British tried unsuccessfully to eradicate the Maroons in a conflict that started in 1660 and ended in 1739, with the government conceding that the Maroons could not be defeated and reaching agreements with them instead. Marronage was ideologically based on the notions of self-sufficiency and cultural autonomy, and to this day the Maroons of Jamaica serve as a model for resistance, rebellion, and survival (Brathwaite 1994, Gottlieb 2000).

Jamaican Maroons trace their history to various groups, including explorers, livestock managers, and militia brought to Jamaica from Africa and the Iberian Peninsula by the Spanish in the 1550s (Brandon 2004). They either managed to escape Spanish slavery between 1517 and 1660, or were freed when the Spanish withdrew from the island. Archaeological evidence identifies Maroon settlements in the east that consisted of groups living in the inaccessible areas of the Blue Mountains during the Spanish occupation. Peoples from different backgrounds, speaking different languages and practicing different customs and traditions, successfully integrated. The growth, development, and survival of the Maroons demonstrates a cultural link between Taíno and African societies and the Spanish on one hand, and the English on the other. For example, contemporary fishing and trap-setting techniques and traditional herbal medicines reflect Taíno, African, and European sources (Agorsah 1994).

Maroon leaders possessed a profound understanding of the supernatural and its manipulation for good or evil for the benefit of the community. The practice of

Obeah was referred to by the Maroons as "Science." Nanny, leader of the Windward Maroons, was said to possess the full skills of Science in addition to being celebrated as a great warrior (Campbell 1990).

Division of labor was observed, in which men were predominantly employed in warfare and hunting. Men not proficient as fighters were employed in clearing ground for women to plant and were generally held in low esteem. In what appears to follow African tradition, women planted and cultivated as agriculturists and also raised domestic animals. Crops included plantains, bananas, cocoa, sweet corn, pineapple, cassava, and sugar cane. Salt making was an important occupation. The Windward Maroons reportedly made salt derived from plant ash, an African tradition. They used the ash salt to first cure and then smoke wild hog meat (Campbell 1990).

Agriculturally, the Maroons followed the traditional "slash-and-burn" method, an effective method of clearing land that dates from pre-industrial societies. After a period of cultivation, land was left fallow for a season or two, and often during this period nutritious roots were allowed to grow. The fallow system is deemed necessary in tropical agriculture, which mainly suffers from shallow topsoil and low energy content. To this day, Maroons graze their cattle on communal pastures. Commercial activities increased after the signing of the peace treaties with the British, particularly the cultivation and manufacturing of tobacco. Tobacco, the native plant of the Taíno people, from which they made intoxicating drinks and which they inhaled for hallucinatory spiritual practices, became an important cash crop for export to Europe. The Maroons dried the leaves and twisted them into ropes and subsequently sold them to the estates and others within Jamaica. This kind of tobacco was finely sliced and used in pipes (Campbell 1990).

Maroons also sold the products of their hunting, particularly jerk pork, in the eastern and south-eastern areas of the country. This "original" jerk pork was a spicy, carefully prepared delicacy, barbecued over selected woods. The boar was shot or pierced through with lances, cut open, the bones taken out, the flesh gashed on the inside into the skin and filled with salt, and exposed to the sun, a process referred to as jerking. During periods of salt shortage, Maroons may have substituted spices, including hot chili peppers, for salt (Campbell 1990).

The Indians and Chinese

The TK contributions from Asian groups in Jamaica, who arrived during the colonial period, cannot always be easily discerned, as they have been fairly well assimilated into Jamaican society. An estimated 36,000 East Indians came to British Jamaica as indentured workers between 1845 and 1915 (Table 5.1). Today, the number of Indian Jamaicans is approximately 74,000. In contrast to their relatively small numbers, Indian Jamaicans have had a significant impact on TK. At least 75 botanical species introduced to Jamaica are native to India, although many were already on the island prior to their arrival, brought in by the Europeans and Africans, such as tamarind and mango. However, Indian Jamaicans are credited with the introduction of the Bombay variety of mango. Traditional Indian foods that are a famous part of Jamaican national cuisine include curry goat, roti, and perhaps callaloo

(*Amaranthus viridis* L.), although the consumption of amaranths as leafy vegetables is also popular in West Africa and may be a continuation of this African tradition. Another native Southeast Asian species that features in Jamaican cuisine is turmeric (*Curcuma longa* L.), referred to locally as *tambrik* by African and Indian Jamaicans alike (Cassidy and Le Page 2002, Payne-Jackson and Alleyne 2004).

Between 1854 and 1884, one thousand Chinese indentured workers came to Jamaica. By 1925, the Chinese Jamaican population had grown to 3,366 (Table 5.1). Most of the Chinese who came before 1980 were Hakka people originating from a group of villages within 20 miles of each other, from an area just north of Hong Kong (CBAJ 2017). By 1991, Chinese Jamaicans numbered 5,372. At the end of their contracts, most former indentured workers switched from agriculture to open retail stores, laundries, and restaurants, and later moved into banking and other professions. This move away from working the soil probably explains why there are no reported direct Chinese plant introductions in Jamaica. Instead, plant species of Chinese origin were introduced to Jamaica via other Caribbean countries or Europe (Payne-Jackson and Alleyne 2004, CABJ 2017).

JAMAICAN ETHNOBOTANY IN THE SCHOLARLY LITERATURE

Jamaican ethnobotany has had a long-standing appeal to scholars since the 18th century, as is evident from natural history accounts by early naturalists, physicians, and botanists such as Sloane (1707–1725), Browne (1756), and Long (1774). This appeal has continued into the 19th and 20th centuries (Lunan 1814, Fawcett 1891, Beckwith 1927, Steggerda 1929, Asprey and Thornton 1953, 1954, 1955a, b, Robertson 1982, Barker and Spence 1988), and has culminated in present-day treatises of Jamaica's useful and medicinal plants, as well as papers, reviews, and books on TK associated with agrobiodiversity and community health (Rashford 1984, 1985, 1988, 1989, 1991, 1995, 1997, 2001, Payne-Jackson and Alleyne 2004, Delgoda et al. 2004, 2010, Mitchell and Ahmad 2006, Beckford and Barker 2007, Warner 2007, Higman 2008, Beckford 2009, Harris 2010, Mitchell 2011, Picking et al. 2011, 2015, Sander and Vandebroek 2016, Vandebroek and Picking 2016). However, given Jamaica's rich biological and cultural heritage, and the dynamic response of TK and LK systems to social and environmental change, continued documentation and valuation of these systems is needed and urgent, especially in the face of challenges such as climate change and loss of biological diversity that disproportionally affect island nations.

MODERN APPLICATIONS OF TK AND LK IN JAMAICA

Agriculture and Diet

Jamaica has been described as the garden of the Caribbean, benefitting significantly from both its diverse native flora and rich exotic flora introduced by the Taíno,

Europeans, Africans, Asians, and others, giving rise to a vegetation that is uniquely Jamaican (Rashford 1991). Due to a tropical climate and rich fertile soil, nearly half of Jamaica's land mass is devoted to agriculture (Harris 2012). Even though the agricultural sector (including fisheries and aquaculture) supports 20% of the population and remains one of the most labor-intensive, it represents only around 6% of the country's GDP (Arias et al. 2013, Selvaraju 2013).

Jamaica's agricultural economy is dualistic and consists of large-scale commercial plantations of monocultures for exportation (bananas, citrus, coffee, sugar cane, pimento, and nutmeg), and smallholder farms for household subsistence and the domestic market (Selvaraju 2013). According to the World Bank, in 2007, the total agricultural land area in Jamaica consisted of 326,000 ha, of which more than 60% was allocated to farm activities (crops and pasture). As a legacy of slavery, there exist stark inequalities in the size and land quality of small farms as compared to large plantations. In 2007, the number of small farms (with a size of less than 5 ha) was much higher than that of large plantations (measuring 5 to 200+ ha), representing 86% versus 2% of all farms, with the remaining 12% being worked by landless farmers. However, large plantations dominated the landscape, leaving a mere 2% of landowners in control of 60% of the island's most fertile soil (Arias et al. 2013).

Maroons and other free African-Jamaicans, who became Jamaican subsistence farmers, were forced to develop agriculture in the less fertile hills of the island, giving rise to the name "hillside farming" (Harris 2012) (Figure 5.1). Traditional

Figure 5.1 Hillside farming in present day Jamaica, Portland parish. (Photo credit I. Vandebroek.)

farming techniques have persisted since the time of slavery and play a critical role in maintaining Jamaica's food security. However, negative attitudes continue to exist about the validity and relevance of small-scale farmers and their TK, and they are frequently overlooked by the government, international funding agencies, and research and development programs (Beckford et al. 2007). In stark contrast, increasing climatic challenges, such as drought and flooding, have led to a global resurgence of interest in local solutions (Beckford and Barker 2007, Mistry et al. 2009, Vandebroek et al. 2011).

Currently, Jamaican subsistence farmers continue to use mound farming techniques similar to those of Taíno *conucos* (farm plots) for the cultivation of cassava, yams, and other crops. Likewise, they still produce cassava products, including breads (called bammy) derived from Taíno culture (Atkinson 2006). Other authors have noted that traditional farming methods observed throughout Jamaican history are part of an "ecological inheritance" from African ancestors. This rich body of agricultural knowledge, which was originally developed on the African continent, steadily became incorporated into daily Jamaican farming experiences through centuries of forced agricultural labor within the colonial slave system (Harris 2012).

Until the 1960s, yams (*Dioscorea* spp.) were cultivated for the domestic market and their own consumption by subsistence farmers. Since the 1960s, yams have also become an important export crop. It has been proposed that yam-growing areas of Jamaica can contribute significantly to the country's food security and rural livelihoods through its exploitation as a value-added crop (Beckford et al. 2011).

Subsistence farmers in Jamaica develop "food forests" that consist of intercropping different levels of useful plants, including tall trees such as coconut and breadfruit; bushes such as cocoa; and ground level plants such as dasheen, pumpkin and squashes, and medicinal herbs and spices (e.g. thyme, scallions) (Figure 5.2). These food forests are considered a cultural adaptation of the African practice of kitchen compound farming (Barker and Spence 1988). In Jamaica, the location of the food forest close to the home is called the "yard" or "kitchen garden." In addition, farm plots situated farther away from the house also represent complex food forests. A study in Portland parish, in the northeast of the island, showed the complementary functionality of plant species cultivated, managed, or tolerated in yards versus farm plots. Whereas yards contained a proportionally higher number of medicinal plants that could be extracted for immediate consumption, farm plots contained significantly more timber tree species that served as a saving fund to generate extra income in times of need (Sander and Vandebroek 2016).

The types of foods eaten by Jamaicans, and their attitudes to them, form a critical aspect of national development and identity. Prior to independence in 1962, foods that were seen as quintessentially Jamaican were noticeably absent, and even today no official national dish exists. Informally, ackee and saltfish is seen as a national dish (Figure 5.3), and while ackee is the National Fruit of Jamaica, saltfish has to be imported (Higman 2008). Amy Bailey, educator and social activist in the 1930s, wrote an article in the national newspaper, titled, "Jamaicanising Jamaica: Yams, cocos, plantains, bananas and bush tea." She believed that many "high society" Jamaicans held negative attitudes to locally grown foods, preferring instead to eat

Figure 5.2 Jamaican food forest with different levels of intercropping in Portland parish. (Photo credit I. Vandebroek.)

Figure 5.3 Jamaica's informal national dish, ackee (*Blighia sapida*) and saltfish, served with roasted-then-fried breadfruit (*Artocarpus altilis*) and a glass of sorrel juice (made with the red, fleshy calyces of *Hibiscus sabdariffa*). (Photo credit I. Vandebroek.)

imported, foreign items (Bailey 1939). From the late 19th century, an increasing number of foods thought to be characteristically Jamaican were in fact naturalized or imported, rather than native. However, what often distinguished them as "uniquely Jamaican" was the manner in which they were prepared or combined. This struggle between local and foreign foods continues today, and there have been a number of attempts by the government to promote locally grown foods through public awareness campaigns such as "Eat what you grow and grow what you eat" (Higman 2008). This apparent disconnect between certain levels of Jamaican society and Jamaican foods and agriculture may have its roots in the way in which forced migration and slavery altered people's relationship with the soil. The indignity of forced agricultural labor is unlikely to encourage a desire for a close relationship with the soil, and many Jamaicans today see farming as "backward" and something to fall back on when times are hard (Harrison 2001, DeLoughrey 2011).

Folk Medicine or Traditional Medicine

Following the abolition of slavery in 1834, the majority of the population, many impoverished, continued to rely on their folk medicine heritage, passed on to them by their enslaved, Maroon, and free black ancestors (Carney 2003). Payne-Jackson and Alleyne (2004) have identified two distinct healthcare systems in Jamaica: Western (or biomedicine) and folk medicine (here used interchangeably with traditional medicine). Jamaican folk medicine has evolved into a generalized TK system resulting from the fusion of indigenous (Taíno), African, European, and Asian TKs, although it has been argued that the African influence predominates (Payne-Jackson and Alleyne 2004). The unique formation of Afro-Jamaican folk medicine is best viewed through an understanding of the process of creolization and the historical context of plantation slavery (Laguerre 1987, Hall 2015).

Many similarities are still evident today between Jamaican folk medicine and the different cultural knowledge systems that have influenced it. For example, there exist clear links between modern West African folk medicine and religion, which are reflected in today's Jamaican folk medicine (Payne-Jackson and Alleyne 2004). Obeah, Obi, or "Science" are terms used to refer to practices of traditional African religions still practiced in Jamaica today. However, such practices are generally viewed negatively, often referred to in a pejorative manner as black magic, witchcraft, and evil, and are against the law. The Maroons, amongst others, take pride in retaining many aspects of African religious practices, and some Jamaicans believe that Obeah, in fact, represents the positive and noble application of African spiritual knowledge, used for healing, spiritual protection, and addressing the problems of daily life (Gottlieb 2000, Bilby 2012, MOJ 2017).

The modern application of Jamaican folk medicine takes a number of forms, often as first-line self-medication with medicinal plants and consumption of "bush teas," with 73% of Jamaicans using medicinal plants and foods to maintain health and treat illness (Picking et al. 2011). At least 334 medicinal plants have been recorded in Jamaica to date (Mitchell and Ahmad 2006), but the actual number is likely higher. People may also seek the guidance of a physician to obtain or confirm a diagnosis and subsequently combine medicinal plants and pharmaceutical drugs, either at the same time or consecutively. This combined or concomitant use of drugs and plant medicines is often undertaken without the knowledge of the physician (Payne-Jackson and Alleyne 2004, Picking et al. 2011). When resolution of a particular health problem is not achieved, people often seek the treatment of a traditional practitioner. Three major divisions of traditional practitioners have been distinguished in Jamaica: (1) Natural practitioners, such as bush doctors (herbalists) and lay midwives; (2) spiritual healers, such as church healers, psychic mothers, revealers, and balm-yard healers; and (3) occult healers, such as Obeah-(wo)men, Science-(wo)men, and sorcerers (Payne-Jackson and Alleyne 2004).

Development of the National Nutraceutical Industry and Intellectual Property Rights (IPR)

Despite long-term legal restrictions, two pioneering Jamaican scientists, Professor Manley West and Dr. Albert Lockhart, successfully developed and commercialized three medicinal ganja products, based on TK, in the 1970s and 1980s. These products, Canasol, Asmasol, and Canavert, are used to treat glaucoma, asthma, and seasickness, respectively (Senior 2003, Abel 2014, JIS 2017b). The debate surrounding ganja decriminalization has been long, contentious, and controversial (Abel 2014). In 2015, Jamaican legislators, in a groundbreaking decision, amended the Dangerous Drugs Act and decriminalized sacramental use of the plant by members of the Rastafarian faith, and personal use in small amounts. They also established the Cannabis Licensing Authority (CLA) to support and regulate the development of a lawful medicinal cannabis industry (MOJ 2015, CLA 2017).

The development of a medicinal cannabis, or ganja, industry forms part of a broader goal to establish a national nutraceutical industry. The term nutraceuticals, in a Jamaican context, refers predominantly to medicinal or "healing" plants (NCST 2017). Notwithstanding the complex legal issues surrounding medical ganja, the Jamaican government in 2015 made a commitment to establish a regulatory framework for nutraceuticals through amendments to the existing Food and Drugs Act. The proposed changes seek to incorporate the classification and official registration of nutraceutical products and to align the industry with international standards (NCST 2017). Under current legislation, medicinal plants and other forms of nutraceuticals (e.g. functional foods, vitamins, and minerals) are classified as either food or pharmaceutical drugs. The proposed amendments seek to establish a framework similar to that in Canada, where products are classified as Natural Health Products (NHPs) and regulated as medicines that can make health claims. The levels of evidence required to support claims are based on the proposed health claims and the risk profile of the product. Licenses issued for Canadian NHPs fall into two main categories, those making modern health claims, and those with a documented history of use as traditional medicines (Smith 2014, NCST 2017, Picking 2017).

The nascent medicinal ganja "industry" provides an interesting case study for three evolving legislative areas: regulation, IPR, and access and benefit sharing (ABS). Jamaica, in 2017, is reviewing its current intellectual property (IP) legislation, with the stated goal of developing a national strategy and policy for the IP protection of TK and traditional cultural expressions (folklore) (JIPO 2017). In addition, Jamaica has begun a consultation process to develop and implement a national framework for ABS of genetic resources and associated TK in accordance with provisions of the Convention on Biological Diversity (CBD) and the Nagoya Protocol. Amongst the issues that spark ongoing debate is that TK in Jamaica is diffuse and not easily attributable to clearly identified groups or communities, which has been observed in other countries as well. The stated goal of the government is to become a signatory of the Nagoya Protocol on ABS and to have the agreement ratified by 2020.

Bush Teas and Root Tonics

The consumption of "bush tea," made from plants growing in kitchen gardens around the house, as a beverage in the morning and evening, is such a common Jamaican practice, especially in rural communities, that it has gained its own entry in the "Dictionary of Jamaican English" (Cassidy and Le Page 2002) (Figure 5.4). Apart from its appreciation as a hot beverage, bush tea is consumed to prevent gas (which according to Jamaicans invariably develops during digestion), as a prophylactic, and as a blood cleanser (Sobo 1993). Specific bush teas with individual plants or plant mixtures are also prepared for specific illnesses (Picking et al. 2015), or for spiritual problems (Payne-Jackson and Alleyne 2004). Healthcare professionals are often not supportive of the use of bush tea (Sobo 1993), a fact borne out by only 17% of Jamaicans, on average, reporting that they discuss such consumption with their physician or other healthcare professional (Delgoda et al. 2004, 2010, Picking et al. 2011).

Figure 5.4 Bush tea preparation of cerasee (*Momordica charantia* L.), as a morning beverage, or to treat skin rash that, according to cultural beliefs, should be treated by cleansing the blood (Vandebroek and Picking 2016). (Photo credit I. Vandebroek.)

Jamaican "root tonics," in short "roots," are popular fermented formulations of roots, bark, and other parts of several plant species widely used for general health and stamina, for reproductive and mental health, to cleanse the blood, and as aphrodisiacs (Figure 5.5). They are consumed to keep the body strong and healthy rather than as medicine (even though some people also prepare root tonics for specific illnesses) and represent an important aspect of Jamaican cultural heritage (Mitchell 2011). Nowadays, the preparation and use of root tonics are common across both urban and rural residents, Maroons, and Rastafari in Jamaica. Such is their popularity that they are commercialized across the Jamaican diaspora in New York City, Toronto, and London, and even sought after online, especially for their aphrodisiac properties. In New York City, root tonics are either imported from Jamaica or prepared locally by Jamaican immigrants with imported raw plant material. In stark contrast to their popularity as a symbol of Jamaican cultural heritage, root tonics remain understudied from a scientific perspective. So far, only one paper has reviewed the plant composition of Jamaican root tonics, in which

Figure 5.5 Commercially available Jamaican root tonics in New York City (above) and home-made root tonic preparation in rural Jamaica. (Photo credit I. Vandebroek.)

plant data was mainly obtained from labels of listed ingredients on commercial products (Mitchell 2011). Data from the same author contributed to a comparison across the Caribbean and Africa for the use of plant mixtures as aphrodisiacs (van Andel et al. 2012).

The origin and history of Jamaican root tonics remain obscure. Hans Sloane, one of the earliest Anglo-Irish naturalists to report on the Jamaican flora in the 18th century, makes no mention of root tonics in his influential work. The *Dictionary of Jamaican English* (Cassidy and Le Page 2002), a comprehensive and authoritative scholarly work that represents a lexicon of Jamaican Patois recorded since 1655, includes many common Jamaican plant names and their botanical identifications but no entries for "root tonic," "tonic," or "roots," which seems surprising given the general representation of root tonics within Jamaican culture. A first review of the literature, using historical explorations of naturalists, Caribbean folk medicine, plantation or slave medicine, anthropological studies of health and the body, or treatises of different cultural groups that contribute to Jamaican society, has not yet elucidated when or by whom this cultural practice was developed in Jamaica. Further research is needed to unravel possible origins and cultural influences, including Taíno, African, Maroon, European, or others.

Conservation of Biocultural Diversity

TK of place-based biodiversity has been called the "oldest scientific tradition on earth" (Wilder et al. 2016). Jamaican TK offers invaluable opportunities for community-guided conservation of useful plant diversity and cultural heritage (Vandebroek et al. 2011). Combining a high degree of species endemism (plants occurring only there and nowhere else) with high levels of deforestation (Anadon-Irizarry et al. 2012), the Caribbean region is a plant diversity hotspot critical to conservation. Jamaica is rapidly losing its forest cover (Tole 2001). Apart from an inherent ecological value, forests and forest products are essential to local livelihoods for construction materials, bush teas, plant medicines, wild vegetables, and fuelwood. Whereas some authors consider subsistence farmers to be the main drivers of deforestation in Jamaica (Tole 2001), others have put forward a more nuanced perspective (Weis 2000) or have called attention to the biodiversity-promoting role of smallholder farming (Sander and Vandebroek 2016).

Neglected and underutilized species (NUS) are useful wild or semi-domesticated varieties and non-timber forest species, and agricultural species that are not amongst the major staple crops (often called "orphan crops") (Padulosi et al. 2013). NUS are culturally important, adapted to local environments, managed by TK systems, in general not widely traded as commodities, often nutritionally rich, but they are ignored by plant breeders, agricultural researchers, and policymakers, and often considered "poor man's food" (Padulosi et al. 2013). NUS abound in Jamaica, with examples such as breadfruit (*Artocarpus altilis* (Parkinson) Fosberg), jackfruit (*Artocarpus heterophyllus* Lam.), June plum (*Spondias dulcis* Parkinson), noni (*Morinda citrifolia* L.), moringa (*Moringa oleifera* Lam.), cho cho (*Sechium edule* (Jacq.) Sw.), gungo peas (*Cajanus cajan* (L.) Millsp.), yams (*Dioscorea* spp.), dasheen (*Colocasia esculenta* (L.) Schott), coco (*Xanthosoma sagittifolium* (L.) Schott), and many others (Padulosi et al. 2013, Sander and Vandebroek 2016). These species feature heavily in Jamaican TK and traditional diets, and offer tremendous opportunities for food security, resilience of food systems, income generation, reaffirmation of cultural identity, and empowerment of traditional lifestyles (Padulosi et al. 2013).

TK systems are embedded in local languages (Wilder et al. 2016). The term "biocultural diversity" was coined to demonstrate the intimate link between biological, cultural, and linguistic diversity (Maffi 2005, Pretty et al. 2009). Jamaican Patois (or Patwa), the Creole language spoken by the majority of Jamaicans, is the vehicle for cultural life and reflects a deep understanding of the natural world (Cassidy and Le Page 2002). Patois has been disparaged as a "fragmented language of a fragmented people", with English considered the language of prestige and power since the British seized control of Jamaica in 1655 (Beckford Wassink 1999). However, research has shown that the loss of native language speakers and knowledge-keepers can directly and negatively affect useful plant diversity (Salick et al. 1997). The combined loss of TK and biocultural diversity has long-term dramatic consequences for the future, not only in the loss of potential species for food security, health, and climate change but ultimately for human resilience (Maffi 2005).

FUTURE OPPORTUNITIES AND RECOMMENDATIONS

Jamaica's rich TK and LK systems, consisting of intertwined elements of biological, cultural, and linguistic diversity, merits appreciation, protection, and deeper understanding – as well as sustainable development for the benefit of communities across the island. The widespread popularity of Jamaican bush teas and root tonics illustrates the multitude of ways in which TK can guide and expand future sustainable development, encompassing ecotourism, gastronomy, conservation, education and outreach programs, healthcare, business development, IPR legislation, and the nutraceutical industry.

In order to better understand the multiple dimensions of Jamaican biocultural diversity, counteract its loss, and recognize its value, an equitable and dynamic collaboration is needed across multiple sectors, including social and natural scientists, policy and decision makers, the private sector, educators, and local communities (Wilder et al. 2016). An example of such an approach entails incorporating aspects of traditional medicine into medical training curricula to address knowledge gaps, promote cultural sensitivity, and improve physician-patient communication (Vandebroek 2013). Another example is addressing the stigma that still exists against African-based religious practices, folk medicine, and small-scale farming. Repealing the Obeah Act would be a powerful symbolic gesture.

Jamaica, like many other Caribbean island nations, is a food-deficit country, relying on importation of many food products (Beckford et al. 2007). Today, in addition to problems in meeting local food requirements, overseas demands for Jamaican produce and culinary by-products exist, which are consistently not being met. Much of this export demand comes from expatriate Jamaican communities living abroad, with more Jamaicans now living outside Jamaica than on the island (Senior 2003). Researching, promoting, conserving, and sustainably using NUS that are an integral component of small-scale farming systems can strengthen Jamaica's food supply, buffer economic and social shocks, provide a safety net during periods of stress and after disasters, and help make agricultural production systems more resilient to climate change (Padulosi et al. 2013). In addition, these NUS are grown in complex agroforestry systems that not only maintain useful plant diversity but also protect the wild native flora that coexists within the boundaries of these systems. Finally, recognition of NUS and development of on-farm NUS conservation programs can empower local communities to reaffirm their cultural identity. Key priorities in developing NUS include improving access to markets, adding value, and stimulating demand. Chefs, restaurants, and the audiovisual and virtual media can all play a role in promoting the gastronomy of Jamaican NUS and food systems based on TK. It is essential that small-scale farmers are stakeholders in these national, regional, and international networks and collaborative platforms that address the strategic development of NUS (Padulosi et al. 2013).

In addition, opportunities exist for the transition of traditional farmers from the illicit growing, processing, and exporting of "recreational" ganja to becoming equitable partners in a medicinal ganja industry. Such an industry holds the promise of improved livelihoods and wealth creation across the full spectrum of Jamaican society, from traditional farmers and holders of TK to scientists, clinicians, and the business sector. However, there are major concerns that holders of TK relating to the farming and processing of ganja will be marginalized and priced out of the industry, as the requirements put in place for the issuing of licenses will require significant capital investment. One option, within the framework of the Nagoya Protocol, would be to establish a national trust fund that supports and provides access to resources and funding for TK holders in Jamaica.

REFERENCES

Abel WD (2014) Medical marijuana in Jamaica. Paper presented at Science Research Council 3rd Biennial Science & Technology Conference, Kingston, 10/11/2014. http://www.src.gov.jm/wp-content/uploads/2014/11/SRC.pdf.

Abel WD, Sewell C, Eldemire-Shearer D (2011) Decriminalization of marijuana: Is this a realistic public mental health policy for Jamaica? *West Indian Med J* 60(3):367–370.

Agorsah EK (ed.) (1994) *Maroon Heritage. Archaeological, Ethnographic and Historical Perspectives.* Canoe Press, Kingston, Jamaica.

Anadon-Irizarry V, Wege DC, Upgren A, Young R, Boom B, Leon YM et al. (2012) Sites for priority biodiversity conservation in the Caribbean Islands biodiversity hotspot. *J Threat Taxa* 4(8):2806–2844.

Arias D, Gurria M, Pena H, Brown-Knowlton M, Boyce R, Smikle C. (2013) Agricultural support policies and programs in Jamaica 2006–2011. LCSSD occasional paper series. World Bank, Washington, DC. http://hdl.handle.net/10986/18998.

Asprey GF, Thornton P (1953) Medicinal plants of Jamaica part I. *West Indian Med J* 2:233–252.

Asprey GF, Thornton P (1954) Medicinal plants of Jamaica part II. *West Indian Med J* 3:17–41.

Asprey GF, Thornton P (1955a) Medicinal plants of Jamaica part III. *West Indian Med J* 4:69–82.

Asprey GF, Thornton P (1955b) Medicinal plants of Jamaica part IV. *West Indian Med J* 4:145–168.

Atkinson L (2006) The exploitation and transformation of Jamaica's natural vegetation. In: Atkinson L (ed.) *The Earliest Inhabitants: The Dynamics of the Jamaican Taíno.* University of the West Indies Press, Kingston, Jamaica, pp. 97–112.

Bailey A (1939) Jamaicanising Jamaica: Yams, cocos, plantains, bananas and bush tea. *Kingston Gleaner,* Aug 26th 1939.

Barker D, Spence B (1988) Afro-Caribbean agriculture: A Jamaican Maroon community in transition. *Geogr J* 154(2):198–208.

Beckford C, Barker D (2007) The role and value of local knowledge in Jamaican agriculture: Adaptation and change in small-scale farming. *Geogr J* 173(2):118–128.

Beckford C, Barker D, Bailey S (2007) Adaptation, innovation and domestic food production in Jamaica: Some examples of survival strategies of small-scale farmers. *Singap J Trop Geogr* 28:273–286.

Beckford C, Campbell D, Barker D (2011) Sustainable food production systems and food security: Economic and environmental imperatives in yam cultivation in Trelawny, Jamaica. *Sustainability* 3:541–561.

Beckford CL (2009) Sustainable agriculture and innovation adoption in a tropical small-scale food production system: The case of yam minisetts in Jamaica. *Sustainability* 1:81–96.

Beckford Wassink A (1999) Historical low prestige and seeds of change: Attitudes toward Jamaican creole. *Lang Soc* 28(1):57–92.

Beckwith MW (1927) *Notes on Jamaican ethnobotany.* Publication 8. Vassar College Folklore Foundation, New York, NY.

Bilby K (2012) An (un)natural mystic in the air: Images of obeah in Caribbean song. In: Paton D, Forde M (eds) *Obeah and Other Powers, the Politics of Caribbean Religion and Healing.* Duke University Press Books, Durham, NC, pp. 45–79.

Brandon G (2004) Jamaican maroons. In: Ember CR, Ember M (eds) *Encyclopedia of Medical Anthropology. Health and Illness in the World's Cultures, vol II.* Springer, New York, NY, pp. 754–764.

Brathwaite K (1994) Nanny, Palmares & the Caribbean maroon connexion. In: Agorsah EK (ed.) *Maroon Heritage: Archaeological, Ethnographic and Historical Perspectives, Jamaica.* Canoe Press, Kingston, Jamaica, pp. 119–138.

Browne P (1756) *The Civil and Natural History of Jamaica.* Printed by T. Osborne and J. Shipton for the author, London, UK.

Campbell MC (1990) *The Maroons of Jamaica 1655: A History of Resistance, Collaboration & Betrayal.* Africa World Press, Inc, Trenton, NJ.

Cannabis Licensing Authority (CLA) (2017) Cannabis Licensing Authority, Jamaica. http://cla.org.jm/. Accessed 15 November 2017.

Carney JA (2003) African traditional plant knowledge in the circum-Caribbean region. *J Ethnobiol* 23(2):167–185.

Carney JA, Rosomoff RN (2009) *In the Shadow of Slavery: Africa's Botanical Legacy in the Atlantic World.* University of California Press, Los Angeles, CA.

Cassidy FG, Le Page RB (2002) *Dictionary of Jamaican English,* 2nd edn. University of the West Indies Press, Kingston, Jamaica.

Chinese Benevolent Association of Jamaica (CBAJ) (2017) Timeline celebrating the 150th anniversary of the arrival of the Chinese in Jamaica 1854–2004. http://www.cbaj

amaica.com/yahoo_site_admin/assets/docs/CBA_Timeline.257142138.pdf. Accessed 15 November 2017.

Deason ML, Salas A, Newman SP, Macaulay VA, St. A. Morrison EY, Pitsiladis YP (2012) Interdisciplinary approach to the demography of Jamaica. *BMC Evol Biol* 12:24. https://doi.org/10.1186/1471-2148-12-24.

Delgoda R, Ellington C, Barrett S, Gordon N, Clarke N, Younger N (2004) The practice of polypharmacy involving herbal and prescription medicines in the treatment of diabetes mellitus, hypertension and gastrointestinal disorders in Jamaica. *West Indian Med J* 53:400–405.

Delgoda R, Younger N, Barrett C, Braithwaite J, Davis D (2010) The prevalence of herbs used in conjunction with conventional medicines in Jamaica. *Complement Ther Med* 18:13–20.

DeLoughrey E (2011) Yam, roots, and rot: Allegories of the provision grounds. *Small Axe* 15(1):58–75.

Fawcett W (1891) *Economic Plants. An Index to Economic Products of the Vegetable Kingdom in Jamaica*. Government Printing Establishment, Kingston, Jamaica.

Forte MC (2004) Extinction: The historical trope of anti-indigeneity in the Caribbean. Paper presented at: Atlantic History: Soundings. 10th Anniversary Conference of the Atlantic History Seminar. Cambridge, MA: Harvard University, August 10–13, 2005.

Franco JL (1978) The Slave Trade in the Caribbean and Latin America. The African Slave Trade from the Fifteenth to the Nineteenth Century. UNESCO, Port-au-Prince.

Gottlieb KL (2000) *The Mother of us all: A History of Queen Nanny Leader of the Windward Jamaican Maroons*. Africa World Press, Inc, Trenton, NJ and Eritrea.

Grigg DB (1974) *The Agricultural Systems of the World: An Evolutionary Approach*, vol. 5. Cambridge University Press, Cambridge, UK.

Hall S (2015) Créolité and the process of creolization. In: Rodriguez EG, Tate SA (eds) *Creolizing Europe: Legacies and Transformations*. Liverpool University Press, Liverpool, UK.

Harris I (2010) *Healing Herbs of Jamaica*. AhHa Press, Royal Palm Beach, FL.

Harris LZ (2012) Roots of history, seeds of change: Women organic farmers & environmental health in Jamaica. Dissertation, University of Guelph.

Harrison RP (2001) Hic Jacet. *Crit Inq* 27(3):393–407.

Hart R (1985) *Slaves Who Abolished Slavery: Blacks in Rebellion*, vol. 2. University of the West Indies, Kingston, Jamaica.

Higman BW (2008) *Jamaican Food: History, Biology, Culture*. University of the West Indies Press, Kingston, Jamaica.

Jamaica Information Service (JIS) (2017a) National symbols. http://jis.gov.jm/symbols/jamaican-coat-of-arms/. Accessed 24 May 2017.

Jamaica Information Service (JIS) (2017b) Celebrating Jamaican inventors and innovators. http://jis.gov.jm/media/JA-Inventors-Innov-22-12-15.pdf. Accessed 15 November 2017.

Jamaica Intellectual Property Office (JIPO) (2017) Traditional knowledge, traditional cultural expressions and genetic resources. http://www.jipo.gov.jm/node/90. Accessed 15 November 2017.

Keegan WF (2000) The Caribbean, including North South America and Lowland Central America: Early history. In: Kiple KF, Ornelas KC (eds) *The Cambridge World History of Food*, vol 2. Cambridge University Press, Cambridge, UK, pp. 1260–1278.

Laguerre MS (1987) *Afro-Caribbean Folk Medicine*. Bergin & Garvey Publishers, Inc, South Hadley, MA.

Long E (1774) The History of Jamaica or, General Survey of the Ancient and Modern State of the Island: With Reflections on its Situation, Settlements, Inhabitants, Climate, Products, Commerce, Laws, and Government. Printed for Lowndes T in Fleet-Street, London, UK.

Lunan J (1814) Hortus Jamaicensis, or a Botanical Description, (according to the Linnaean System) and an Account of the Virtues of its Indigenous Plants Hitherto Known, as also of the Most Useful Exotics. Printed at the office of St Jago de la Vega Gazette, Jamaica.

Maffi L (2005) Linguistic, cultural, and biological diversity. *Annu Rev Anthropol* 34:599–617.

Ministry of Justice (MOJ) (2015) Dangerous drugs (Amendment) Act 2015. http://moj.gov.jm/sites/default/files/Dangerous%20Drugs%20Amendment%20Act%202015%20Fact%20Sheet_0.pdf. Accessed 15 November 2017.

Ministry of Justice (MOJ) (2017) The Obeah Act. http://moj.gov.jm/laws/obeah-act. Accessed 2 November 2017.

Mistry J, Berardi A, McGregor D (2009) Natural resource management and development discourses in the Caribbean: Reflections on the Guyanese and Jamaican experience. *Third World Q* 30(5):969–989.

Mitchell SA (2011) The Jamaican root tonics: A botanical reference. *Focus Altern Complement Ther* 16:271–280.

Mitchell SA, Ahmad A (2006) A review of medicinal plant research at The University of the West Indies, Jamaica, 1948–2001. *West Indian Med J* 55:243–269.

National Commission on Science & Technology (NCST) (2017) The establishment of the National Nutraceutical Industry (NNI). http://ncst.gov.jm/programmes/nutraceuticals-and-functional-foods. Accessed 15 November 2017.

Padulosi S, Thompson J, Rudebjer P (2013) Fighting poverty, hunger and malnutrition with neglected and underutilized species (NUS): Needs, challenges and the way forward. *Bioversity International*. https://www.bioversityinternational.org/uploads/tx_news/Fighting_poverty__hunger_and_malnutrition_with_neglected_and_underutilized_species__NUS__1671_01.pdf. Accessed 19 November 2017.

Parry JH (1955) Plantation and provision ground: An historical sketch of the introduction of food crops into Jamaica. *Rev Hist Am* 39:1–20.

Payne-Jackson A, Alleyne M (2004) *Jamaican Folk Medicine: A Source of Healing*. University of the West Indies Press, Kingston, Jamaica.

Picking D (2017) The global regulatory framework for medicinal plants. In: Badal S, Delgoda R (eds) *Pharmacognosy: Fundamentals, Applications and Strategy*. Elsevier, London, UK, pp. 663–675.

Picking D, Delgoda R, Younger N, Germosen-Robineau L, Boulogne I, Mitchell S (2015) TRAMIL ethnomedicinal survey in Jamaica. *J Ethnopharmacol* 169:314–327.

Picking D, Younger N, Mitchell S, Delgoda R (2011) The prevalence of herbal medicine home use and concomitant use with pharmaceutical medicines in Jamaica. *J Ethnopharmacol* 137:305–311.

Pretty J, Adams B, Berkes F, de Athayde S, Dudley N, Hunn E et al. (2009) The intersections of biological diversity and cultural diversity: Towards integration. *Conserv Soc* 7:100–112.

Rashford J (1984) Plants, spirits and the meaning of "John" in Jamaica. *Jam J* 17: 62–70.

Rashford J (1985) The cotton tree and the spiritual realm in Jamaica. *Jam J* 18: 49–57.

Rashford J (1988) Packy tree, spirits and duppy birds. *Jam J* 21: 2–11.

Rashford J (1989) Leaves of fire: Jamaica's crotons. *Jam J* 21: 19–25.

Rashford J (1991) Arawak, Spanish and African contributions to Jamaica's settlement vegetation. *Jam J* 24(3):17–23.

Rashford J (1995) The past and present uses of bamboo in Jamaica. *Econ Bot* 49: 395–405.

Rashford J (1997) Africa's baobab tree in Jamaica: A further comment. *Jam J* 26: 51–58.

Rashford J (2001) Those that do not smile will kill me: The Ethnobotany of the ackee in Jamaica. *Econ Bot* 55: 190–211.

Robertson D (1982) *Jamaican Herbs: Nutritional and Medicinal Values.* Jamaican Herbs Ltd, Kingston, Jamaica.

Rouse I (1992) *The Taíno: Rise and Decline of the People Who Greeted Columbus.* Yale University Press, New York, NY.

Rubin V, Comitas L (1975) *Ganja in Jamaica: The Effects of Marijuana Use.* Anchor Books, New York, NY.

Salick J, Cellinese N, Knapp S (1997) Indigenous diversity of cassava: Generation, maintenance, use and loss among the Amuesha, Peruvian Upper Amazon. *Econ Bot* 51(1):6–19.

Sander L, Vandebroek I (2016) Small-scale farmers as stewards of useful plant diversity: A case study in Portland parish, Jamaica. *Econ Bot* 70:303–319.

Saunders N (2006) Zemís, trees and symbolic landscapes: Three Taíno carvings from Jamaica. In: Atkinson L (ed.) *The Earliest Inhabitants. The Dynamics of the Jamaican Taíno.* University of the West Indies Press, Kingston, Jamaica, pp. 187–198.

Schultes RE, Hofmann A (1992) *Plants of the Gods: Their Sacred, Healing and Hallucinogenic Powers.* Healing Arts Press, Rochester, VT.

Selvaraju R (2013) Climate change and agriculture in Jamaica: Agricultural sector support analysis. Food and Agricultural Organization of the United Nations, Rome, Italy. http://www.fao.org/3/a-i3417e.pdf. Accessed 14 November 2017.

Senior O (2003) *Encyclopedia of Jamaican Heritage.* Twin Guinep Publishers Ltd, Kingston, Jamaica.

Sloane, H. (1707–25) A voyage to the islands Madera, Barbados, Nieves, S. Christophers and Jamaica with the natural history of the herbs and trees, four-footed beasts, fishes, birds, insects, reptiles, &c. of the last of those islands; to which is prefix'd an intro., wherein is an account of the inhabitants, air, waters, diseases, trade, &c. of that place, with some relations concerning the neighbouring continent, and islands of America. Printed by B.M. for the author, London, UK.

Smith A, Jogalekar S, Gibson A (2014) Regulation of natural health products in Canada. *J Ethnopharmacol* 158:507–510.

Sobo EJ (1993) *One Blood: The Jamaican Body.* State University of New York Press, Albany, NY.

Steggerda M (1929) Plants of Jamaica used by natives for medicinal purposes. *Am Anthropol* 31:431–434.

Tole L (2001) Jamaica's disappearing forests: Physical and human aspects. *Environ Manage* 28:455–467.

Understanding Slavery Initiative (USI) (2017) Understanding slavery initiative. http://www.understandingslavery.com. Accessed 11 May 2017.

van Andel T, Mitchell S, Volpato G, Vandebroek I, Swier J, Ruysschaert S et al. (2012) In search of the perfect aphrodisiac: Parallel use of bitter tonics in West Africa and the Caribbean. *J Ethnopharmacol* 143:840–850.

Vandebroek I (2013) Intercultural health and ethnobotany: How to improve healthcare for underserved and minority communities? *J Ethnopharmacol* 148:746–754.

Vandebroek I, Picking D (2016) *Popular Medicinal Plants in Portland and Kingston, Jamaica.* Pear Tree Press, Kingston, Jamaica.

Vandebroek I, Reyes-García V, de Albuquerque UP, Bussmann R, Pieroni A (2011) Local knowledge: Who cares? *J Ethnobiol Ethnomed* 7:35. https://doi.org/10.1186/1746-4269-7-35.

Vega B (1996) *Las frutas de los Taínos*. Amigo del Hogar, Santo Domingo.
Veloz Maggiolo M (1997) The daily life of the Taíno people. In: Bercht F, Brodsky E, Farmer JA, Taylor D (eds) *Taíno: Pre-Columbian Art and Culture from the Caribbean*. Monacelli Press, New York, NY, pp. 34–45.
Warner M (2007) *Herbal Plants of Jamaica*. MacMillan Education, Oxford, UK.
Weis T (2000) Beyond peasant deforestation: Environment and development in rural Jamaica. *Glob Environ Change* 10:299–305.
Wilder BT, O'Meara C, Monti L, Nabhan GP (2016) The importance of indigenous knowledge in curbing the loss of language and biodiversity. *BioScience* 66(6):499–509.
Wilson SM (1997). *The Indigenous People of the Caribbean*. The University Press of Florida, Gainesville, FL.

CHAPTER 6

Indigenous Knowledge Systems: Practices in Modern-Day China

Weiyang Chen

CONTENTS

ABSTRACT

Background Traditional Chinese Medicine (TCM) is rooted in the ancient philosophy of Taoism. The theory of TCM takes a holistic view of the human body by seeing it as integral with its social and natural environment. As an important part of the Chinese culture, TCM has evolved into a unique and complete medical system during the long history of China. TCM practice includes herbal medicine, acupuncture, massage (tui na), exercise (qigong), and dietary therapy. Today, TCM is widely practised both in China and around the world.

Relevance TCM has come to be respected as a credible treatment modality in many countries around the world. It is the subject of active basic and practice research, which appears to validate much of the underlying principles. All this is owed to a clear vision and strategy by the Chinese government to systematize, modernize, and promote TCM to be on a par with Western medicine. It is therefore clear from this that the mainstreaming of traditional knowledge and practices can only succeed with the active support of government policies and political will.

INTRODUCTION

"Indigenous knowledge systems" refers to knowledge systems that are unique to a given culture or society (Ellen and Harris, 1996). These systems are separate and different from the "modern scientific system" which they use to cope with various daily situations. The "modern scientific system" includes knowledge generated by universities, research institutions, and private firms. Thus indigenous knowledge systems are the basis for local-level decision-making and include information, knowledge, practices, and rituals pertaining to agriculture, healthcare, food preparation, education, and natural resource management within indigenous communities and cultures (Tharakan, 2015). Indigenous knowledge is locally embedded knowledge that is context-specific, accumulated over time, and unique to a given culture, society, or local community (Sillitoe and Marzano, 2009). It is learned by repetition and often transmitted orally or through demonstration and imitation (Rao, 2006). Societies the world over possess rich sets of experiences, understanding, and explanations. For rural and indigenous peoples, local knowledge is a cornerstone to their decision-making in day-to-day life.

Traditional Chinese Medicine (TCM) is rooted in Taoist philosophy. It takes a holistic view that sees the human body as integrated with its social and natural environments (Chi, 1994; Lu et al., 2004). TCM includes various forms of herbal medicine, acupuncture, massage (tui na), exercise (qigong), and dietary therapy (Hesketh et al., 1997). China is the only country in the world where Western medicine and traditional medicine are practised alongside each other at every level of the healthcare system. TCM now accounts for around 40% of all healthcare delivered in China. Starting in the 1950s, the principles of TCM were standardized in the People's Republic of China. Standardization included attempts to integrate these principles with modern notions of anatomy and pathology. In addition, the Chinese government promoted a systematized form of TCM (Levinovitz, 2013) and gave a high priority to TCM as part of communist healthcare reform. TCM was also widely promoted in the mass media and has gradually gained widespread international recognition. In 1986, the government set up the State Traditional Chinese Medicine Administration, which has a key role in formalizing education programmes in the field of traditional medicine. Thus, in China, traditional medicine is held in high regard, and there has been a policy to promote its use and blend its benefits with those of modern Western medicine. In 2008, China's Minister of Health identified several key priorities for TCM development. These included:

- Increasing policy support for traditional Chinese medicine.
- Strengthening research on key traditional Chinese medicine issues and building capacity for traditional Chinese medicine research.

- Training prominent traditional Chinese medicine doctors and establishing well-known traditional Chinese medicine hospitals and departments.
- Improving and adapting traditional Chinese medicine services to meet public need.
- Increasing access to, and improving the quality of, traditional Chinese medicine services in rural and urban communities.
- Strengthening international cooperation and communication on traditional Chinese medicine (Umakanta, 2012). In 2016, China's State Council Information Office issued a white paper on the development of TCM in China, meaning that the government set great store by TCM and rendered vigorous support to its development.

TRADITIONAL CHINESE MEDICINE

The doctrines of Chinese medicine are rooted in books such as the *Yellow Emperor's Inner Canon* and the *Treatise on Cold Damage*, as well as in cosmological notions such as "Yin yang" and the "Five phases". Yin and yang are polar opposites and complements, whose relationship determines the nature of existence and change. They represent two abstract and complementary aspects that every phenomenon in the universe can be divided into (Zhang et al., 2000). The concept can also be applicable to the human body. The upper part of the body and the back are assigned to yang, while the lower part of the body is believed to have the yin character. This characterization also extends to the various body functions, and more importantly, to disease symptoms (e.g., cold and heat sensations are assumed to be yin and yang symptoms, respectively). Another key concept in TCM, which further illustrates the correlations between different domains of experience, is the "Five Phases" (wu xing) model. "Five Phases" presumes that all phenomena of the universe and nature can be broken down into five elemental qualities – represented by wood, fire, earth, metal, and water (Singh and Ernst, 2008). These five phases constitute the basis of the zàng-fû concept, and thus have great influence regarding the medical approach to the body, health, and disease.

The holistic TCM approach towards achieving harmony includes acupuncture, dietary therapies, tui na and Chinese massage, as well as the use of herbal medicines. Due to its long history of effectively treating certain challenging diseases, comparatively lower costs and reportedly fewer side effects, the role of TCM in health is widely recognized around the globe. Acupuncture, an important and indispensable part of TCM, has played a major role in the healthcare of the Chinese people. Qigong is a special and unique part of TCM. Through physical exercises, it regulates the vital energy (Qi) of the human body, thus achieving health protection and treatment of diseases. Tai Chi Chuan is another way of physical exercise that is a combination of martial arts and the regulation of Qi in the theory of TCM. It integrates the breath, mind, and physical activity, thus making exercisers achieve greater awareness and a sense of inner peace. These modalities may be used separately or in combination to effect a treatment.

Acupuncture and Moxibustion

The Chinese term "zhēn jiū" refers to acupuncture (zhēn) and moxibustion (jiū). They are forms of TCM widely practised in China, Southeast Asia, Europe, and the Americas to strengthen the blood, stimulate the flow of Qi, and maintain general health. The theories of acupuncture and moxibustion hold that the human body acts as a small universe connected by channels, and that by physically stimulating these channels the practitioner can promote the human body's self-regulating functions and bring health to the patient. Acupuncture is the practice of inserting needles into the superficial skin, subcutaneous tissue, and muscles at particular acupuncture points and manipulating them (Ng and Liao, 1997). In TCM, there are as many as 2,000 acupuncture points on the human body that are connected by 12 main meridians. These meridians conduct energy, or Qi, between the surface of the body and its internal organs. Acupuncture is believed to keep the balance between Yin and Yang, thus allowing for the normal flow of Qi throughout the body and restoring health to the mind and body. Qi refers to the "finest matter influence", which flows through the body (Unschuld, 1985). Any potential disruptions of this Qi flow are believed to be responsible for diseases.

Moxibustion consists of burning dried mugwort (*Artemisia vulgaris*) on particular points on the body. Artemisia species are herbs used to produce Moxa, in Chinese called "Ai". There are two types of moxibustion: direct and indirect. In direct moxibustion, a small, cone-shaped amount of moxa is placed on top of an acupuncture point and burned. In indirect moxibustion, a practitioner lights one end of a moxa stick, roughly the shape and size of a cigar, and holds it close to the area being treated for several minutes until the area turns red. Indirect moxibustion is currently the more popular form of care because there is a much lower risk of pain or burning. Burning moxa produces a great deal of smoke and a pungent odor that is often confused with that of cannabis. The purpose of moxibustion is to warm and invigorate the blood, stimulate the flow of Qi, strengthen the kidney Yang, expel wind and disperse cold, and dissolve stagnation. Historically, this therapy had been used to treat menstrual pain. Generally, moxibustion is applied in "cold syndrome," deficient conditions, and chronic diseases (Li, 2013). Moxibustion may be used in combination with acupuncture or separately to effect a treatment.

Acupuncture and moxibustion are now well integrated into the Chinese healthcare system and are practised along with conventional (Western) medicine in China (Xu and Yang, 2009). Since 1999, the Chinese government has promoted the development of Community Health Services Centers as major providers of primary healthcare, which includes the practice of acupuncture and moxibustion. Higher education in acupuncture and moxibustion in China is offered by Chinese medicine universities and by Western medical and non-medical universities. There are presently 46 tertiary Chinese medicine universities, 10 Western medical, and six non-medical universities offering acupuncture and moxibustion programs. More than 90% of the 4,169 Chinese medicine hospitals in China (including integrated Chinese medicine and Western medicine hospitals) now have an acupuncture and moxibustion department (China Association for Acupuncture and Moxibustion, 2012). The legislation regarding Chinese medicine physicians was established in 1999 to outlaw unqualified and unreasonable practices, and delivery of TCM services is now under the management of the State Administration of Traditional Chinese Medicine. A physician applying for a practicing license must pass an examination. The examination contains two components: theoretical and practical. The practical component requires candidates to demonstrate basic proficiency in acupuncture and moxibustion, including needling, moxibustion, cupping, and scalp and auricular acupuncture (Zou and Huang, 2013). In 2002, Continuing Medical Education was established to improve the competency of practitioners and ensure the passing down of knowledge outside of China.

Tui Na (Massage)

The term tui na is composed of two Chinese characters, Tui and Na. Tui can be translated as pressing and dragging, and Na as grasping. Tui na is a form of Chinese manipulative therapy often used in conjunction with acupuncture, moxibustion, fire cupping, Chinese herbalism, tai chi, and qigong (Tao et al., 2016). This Asian bodywork therapy form has been used in China for centuries. In a typical tui na session, the patient remains clothed and sits on a chair. The practitioner will ask a series of questions and then begin treatment. The type of massage delivered by a tui na practitioner can be quite vigorous at times. Practitioners may use herbal compresses, ointments, and heat to enhance these techniques. The practitioner uses his finger, hand, elbow, knee, or foot to apply pressure to muscle or soft tissue (Tao et al., 2016). Tui na is best suited for treating chronic pain and musculoskeletal conditions. Tui na was thought to be especially suitable for use on the elderly population and on infants.

Tui na is a treatment based on a solid theoretical background that includes basic theories, diagnostic methods and syndrome differentiation of TCM, particularly the theory of meridians. Tui na has specific benefits and advantages in a wide range of applications in both prevention and treatment of diseases. It may be used in the treatment of internal and external conditions, traumatic injury, and musculoskeletal, gynecological, obstetric, and pediatric diseases. The physician must first concentrate his mind when performing Tui na, then regulate his breathing, and actuate the Qi and power of his entire body towards his hands, elbow, or other part of the body required for treatment. The flow of Qi and blood is stimulated by applying manipulation on certain points or areas of the patient's body in order to normalize the function of zang fu and balance yin and yang (Chaudhury and Rafei, 2001). Tui na practice involves a range of conventional diagnostic methods, such as imaging, laboratory tests, and orthopedic and neurological assessments. Patient management includes a range of manual techniques, rehabilitative exercises, patient education, and other therapeutic advice. Thus, competence is important. In 1956, the first Tui na school was established in Shanghai, China. Higher education for Tui na is now available in universities and colleges of TCM throughout China. Significant progress has been made in Tui na education and research (Cao, 1992). Over the last 50 years, a number of textbooks on Tui na have been published. In modern China, many hospitals include Tui na as a standard aspect of treatment, with specialization for infants, adults, orthopedics, traumatology, cosmetology, rehabilitation, and sports medicine. Currently, Tui na is taught in many academic traditional universities and used globally (World Health Organization, 2010).

Cupping Therapy

Cupping therapy is a form of alternative medicine in which a local suction is created on the patient's skin to dispel stagnation (stagnant blood and lymph), thereby improving Qi flow, in order to treat respiratory diseases such as the common cold, pneumonia, and bronchitis (State Administration of Traditional Chinese Medicine and Pharmacy, 1997). Through suction, the skin is drawn into the cup by creating a vacuum in the cup placed on the skin over the targeted area. The vacuum can be created either by the heating and subsequent cooling of the air in the cup, or via a mechanical pump. The cup is usually left in place for somewhere between five and fifteen minutes. The suction and negative pressure provided by cupping can loosen muscles, encourage blood flow, and relieve the nervous system. Cupping is used on the back, neck, and shoulders to relieve musculoskeletal complaints. Acupuncturists often use cupping therapy with Tui na or needling. Cupping is not advised in TCM over skin ulcers or to the abdominal or sacral regions of pregnant women (Cheng, 1987).

Cupping therapy can be classified via six main categories (Shaban, 2013; Al-Bedah et al., 2016). The first category distinguishes the "technical types" which include dry, wet, massage, and flash cupping therapies. The second category distinguishes the "power of suction related types", which include light, medium, and strong cupping therapies. The third category distinguishes the "method of suction related types", which include fire, manual suction, and electrical suction cupping therapies. The fourth category is made up of "materials inside cups related types", which include herbal, water, ozone, moxa, needle, and magnetic cupping therapies. A fifth category and a sixth category were developed later. The fifth category is made up of treatment types focused on a specific body area, which include facial, abdominal, female, male, and orthopedic cupping therapy. The sixth category is made up of special-purpose cupping types that include sports and aquatic cupping.

In today's China, cupping therapy is commonly used in clinics and hospitals. In the 1950s, the clinical efficacy of cupping therapy was confirmed by the co-research of China and the former Soviet Union, and was established as an official therapeutic practice in hospitals all over China.

Scraping (Gua Sha)

Scraping, or Gua sha, is a folk medicine technique. The term gua sha is comprised of two Chinese words: gua, meaning to rub or scrape, and sha, a type of stagnant energy that causes excess heat to build within the body. Gua sha involves repeated pressed strokes over lubricated skin with a smooth edged and blunt instrument. Skin is typically lubricated with massage oil or balm, and commonly using pieces of smooth jade, bone, animal tusks, horns, smooth stones, or even a simple metal cap with a blunt rounded edge. The smooth edge is placed against the oiled skin surface, pressed down firmly, and then moved down the muscles or along the pathway of the acupuncture meridians along the surface of the skin in order to release obstruction and toxins that are trapped at the surface of the skin. The scraping is done until red spots appear and then bruising covers the treatment area. Gua sha is usually applied to the back, buttocks, and posterior surfaces of the neck, shoulders, arms, and legs. It is occasionally applied to the chest or abdomen. Practitioners believe that gua sha releases unhealthy bodily matter from blood stasis within open sores, tired, stiff, or injured muscle areas to stimulate new oxygenated blood flow to the areas, thus promoting metabolic cell repair, regeneration, healing, and recovery (GuaSha Treatment of Disease, 2009).

The principle of gua sha is based on 12 meridians and eight extra-meridians. The local skin redness and bruising caused by gua sha is considered to release blocked energy and heat by bringing heat to the skin surface. Gua sha also increases lymphatic drainage.

In China, gua sha is widely available in national and public hospitals and in private massage shops, and is used to treat joint and muscle pain. Because of the reasonable price, many patients in China and other parts of Asia also use gua sha as a form of preventive medicine and a first line of defense against illness. Gua sha has been used to treat a wide range of both acute and chronic health conditions, including headache, fever, digestive disorders, asthma, and respiratory infections, as well as women's health issues, insomnia, and general fatigue (Roizman, 2016).

Qigong

The term qigong is composed of two Chinese characters, "qi" and "gong", where "qi" can be translated roughly into English as "vital energy", and "gong" as "method" or "performance". Qigong is a holistic system of coordinated body postures, including movement, breathing, and meditation, that is used in the belief that it promotes health, spirituality, and martial arts training (Xie, 1994). Qigong is based on Chinese philosophy. The Chinese sage Lao Tzu first propagated this form of therapy. It works by creating a balance between the body, spirit, and mind of the patient. Using a series of breathing exercises, visual methods, gradual movements, and unmoving postures, the therapy attempts to stimulate the body's meridians and promote healing. It is known to induce the awakening of the body's innate mechanisms to heal itself. Qigong is commonly classified into two foundational categories: (1) dynamic or active qigong (dong gong) – with slow, flowing movement, and (2) meditative or passive qigong (jing gong) – with still positions and inner movement of the breath. From a therapeutic perspective, qigong can be classified into two systems: (1) internal qigong, which focuses on self-care and self-cultivation, and (2) external qigong, which involves treatment by a therapist who directs or transmits qi (Micozzi, 2010). Qi plays an important role in the vital processes of the human body. Regulation of the flow of qi can be used as a method to preserve health and prevent diseases (Cohen, 1999).

There are 75 ancient forms of qigong that can be found in ancient literature, and 56 common or contemporary forms have been described in a qigong compendium (Ma, 1992). People who recovered from their illnesses after qigong practice also developed many contemporary forms.

Qigong is now practised throughout China and worldwide for recreation, exercise, and relaxation, preventive medicine and self-healing, alternative medicine, meditation and self-cultivation, and training for martial arts. Research concerning qigong has been conducted for a wide range of medical conditions, including hypertension, pain, and cancer.

Herbal Medicine

Chinese medicine involves the use of herbal medicine, animal parts, and minerals. There are roughly 13,000 medicines used in China and over 100,000 medicinal recipes recorded in the ancient literature (Chen and Yu, 1999): thus, medicinal herbs play a major role in health services. The substances TCM practitioners most commonly use can come from different leaves, roots, stems, flowers, and seeds, such as cinnamon bark, ginger, ginseng, licorice, and rhubarb.

The concepts used to describe the pharmacology of Chinese medicines include property; taste or flavor; actions of ascending or descending, floating and sinking; meridian tropism; and toxicity (Chaudhury and Rafei, 2001). Properties include cold, hot, warm, and cool. These properties are based on their therapeutic effects: For example, herbs effective for the treatment of heat syndromes are endowed with cold or cool properties, while those effective for cold syndromes are herbs with warm or hot properties. Taste includes the sour, bitter, sweet, pungent, and salty taste of herbs. They are the expression of the feature of the herb's actions; thus, the taste does not necessarily mean the real taste of the herb. Action of ascending, descending, floating, and sinking involves the direction of an herb's action. The ascending and floating herbs have an upward and outward effect, while descending and sinking herbs have a downward and inward effect. The action of an herb can be changed by processing or by combined use with other herbs. Meridian tropism theory is the classification of herbs according to the meridians on which their therapeutic action is differentially manifested. In addition, meridian tropism also refers to the ability of some herbs to guide or lead other herbs included in the prescription to the targeted meridians or organs. For example, *Radix bupleuri* (Chai Hu) can relieve pain in the hypochondriac region, which is believed in Chinese medicine to be the result of stagnation of liver qi. As the liver meridian crosses through the hypochondriac region, it can be deduced that this herb acts on the liver meridian. An herb can act on more than one meridian. Toxicity refers to the harmful effects produced by herbs on the human body. Toxic herbs are usually marked by "highly toxic", "moderately toxic" or "slightly toxic" to indicate the different degrees of toxicity (Chaudhury and Rafei, 2001).

There are three types of Chinese medicine preparations currently being used. (1) Preparations originating from an individual Chinese medicine, its active fraction or total extract, (2) Preparations made of pure compounds derived from a medicine or analogues produced by structural modification, (3) Preparations from a composite prescription of TCM, possibly used in a traditional dosage form (pill, powder, etc.) or in a modern dosage form (tablet, granule, etc.) (Chen and Yu, 1999).

Many treatments in Chinese medicine make sense within both the biomedical model of the body and TCM ideas about the body. Artemisia, for example, is traditionally used in treating symptoms such as fever and now is globally used in treating malaria. Artemisinin, a compound extracted from *Artemisia annua*, is employed against malaria (Hsu, 2006), and a polysaccharide peptide isolated from *Coriolus versicolor* is used for cancer treatment in China (Chan and Yeung, 2006; Wong et al., 2005).

Food Therapy

Chinese food therapy is a mode of dieting rooted in Chinese beliefs concerning the effects of food on the human body and is centered on concepts such as eating in moderation. Its basic precepts are a mix of folk views and concepts drawn from traditional Chinese medicine (Whang, 1981). The main tenets central to food therapy include "curing and nourishing come from the same source" and "when you eat, satisfy only seven-tenths of your hunger". (Anderson, 2013). Four principles of TCM food therapy are light eating, balancing the "hot" and "cold" nature of foods, maintaining the harmony of the five flavors of food, and maintaining consistency between dietary intake and different health conditions (Zou, 2016). In Chinese nutrition, a balanced diet includes all five tastes—spicy (warming), sour (cooling), bitter (cooling), sweet (strengthening), and salty (cooling). Each of these "flavors" purportedly has specific effects on particular organs. The sour flavor has "constriction and emollient effects" and "can emolliate the liver and control diarrhea and perspiration", whereas "bitter" food can "purge the heart 'fire'", reduce excessive fluids, induce diarrhea, and reinforce the heart "Yin" (Zou, 2016). Foods that have a particular taste tend to have particular properties. There are no forbidden foods or "one size fits all" diets in Chinese nutrition. In TCM, nutrition is considered the first line of defense in health matters.

Food therapy has long been a common approach to health among Chinese people both in China and overseas, and was popularized for Western readers in the 1990s with the publication of books like *The Tao of Healthy Eating* (Flaws, 1995) and *The Wisdom of the Chinese Kitchen* (Young, 1999).

CONCLUSION

Indigenous knowledge systems comprise broad and comprehensive knowledge systems that address societal issues in various fields important to human survival and the quality of life. The extensive history of indigenous knowledge systems and practices in China such as TCM provide a rich resource and a history of engagement, success, and failure that could beneficially inform communities in their search for an improved quality of life. With a history of over 2,000 years, the TCM practice offers natural, safe, and effective therapies and cures for many diseases with much fewer side effects. The vast system of TCM has been gradually incorporated into modern medicine, and the knowledge of TCM can beneficially affect the direction of future medical development and is simple enough to fit changing times and varied cultures.

REFERENCES

Al-Bedah, A.M., Aboushanab, T.S., Alqaed, M.S., Qureshi, N.A., Suhaibani, I., Ibrahim, G., Khalil, M. 2016. Classification of cupping therapy: A tool for modernization and standardization. *Journal of Complementary and Alternative Medical Research* 1:1–10.

Anderson, E.N. 2013. Folk nutritional therapy in modern China. In: *Chinese Medicine and Healing: An Illustrated History Hinrichs*, eds Hinrichs, T.J. and Barnes, L.L. The Belknap Press of Harvard University Press, pp. 259–260.

Cao, R.F. 1992. *Zhong yi Tuina xue [Tuina therapy]*. Ren Min Wei Sheng Chu Ban She, Beijing, China.

Chan, S.L., Yeung, J.H.K. 2006. Effects of polysaccharide peptide (PSP) from Coriolus versicolor on the pharmacokinetics of cyclophosphamide in the rat and cytotoxicity in HepG2 cells. *Food and Chemical Toxicology* 44:689–694.

Chaudhury, R.R., Rafei, U.M. 2001. *Traditional Medicine in Asia*. WHO Regional Publications.

Chen, K.J., Yu, B. 1999. Certain progress of clinical research on Chinese integrative medicine. *Chinese Medical Journal* 112:934–937.

Cheng, X. 1987. *Chinese Acupuncture and Moxibustion* (Revised Edition). Foreign Languages Press, Beijing, China, p. 370.

Chi, C. 1994. Integrating traditional Chinese medicine into modern health care systems: Examining the role of the Chinese medicine in Taiwan. *Social Science and Medicine* 39:307–321.

China Association for Acupuncture and Moxibustion. 2012. Report on advances in acupuncture and moxibustion.

Cohen, K.S. 1999. *The Way of Qigong: The Art and Science of Chinese Energy Healing*. Random House of Canada, Toronto, Canada.

Ellen, R., Harris, H. 1996. Concepts of indigenous environmental knowledge in scientific and development studies literature: A critical assessment. East-west Environmental Linkages Network Workshop 3, Canterbury.

Flaws, B. 1995. The Tao of Healthy Eating: Dietary Wisdom According to the Traditional Chinese Medicine. Blue Poppy Press, Boulder, CO.

GuaSha Treatment of Disease. 2009. Tcmwell.com. Retrieved 2011-05-17.

Hesketh, T., Zhu, W.X., Asiab, E. 1997. Health in China: Traditional Chinese medicine: One country, two systems. *British Medical Journal* 315:115–117.

Hsu, E. 2006. Reflections on the "discovery" antimalarial qinghao. *British Journal of Clinical Pharmacology* 61:666–670.

Levinovitz, A. 2013. "Chairman Mao Invented Traditional Chinese Medicine". *Slate*.

Li, Z.G. 2013. English Translation of Traditional Chinese Medicine: Theory and Practice. 上海三联书店.

Lu, A., Jia, H., Xiao, J., Lu, Q. 2004. Theory of traditional Chinese medicine and therapeutic method of diseases. *World Journal of Gastroenterology* 10:1854–1856.

Ma, J.R. 1992. *Practical Qigong for Traditional Chinese Medicine*. Shanghai Scientific and Technical Publishers, Shanghai, China.

Micozzi, M.S. 2010. *Fundamentals of Complementary and Alternative Medicine*. Elsevier Health Sciences, Kindle Edition.

Ng, L.K.Y., Liao, S.J. 1997. Acupuncture: Ancient Chinese and Modern Western. A comparative inquiry. *Journal of Alternative and Complementary Medicine* 3:11–23.

Rao, S. 2006. Indigenous knowledge organization: An Indian scenario. *International Journal of Information Management* 26:224–233.

Roizman, T. 2016. What is Gua Sha Therapy and how does it work? https://trainedto.com/tcm/what-is-gua-sha-therapy-how-it-works.

Shaban, T. 2013. *Cupping Therapy Encyclopedia*. CreateSpace Publishing.

Sillitoe, P., Marzano, M. 2009. Future of indigenous knowledge research in development. *Futures* 41:13–23.

Singh, S, Ernst, E. 2008. *Trick or Treatment: Alternative Medicine on Trial*. Bantam, London, UK.

State Administration of Traditional Chinese Medicine and Pharmacy. 1997. *Advanced Textbook on Traditional Chinese Medicine and Pharmacology*. New World Press, Beijing, China.

Tao, W.W., Jiang, H., Tao, X.M., Jiang, P., Sha, L.Y., Sun, X.C. 2016. Effects of Acupuncture, Tuina, Tai Chi, Qigong, and Traditional Chinese medicine five element music therapy on symptom management and quality of life for cancer patients: A meta-analysis. *Journal of Pain and Symptom Management* 51:728–747.

Tharakan, J. 2015. Indigenous knowledge systems-a rich appropriate technology resource. *African Journal of Science, Technology, Innovation and Development* 1:52–57.

Umakanta, S. 2012. *Clinical Research in China. Clinical Research in Asia, Opportunity & Chanllege*. Woodhead Publishing Series in Biomedicine.

Unschuld, P. 1985. *Medicine in China: A History of Ideas*. University of California Press, Berkeley, CA.

Whang, J. 1981. Chinese traditional food therapy. *Journal of the American Dietetic Association* 78:55–57.

Wong, C.K., Bao, Y.X., Wong, E.L.Y., Leung, P.C., Fung, K.P., Lam, C.W.K. 2005. Immunomodulatory activities of Yunzhi and Danshen in post-treatment breast cancer patients. *The American Journal of Chinese Medicine* 33:381–395.

World Health Organization. 2010. Benchmarks for Training in Traditional/Complementary and Alternative Medicine-Benchmarks for Training in Tuina. WHO, Geneva, Switzerland.

Xie, Z.F. 1994. *Classified Dictionary of Traditional Chinese Medicine*. New World Press, Beijing, China.

Xu, J., Yang, Y. 2009. Traditional Chinese medicine in the Chinese health care system. *Health Policy* 90:133–139.

Young, G. 1999. *The Wisdom of the Chinese Kitchen: Classic Family Recipes for Celebration and Healing*. Simon and Schuster, New York, NY.

Zhang, H, Wang, N, Zheng, L. 2000. The influence of acceding to WTO on progress of modernization of Chinese materia medica in the 21st century. *China Pharmacy* 11:51–53.

Zou, H., Huang, Y. 2013. Optimization of the teaching content of acupuncture and moxibustion based on traditional Chinese medicine and Western medicine licensure examination. *Health Vocational Education* 31:60–62.

Zou, P. 2016. Traditional Chinese medicine, food therapy, and hypertension control: A narrative review of Chinese literature. *The American Journal of Chinese Medicine* 44:1579–1594.

CHAPTER 7

Kampo Medicine: A Different Model for Integrating Health Care Practices

Naotoshi Shibahara

CONTENTS

ABSTRACT

Background Kampo medicine, though evolved from Traditional Chinese Medicine, has taken a different path. While Kampo formulae are based on the ancient books of classical Chinese medicine, the dosage applied in daily practice by Japanese Kampo doctors differs from that used in China. There are other distinctions; Kampo medicine relies less on deep philosophies and ideologies, and it is practised by Western (conventional) medicine doctors.

Today Kampo medicine is at the centre of a thriving billion dollar-industry organized along the lines of pharmaceutical medicine, including the incorporation of a pharmacopoeia and the principles of Good Manufacturing Practice. This chapter describes the history of Kampo medicine, its evolution into inclusion into the medical curriculum in Japan, and the theoretical underpinnings of its practice. It ends with some clinical applications showing its utility.

Relevance Kampo medicine shows how traditional medicine can be harnessed for modern ills. It is unique in integrating conventional and traditional, unlike healing systems in which the practitioners of the two disciplines are generally separated. This is something that other healing traditions can learn from.

INTRODUCTION

A visitor to Japan is struck by how modern the country is – being a member of the G7 and hence one of the most industrialized countries in the world – and yet how it is also deeply rooted in its culture. There is little room for Western foods, for instance, as the Japanese eat traditional foods indigenous to the Nippon islands. Modern Japan therefore illustrates that truism that modernity does not mean westernization (Buntrock, 1996; Lal, 2000), and that modernity and tradition can be comfortable bedfellows. Modernization has been seen as something which non-Western societies can only achieve by abandoning their traditional cultures (Galland & Lemel, 2008), a notion which many of the countries in Asia, specifically China and Japan, appear to contradict.

While much is known about Traditional Chinese Medicine (TCM), there is little in the English literature on Kampo medicine, which can best be described as TCM acculturated into Japanese healing systems. The word "Kampo" is a compound of two words: "Kan (Han)", the old name for the Han Dynasty (206 BC–220 AD) in

China, and "Po", meaning "medicine" or "way of treatment". The practices were codified in the basic handbook called *Shang han lun* (Watanabe et al., 2011).

Chinese herbal medicine was adopted in Japan around the 5th century. It underwent a transformation in the crucible of Japanese culture. In particular, the reformations in the 16th and 17th centuries set its development along an evolutionary trajectory that significantly diverged from medicine in China. Kampo formulae are based on the ancient books of classical Chinese medicine. The dosage applied in daily practice by Japanese Kampo doctors differs from that used in China. It has also been observed that Kampo medicine relies less on deep philosophies and ideologies (e.g. Taoism) and is more simplified in order to be practical, with a focus on diagnostic methods (Terasawa, 2004a). This simplification is a result of the influence of, among others, Yoshimasu Todo in the 18th century, who advocated reliance on clinical observations obtained from physical examination of the patient (Fujikawa, 1979). In addition, Kampo uses a limited number of traditional formulations, while TCM utilizes thousands of crude drugs in an ever-increasing number of formulations.

During the Edo Era (1603–1868) and until the Meiji Restoration (1868) TCM (Kampo) was the official medicine practised throughout Japan. During the Meiji era (1868-1912), modern Western medicine was introduced by the Germans to Japan. Previous encounters with Western medicine had started with the arrival of Portuguese doctors and Christian missionaries in the 1530s (Hiwat, 2016). However, these had been subsequently banned from the islands in the 1600s.

In 1879, with the promulgation of the "Medical License Examination Regulations", only candidates who had studied Western medicine were eligible for a license to practice medicine. Most of the anatomy textbooks written by Japanese anatomists followed a format based on that of German anatomy textbooks of the time (and so medical students had to study German as a language) (Shimada, 2007). As a result, Western Germanic medicine broadened rapidly, and by 1945 the practice of Kampo had nearly disappeared and the knowledge of Kampo medicine had been forgotten. After 1945, some efforts to re-establish the practice of Kampo were made, and in 1976 the system of Kampo medicine and Kampo formulae was officially integrated into the Japanese medical healthcare system. According to statistics in 2011, 89% of Japanese clinicians used Kampo formulae in their daily practice either as the sole source of therapy or in combination with modern drugs (JKMA, 2011). A study by Moschik et al. (2012) found that usage varied by specialty, with physicians/internists and gynecologists using up to 25 or more Kampo medicines, while orthopedic surgeons, pediatricians, and psychiatrists generally used only 1–4 formulae.

Kampo medicine today is at the centre of a thriving industry valued at about 151.9 billion yen (USD1.35 billion) in 2014 (JKMA, 2014). It is organized along the lines of pharmaceutical medicine, including the incorporation of a pharmacopoeia and the principles of Good Manufacturing Practice. The evolution has seen the emergence of standardized, modern, ready-to-use forms mainly as spray-dried granular extracts rather than decoctions. Some Kampo products have been registered as investigational new drugs (IND) by the US Food and Drug Administration (Watanabe et al., 2011). These are things that many traditional healing systems outside of Asia have failed to do.

What makes Kampo medicine different from other traditional healing systems is that it is practised only by medical doctors trained in conventional (or so-called Western) medicine (Moschik et al., 2012). Therefore, it is a unique case study of wholly integrated medicine. Thus, in Japan, Kampo medicine is both conventional and traditional, unlike healing systems in China, India, or Africa, where the practitioners of both disciplines are generally separated.

This integrative approach of Kampo has led to high acceptance of the practice – 248 medicines are codified in the pharmacopoeia and are reimbursable by medical insurance companies. Another consequence is that the practice seems to be empirically based – though individualized to specific patient factors – and eschews the inclusion of spirituality or medico-religious rituals. This is unusual for a traditional healing system.

It should be noted that there are also more traditional healing modalities (or folk remedies) operating outside Kampo in Japan. These include hot springs baths (onsen), herbalism, massage therapy, and acupuncture.

PRESENT SITUATION OF KAMPO FORMULATION IN JAPAN

The development of public health services and the development of excellent antibiotics have almost eradicated fatal infectious diseases in Japan. Now non-specific, constitutional, or psychosomatic diseases, such as arteriosclerotic diseases, autoimmune disorders, allergic diseases, malignant neoplasms, and degenerative diseases of the nervous system, have become the most pressing of medical problems. These diseases are often caused by an interaction of multiple factors, some of them still unknown. Since Western medicine, which focuses on eliminating the cause, is not good for diseases in which multiple factors are the cause, Kampo medicine has attracted attention. Kampo medicines are said to be philosophical and experimental, thus unlike conventional medicine where a single drug is often symptom-specific; in Kampo, one formulation (a mixture of multiple crude drugs) can be used for multiple symptoms

Kampo medicines are sold as medicines, separate from herbal products used as health foods. In the early 1970s, a list of 210 Kampo formulae suitable for over-the-counter (OTC) status was created from approximately 700 formulae in reference texts by experts and became known as the "210 Kampo Formulae". To date, 294 formulae are listed (Maegawa et al., 2014). These medicines have a history of use spanning many centuries, and there is no requirement for pre-clinical or clinical data under *The Approval Standards for OTC Drugs in Japan*.

In regards to reviews, the Pharmaceuticals and Medical Devices Agency (PMDA), which is the Japanese regulatory agency working together with the Ministry of Health (MHLW), must confirm that the active ingredients, contents, dosage and administration, and indications, all comply with *The Approval Standards* and that the standards and test methods used are appropriate to ensure product quality.

Four ethical (i.e., prescription) Kampo formulations have been covered by governmental insurance since 1972, as a result of the gradual acceptance of the role of Kampo medicine. After that, Kampo formulations that could be prescribed under governmental insurance gradually increased. Currently, medical doctors can prescribe 241 crude drugs as components of decoctions, 148 Kampo formulation extracts, and five crude drug preparations under the National Insurance System. Furthermore, only in Japan can medical doctors prescribe Kampo formulations along with Western drugs. Because of this, more than 80% of medical doctors prescribe Kampo formulations.

On the other hand, many OTC Kampo formulations are used in addition to the Kampo prescription formulations. The benefits of OTC Kampo formulations include convenience to patients, better self-management of minor illnesses, and a reduction in government medical costs. Therefore, OTC Kampo formulations are playing an important role in self-medication.

In form, Kampo medicines are extracts or dry powders made from a mixture of naturally derived herbal medicines and formulated into decoctions, powders, pills, or ointments (Maegawa et al., 2014). Some of the commonly used natural products are ginseng, rhubarb, glycyrrhizia (liquorice), peony, apricot and peach kernels, capsicum, and aloe. China is the main source of the raw materials. Some materials are sourced locally in Japan or from Korea, and others from far afield, in Sudan (senna) and South Africa (aloe).

EDUCATIONAL SYSTEM OF KAMPO MEDICINE IN JAPAN

One of the characteristic features of Kampo medicine in Japan is its prescription by medical doctors (Moschik et al., 2012). Since the end of the 19th century (the Meiji period), all medical students had to be educated according to Western medical curricula (Shimada, 2007). Kampo medicine was largely excluded from the medical education system until the 1960s. This resulted in many doctors of that era having little or no knowledge of Kampo medicine. The situation changed in the 1970s, when components of Kampo medicine began to be included in the curriculum. In 2001, the Japanese Ministry of Education, Culture, Sports, Science and Technology revised Japanese medical education to include Kampo as an essential part of the curriculum. Consequently, by 2007, Kampo medicine was incorporated into medical education in all 80 Japanese medical schools. However, there is as yet no standardized national curriculum for Kampo medicine offered by all universities.

CONCEPTS OF DISEASE AND HEALING IN KAMPO MEDICINE

Kampo treatment is arrived at based on a proper diagnosis (called *sho*) which is based on symptoms and physical examination (Terasawa, 2004b). Symptoms of importance include the patient's temperature, examining sensation, weakness, or sweating (symptoms which are not widely used in conventional medicine). Physical examination involves abdominal palpation, tongue inspection, and pulse diagnosis. This results in a diagnosis and staging of the disease amount and distribution of *Ki* (vital energy), *ketsu* (blood), and *sui* (body fluid).

Sho in Kampo Medicine

Sho in Kampo is a unified concept of several symptoms that appear in a disease. It is considered as a diagnosis, a state of health, or a disease progression stage similar to the concept of "syndrome" in Western medicine (Terasawa, 2004b). However, although syndrome plays an important role in diagnosing and specifying the disease, it does not directly lead to selection of a treatment method. The same (conventional) diagnosis may lead to different treatments (Watanabe et al., 2011).

The biggest difference between *sho* and "Disease state / diagnosis" (that is, a Western medical diagnostic label) is that *sho* is identified based on each individual's constitutional state. Furthermore, diagnosing *sho* directly influences the treatment method. Diagnosis of *sho* incorporates the two steps in Western medicine, "disease specification" and "treatment instruction," into one action.

Kampo medicine has paid attention to individual differences since its early days. The importance of identifying an individual patient's *sho* is that it allows for the selection of a drug with optimum efficacy while minimizing side effects and adverse events

The process of diagnosis and therapy in the traditional system of Kampo medicine is shown in Figure 7.1. Stage 1 refers to how the physician obtains the necessary information from the patient. Stage 2 combines analysis and evaluation of the information compiled in the first stage. In stage 3, the diagnosis of the patient's condition is made on the basis of the evaluated and analyzed symptoms. Stage 4 refers to the process of choosing the appropriate prescription. Stages 2 and 3 lead to *sho*. The concept of *sho* holistically evaluates the condition of the whole human being (not just the body) at the time of examination and is the basis of the final diagnosis and the choice of the corresponding Kampo formula. Figure 7.2 elucidates the supposed position of a patient's unbalanced condition. The vector (arrow) indicates the corresponding Kampo formula which is able to bring the patient back into his center.

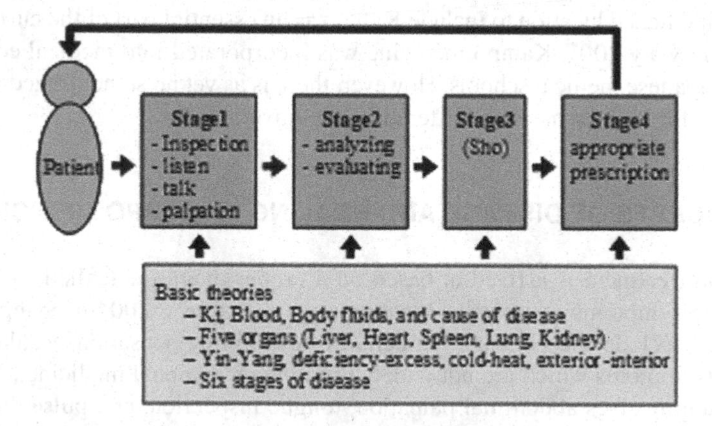

Figure 7.1 Process of diagnosis and treatment in Kampo medicine.

Each of the Kampo formulations possesses a certain dynamic potential to influence the condition of the patient in either a positive or negative manner according to his position prior to treatment (Figure 7.3). Thus it is essential to determine the patient's condition as accurately as possible. In the figure, the dynamic potential of the formula is represented by a vector (arrow). During a patient's clinical course, his condition undergoes certain changes, thus making it necessary to re-evaluate the appropriate formula for his new condition either by altering the dose or by changing the prescription itself.

The concept of *sho* is difficult to understand and has led to the World Health Organization (WHO) proposing the incorporation of the International Classification of Traditional Medicine, China, Japan, Korea (ICTM-CJK) into the International

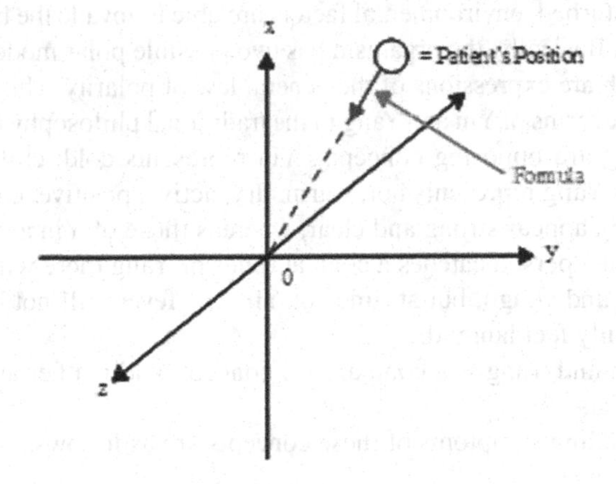

Figure 7.2 Position of imbalance in a patient.

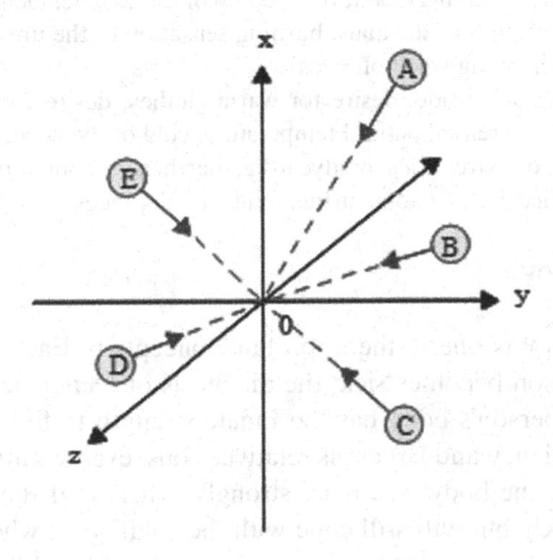

Figure 7.3 Dynamic potential of Kampo medicine to bring the patient to a position of balance.

Classification of Diseases (ICD)-11, while the Japanese government uses a double-coding system in order to maintain both the conventional ICD and that unique to Kampo (Choi & Chang, 2010). This harmonization will facilitate the interpretation of clinical research information.

THE CONCEPTS OF YIN–YANG, EXCESS–DEFICIENCY

Yin–Yang

In Kampo medicine, the living organism is considered to be built on a steady-state balance of various forces like *Ki*, *ketsu*, *sui*, actions of the five organs, and others. If this balance is disturbed, environmental factors are able to invade the body (Sunaga & Watanabe, 2017). Basically, the organism has two possible polar modes of reaction to noxa/insult, which are expressions of the general law of polarity. This law of polarity is described by the terms of Yin and Yang in the traditional philosophy of Eastern Asia.

Yin and Yang are opposing concepts. Yin represents cold, chill, wet, passive, negative, etc. and Yang represents hot, warm, dry, active, positive, etc. For example, symptoms of Yang appear strong and clear, whereas those of Yin are often latent. It is likely that when a person catches a cold, at times of Yang there will be high fever, strong headache, and cough, but at times of Yin, the fever will not be so high, and the patient may only feel languid.

However, Yin and Yang is a comparative concept, and neither absolute Yin nor absolute Yang exists.

The corresponding symptoms of these concepts are as follows:

- Yang: Feeling of heat inside, desire for thin clothes, sweating in the neck and head, desire for cooling and cold water, red face, hyperemic bulbar conjunctiva, increased natural temperature (high-grade fever), red tip of the tongue, tachycardia, diarrhea with burning sensation of the anus, burning sensation of the urethra on urination, hypertonic urine, strong smell of feces
- Yin: Feeling of cold inside, desire for warm clothes, desire for heating and hot water, pale face, decreased natural temperature, cold of the neck, back and lumbar region, coldness of extremities, bradycardia, diarrhea without burning sensation of the anus, pollakisuria, hypotonic urine, faint smell of feces

Excess–Deficiency

Excess–Deficiency is one of the important concepts in East Asian medical traditions. When a person becomes sick, the diagnosis of Deficiency or Excess determines whether the person's body has the innate strength to fight back or not. The relationship of Deficiency and Excess is relative. Thus, even within the same person, if the cold is strong, the body will react strongly, whereas if the cold is weak, the body will react weakly but will still cope with the cold. Also, when disease lingers, a reaction which first appeared strongly may be weakened and the patient may lose the strength to fight the disease.

On the other hand, Excess is sometimes used for classifying physical constitution when the body is in a healthy state. In other words, disease-resistance strength is often based on the defensive physical strength of the person at their normal state. However, it is not necessarily true that a person with Excess has stronger resistance against disease than a person with Deficiency. Often it is the opposite. People who have never been ill can become sick very easily, whereas people who have always been close to doctors and medicines could be strong against disease and even live longer. This means that Excess is not necessarily the better of the two. For example, a balloon that is fully inflated could burst easily but soft balloons will not. But, of course, to retain the function as a balloon, it must have an appropriate amount of air in it. This is the same for human beings, and thus it is important to preserve the balance of Deficiency and Excess.

- Excess: Strong voice, physiological skin color, skin eruption with redness, swelling and pain, severe pain (chest pain, abdominal pain, etc.), tight muscles, constipation with strong smell of feces, non-pitting edema, steer-horn stomach
- Deficiency: Weak voice, fatigue, pathologic skin color, sweating easily without physical exertion, night sweat, constipation with faint smell of feces, pitting edema, ptosis of organs

The terms of Excess and Deficiency are clinically used basically for recognizing a local condition in the organism. Yin and Yang are integral terms, and thus they include Excess and Deficiency. Yin and Yang are also used to describe the overall condition of the whole organism and its reaction to pathogenic challenge.

The position of each formula in relation to the coordinates reflects its inherent dynamic force, which has been empirically evaluated as being able to alter the corresponding condition of a patient, driving it towards the origin of the coordinates (Figure 7.4).

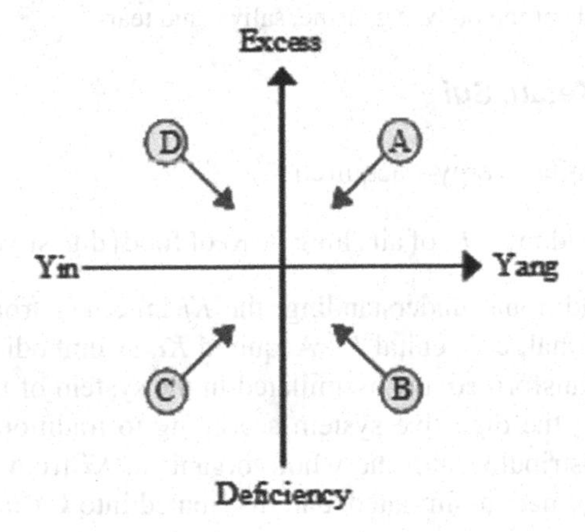

Figure 7.4 Inherent dynamic force of Kampo formula relative to Yin–Yang (Sunaga & Watanabe, 2017).

THE CONCEPTS OF *KI*, *KETSU*, AND *SUI*

Ki, *Ketsu*, *Sui*

In Kampo medicine, it is considered that the homeostasis of the organism is maintained by the three elements of *Ki*, *ketsu*, and *sui*.

(1) *Ki* (in Chinese Chi, Prana in Ayurveda, Vis Vitalis in the traditional European medical system): *Ki* is the elemental energy, which is necessary to life and living, and *Ki* is the invisible and intangible energy recognized only through its action. This elemental energy is the source of all forms of energy encountered on this planet and elsewhere, whether known or yet unknown to humans. According to the traditional medical systems in the Orient, this elemental energy or force (= *Ki*), is subdivided into other forms, such as *Ki* of the nervous system, *Ki* of the kidney, *Ki* of food, and *Ki* of blood.

Ki is also the source of nourishment and stimulation of the circulation of blood and body fluids, and is responsible for protecting the organism against external and internal noxa by way of circulating as Yang-*Ki* in the outer layers of the body (skin).

In the system of Kampo medicine, the human being is regarded as a microcosm reflecting the macrocosm, and hence the treatment aims not only at improving or regaining physical health but also takes the patient's psychic and mental imbalance into account. In this there are similarities with African cosmology, which places the individual as a part/continuum of life from the forebears to those yet to be born (past – present – future continuum) (Benyera, 2016).

(2) *Ketsu*: According to traditional understanding, this refers to the red-stained fluid of the organism.
(3) *Sui*: According to traditional understanding, this term refers to all the clear and not red-stained fluids of the body, e.g. urine, saliva, and tears.

Synthesis of *Ki*, *Ketsu*, *Sui*

$$\begin{aligned} Ki &= \text{congenital energy} + \text{acquired } Ki \\ &= Ki \text{ of kidney} + Ki \text{ of air (lung)} + Ki \text{ of food (digestive system)} \end{aligned}$$

According to traditional understanding, the *Ki* inherited from one's parents is the body's constitutional, congenital *Ki*. Acquired *Ki*, as embodied by the intake of water and food, is transformed and assimilated in the system of the spleen-stomach complex (resembling the digestive system according to traditional understanding), and is from there distributed into the whole organism. *Ki* from the air is taken in via the lungs, transformed, assimilated, and distributed into *ketsu*, *sui* and the whole organism.

Ki Deficiency (*kikyo*)

The concept of *Ki* refers to the energy containing everything and generating everything both in this cosmos as well as in all others. The term *Ki*-deficiency refers to a condition of insufficient volume of *Ki*.

Two processes cause *Ki*-deficiency, as follows:

(1) Disturbance of *Ki*-production:
- Primary deficiency of congenital *Ki*
- *Ki*-deficiency according to dysfunction of kidney
- *Ki*-deficiency according to dysfunction of lung and insufficient *Ki*-intake from air
- *Ki*-deficiency as a result of insufficient intake of *Ki* according to dysfunction of the digestive system (spleen–stomach)

(2) Excessive consumption of *Ki*:
- According to exterior causes
- According to interior causes
- According to progressive disease

Symptoms indicating a state of *Ki*-deficiency are declining psychic power, general malaise, fatigue, organ ptosis, declining sexual desire, impotence and declining level of activity of the whole organism.

Ki-Stagnation (*Ki*-Depression)

The term *Ki*-depression refers to a condition of insufficient circulation of *Ki*. The corresponding symptoms of this state are as follows:

- Head: depression, heavy feeling of the head, disturbance of consciousness
- Throat: feeling of constriction and strangulation, lump sensation in the throat
- Chest: Feeling of distress and accumulation of fluid in the chest, feeling of oppression and difficulty in breathing
- Costal region: tender on pressure, sensation of pain, feeling of constriction
- Abdomen: abdominal distention, sensation of gas retention
- Extremities: numbness with a sensation of swelling

Symptoms also related to a state of *Ki*-stagnation are colic-like pain and sensation of dull pain. All patients with *Ki*-stagnation have various degrees of depression, and complain obstinately. Symptoms show various degrees at time and often change localization.

Ki-Counterflow

Ki-counterflow is a condition due to disharmony of *Ki*-circulation occurring either between the inner and outer layers of the body or between the upper parts and lower parts of the organism. Four basic types of *Ki*-counterflow are distinguished, and the related symptoms are:

- Abdomen: Feeling of distress and constriction moving upwards to the chest, causing nausea and palpitation, and when moving, further proximal headache and restlessness occur, psychic disharmony.
- Chest: A feeling of fullness in the chest accompanied by coughing from the throat to the head, a rising feeling of distress, constriction of the throat, globus hystericus, ruddy face, impatience, aggressiveness, irritability.
- Epigastrium: Uneasy feeling, vomiting of gastric juices, nausea occurring only to a mild degree.
- Extremities: Coldness and pain moving up to the head and brain, spasmodic pain.

Ketsu Deficiency (Deficiency of Blood)

According to traditional understanding, deficiency of *ketsu* means a reduction of blood volume. A short list of possible factors causing this condition:

(1) Insufficient production of blood during growth
(2) Wasting diseases, surgery, malignant neoplasm
(3) Hemorrhage of the digestive tract, hypermenorrhea, hemorrhage of the genitals, bleeding hemorrhoids
(4) Drugs, poison, exposure to radioactive material

Accompanying symptoms of *ketsu* deficiency are insomnia, tachycardia, pale face, loss of weight, vertigo, scintillating scotoma, nail abnormality, alopecia, dystrophy of the skin, muscle cramps, and numbness in the limbs.

Ketsu Stasis (*oketsu*)

The term *ketsu* stasis refers to a state of insufficient blood circulation and blood stasis causing lesions of endothelial cells and hemorrhagic diathesis. Initiating factors of blood stasis are exogenous factors such as cold, heat, moisture, or surgery, and endogenous factors such as stress, lack of physical exercise, or constipation.

Symptoms accompanying *ketsu* stasis are restlessness, insomnia, nervousness, flush, myalgia, lumbago, dysmenorrhea, and hemorrhoids.

Sui Disturbance (Fluid Disturbance)

According to traditional concepts, this term implies disturbance of all non-red fluid of an organism, especially water metabolism. Symptoms of this condition are:

- Imbalance of secretion: Serous rhinitis, serous sputum, enhanced salivation, oliguria, watery diarrhoea.
- Stasis of body fluid: Edema, pleural effusion, ascites, sound of peristaltic movement, sound of fluctuating liquid in the epigastric region, palpitation of the abdominal aorta.

- Subjective symptoms: Palpitation, orthostatic hypotension, vertigo, motion sickness, tinnitus, headache, thirst, nausea, vomiting, morning stiffness, feeling of heaviness in the body, hypermetropia.

Four subtypes of this condition are differentiated:

(1) Whole body type: Edema of whole body, diarrhea, vertigo, pollakisuria
(2) Skin and joint type: Facial edema, swelling of joints, morning stiffness of joints
(3) Chest type: Serous sputum, pleural effusion, palpitation, agony in the chest
(4) Epigastric type: Sound of fluctuating liquid in the epigastric region, nausea, vomiting, diarrhea, sound of peristaltic movement

Transitions between these types are frequently observed.

The Theory of the Five Organs

According to traditional understanding, the five organs of the body are the liver, the heart, the spleen, the lung, and the kidney. These do not correspond precisely in meaning to those organs as anatomically defined, as in Western medicine.

Liver

Functions of the liver: (1) stabilizes and balances function and spiritual activity, (2) metabolism, (3) blood reservoir, supplying the body with essential nutrients, and (4) maintenance of muscle tonus.

Problems with the organ lead to epilepsy, emotional instability, aggressiveness, angry outbursts, malnutrition, growth disorder of hair and nails, and eye fatigue.

Heart

Functions of the heart: (1) maintenance of consciousness, (2) control of states of sleep/wakefulness, and (3) maintenance of blood circulation.

Disorders of the heart result in disturbance of consciousness, loss of consciousness, anxiety, tachycardia, insomnia, glossitis, and hiccup.

Spleen

Functions of the spleen: (1) digestive function, transforming food and water into *Ki*, (2) influencing blood viscosity and prevention of hemorrhage, and (3) influencing development and stabilization of function of muscles.

Spleen disorders result in impatience, depression, fatigability, diminished muscle power, tendency for hemorrhage, loss of appetite, diarrhea, and angular cheilitis.

Lungs

Functions of the lung: (1) intake of air-*Ki*, (2) oxygenizing blood and saturating blood with air-*Ki*, (3) influencing balance of body fluids and acid-base regulation, and (4) influencing functions of the skin and the circulation of Yang-*Ki* in the outer layers of the body.

Consequently, disorders of the lungs result in anxiety, melancholia, fatigability, dyspnea, blocked nasal passage, and excessive hidrosis.

Kidneys

Functions of the kidney are (1) control of growth, formation, and development, (2) influencing formation of bone and teeth, (3) regulating water metabolism and supports lung functions, and (4) influencing memory, decision-making and power of concentration.

Kidney dysfunction results in the deficiency of *Ki*, disturbance of development and growth, impotence, disturbance of calcium-phosphorus equilibrium and bone formation, disturbance of the water and acid-base metabolism, disturbance of urine excretion, vaginitis, and acoustic disturbance.

Five Elements, Cyclic Interrelation of the Five Organs with *Ki*, *Ketsu*, and *Sui*

The theory of the five elements originated around 300 BC in China (Jiuzhang & Lei, 2009). According to this theory, everything in this world is represented by five elements or symbols taken from nature, and various cyclic interrelations are interpreted by this theory (Pachuta, 1991). These five elements are wood, fire, earth, metal, and water. This is similar to the Greek theory of elements.

In Figure 7.5, the five elements, their related organs, and their interactions are described. As an example for elucidating clinical symptoms and their relation to the

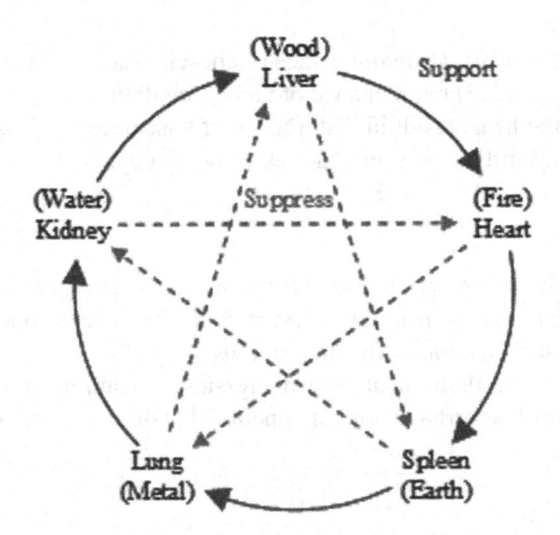

Figure 7.5 How elements and organs are related in East Asian medicine.

five organs and to the theory of the five elements, the following example may be helpful: aggressiveness/outburst of anger (liver in a state of Excess/Excess of wood) enhances the activity of the nervous system (heart/fire) and sedates appetite (spleen/earth), resulting in deficiency of *Ki* and weakening of the organism if this condition prevails.

CURRENT SITUATION OF KAMPO TREATMENT IN JAPAN

Although medical students have to learn Kampo medicine in Japan, there is not enough time for learning about the basic concepts of Kampo medicine. So almost all medical doctors only know that Kampo formulations are useful for various diseases, and so Kampo formulations are selected without utilizing sho. As a result, while there are more than 310,000 doctors in Japan, the number of fellows of the Japan Society for Oriental Medicine (i.e., specialists of Kampo Medicine) is only 2,039. However, Kampo formulations are prescribed for the treatment of many diseases, such as headache, insomnia, dementia, dizziness, allergic rhinitis, common cold, functional dyspepsia, constipation, irritable bowel syndrome, chronic kidney disease, pollakisuria, menstrual disorder, menopausal syndrome, atopic dermatitis, lumbago, knee steoarthritis, rheumatoid arthritis, edema, and muscle cramps. The most common ones are the common cold, constipation, muscle cramps, menopausal syndrome, bronchitis, gastritis, fatigue, and neurosis (Moschik et al., 2012).

Although Kampo formulations should be selected according to *sho*, many medical doctors select Kampo formulations according to disease name. Meanwhile, as a result of the use of Kampo medicine selected according to disease name, much clinical evidence has been accumulated. Below are some of the common conditions which illustrate the clinical utility of Kampo.

Common Cold

The common cold is an easily spread infectious disease of the upper respiratory tract. The symptoms include cough, painful throat, rhinorrhea, and fever. Over 200 different viruses can cause the common cold. The symptoms come from the immune system's response to the infection, not from direct destruction by the viruses themselves. There is no cure for the viral infection, but the symptoms can be treated.

The *Clinical Practice Guidelines on the Treatment of Respiratory Diseases* published by the Japanese Respiratory Society, in the guideline for the usage of Kampo medicine on acute infections, states that appropriate Kampo formulation in the early stage of a cold may provide immediate benefits. Kampo formulations are recommended as first-line therapy for the treatment of colds, except for influenza, as was advised in the *Shang han lun*.

There is some clinical evidence for the treatment of acute respiratory tract infections using Kampo formulations such as *maoto* (Ma Huang Tang) and *kakkonnto*. In a study using influenza virus-infected mice, *maoto* was shown to possess antipyretic activity and to reduce the effects of the virus when administered within 52 hours post-infection (Nagai et al., 2014). In a study with 45 patients and with oseltamivir as a control, *maoto* showed comparable efficacy in reducing fever and influenza symptoms (Saita et al., 2011)

Functional Dyspepsia

If patients have anorexia, nausea, vomiting, and heartburn originating in the upper digestive tract, and malignant tumor and peptic ulcer are ruled out in the diagnosis, these symptoms often indicate a functional gastrointestinal disorder such as functional dyspepsia. Functional gastrointestinal disorders are complex pathologic conditions. Western medicine having a singular site or mechanism of action is often not completely effective. For such pathologic conditions, Kampo medicines considered to have multiple sites of action are as effective as, or more effective than, Western therapies.

Randomized clinical trials have shown *rikkunshito* to be superior in therapeutic effects when compared to the gastroprokinetic agent cisapride (Miyoshi et al., 1991). It has also been reported to be efficacious in various types of dyspepsia, including non-ulcer dyspepsia (NUD) (Tatsuta & Iishi, 1993), post-gastrectomy symptoms (Gochi et al., 1995), and functional dyspepsia (Endo, 2008). *Rikkunshito* improves gastric emptying function and adaptive relaxation as well as stimulation of secretion of ghrelin, which has an appetite-increasing effect.

Because of the compelling clinical evidence for *rikkunshito* for the treatment of dyspepsia, it is now the drug of choice for the treatment of functional dyspepsia in Japan (Tominaga & Arakawa, 2013).

Post-Operative Ileus

Randomized clinical studies with *daikenchuto* on post-operative ileus have produced mixed results (Kono et al., 2011, 2016, Nishino et al., 2018). However, overall *daikenchuto* appears to have positive effects and may significantly prevent the need for surgery in patients with ileus.

Dementia

There are many clinical studies of Kampo formulations for the treatment of dementia. *Yokukansan* is often used for the treatment of behavioral and psychological symptoms of dementia (BPSD) with Alzheimer-type dementia. In a study with 20 patients, *yokukansan* significantly reduced various symptoms such as "delusions", "agitation/aggression", "apathy/indifference", "disinhibition", "irritability/lability", and "aberrant behavior" in the patients (Kimura et al., 2010). *Choto-san* is used for the treatment of vascular dementia and has shown good results in double-blind clinical trials (Terasawa et al., 1997). Components of the formulation have been shown to reduce glutamate-induced neuronal death in cultured cerebellar granule cells (Itoh et al., 1999).

Cancer

Kampo medicines have been found to relieve the side effects of cancer treatment and to improve the quality of life of patients (Iwase et al., 2012). In a study of over 300 physicians, 64% of the respondents cited that they used Kampo medicines to manage palliative cancer patients' symptoms, especially for numbness/hypoesthesia, constipation, poor appetite, weight loss, muscle cramps, and fatigue (Iwase et al., 2012, Mori et al., 2003). The most frequently prescribed medicines were *daiken-chuto, goshajinkigan, rikkunshito, hochuekkito, juzentaihoto*, and *hangeshashinto* (Amitani et al., 2015, Nagata et al., 2016).

CONCLUSION

Kampo medicine shows how traditional medicine can be harnessed for modern ills. Though its development into a modern science may seem to have been accidental (developing out of foreign influence and philosophies [Chinese], being rejected and side-lined for foreign healing modalities [German/Western medicine], re-emerging after the darkest period of Japanese history, and then eventually re-establishing itself as a truly integrative medical discipline), it also seems to reflect how much modern Japan has evolved, incorporating useful foreign influences yet leaning strongly on its own traditions and culture. There is, therefore, much that the world can learn from Japan.

REFERENCES

Amitani, M., Amitani, H., Sloan, R. A., Suzuki, H., Sameshima, N., Asakawa, A., Nerome, Y., Owaki, T., Inui, A. & Hoshino, E. 2015. The translational aspect of complementary and alternative medicine for cancer with particular emphasis on Kampo. *Frontiers in Pharmacology*, 6, 150.

Benyera, E. 2016. Expected yet uncomprehendible: Unpacking death through Nikolas Zakaria's Rufu Chitsidzo. *Gender and Behaviour*, 14(1), 7171–7181.

Buntrock, D. 1996. Without modernity: Japan's challenging modernization. *Architronic*, 5, 1–5.

Choi, S.-H. & Chang, I.-M. 2010. A milestone in codifying the wisdom of traditional oriental medicine: TCM, Kampo, TKM, TVM—WHO international standard terminologies on traditional medicine in the Western Pacific region. *Evidence-Based Complementary and Alternative Medicine*, 7, 303–305.

Endo, G. 2008. Analysis of the rikkunshito efficacy on patients with functional dyspepsia. *Japanese Journal of Medical and Pharmacological Science*, 60, 547–552.

Fujikawa, Y. 1979. *Nihon Igakushi (History of Japanese Medicine)*. Tokyo, Japan, Nagayama Shoten.

Galland, O. & Lemel, Y. 2008. Tradition vs. modernity: The continuing dichotomy of values in European society. *Revue française de sociologie*, 49, 153–186.

Gochi, A., Hirose, S., Sato, K., Hiramatsu, S., Asakura, A., Tokuoka, H., Matsuno, T., Kamikawa, Y. & Orita, K. 1995. The effect of Hange-shashin-to and Rikkunshi-to against the digestive symptoms after gastrectomy. *The Japanese Journal of Gastroenterological Surgery*, 28, 961–965.

Hiwat, F. 2016. Medicine in Edo Japan. Views from different angles on the history of Japanese medicine. Thesis: Leiden University.

Itoh, T., Shimada, Y. & Terasawa, K. 1999. Efficacy of Choto-san on vascular dementia and the protective effect of the hooks and stems of *Uncaria sinensis* on glutamate-induced neuronal death. *Mechanisms of Ageing and Development*, 111(2), 155–173.

Iwase, S., Yamaguchi, T., Miyaji, T., Terawaki, K., Inui, A. & Uezono, Y. 2012. The clinical use of Kampo medicines (traditional Japanese herbal treatments) for controlling cancer patients' symptoms in Japan: A national cross-sectional survey. *BMC Complementary and Alternative Medicine*, 12(1), 222.

Japan International Cooperation Agency. 2005. Japan's Experience in Public Health and Medical Systems. Tokyo, Japan, JICA.

Jiuzhang, M. & Lei, G. (eds) 2009. *A General Introduction to Traditional Chinese Medicine*. Baton Rouge, LA, Taylor & Francis.

JKMA. 2011. Kampoyakusyoho jittaityosa (teiryo) summary report. ⟨http://www.nikkankyo.org/aboutus/investigation/pdf/jittaichousa2011.pdf⟩ (in Japanese).

JKMA. 2014. Kamposeizai no seisandotai. ⟨http://www.nikkankyo.org/publication/movement/h24/all.pdf⟩ (in Japanese).

Kimura, T., Hayashida, H., Furukawa, H.& Takamatsu, J. 2010. Pilot study of pharmacological treatment for frontotemporal dementia: Effect of Yokukansan on behavioral symptoms. *Psychiatry and Clinical Neurosciences*, 64(2), 207–210.

Kono, T., Omiya, Y., Hira, Y., Kaneko, A., Chiba, S., Suzuki, T., Noguchi, M. & Watanabe, T. 2011. Daikenchuto (TU-100) ameliorates colon microvascular dysfunction via endogenous adrenomedullin in Crohn's disease rat model. *Journal of Gastroenterology*, 46(10), 1187.

Kono, T., Shimada, M., Yamamoto, M. & Kase, Y. 2016. Daikenchuto and GI disorders. In: Inui, A. (ed.) *Herbal Medicines. Methods in Pharmacology and Toxicology*. Humana Press, New York, pp. 165–180.

Lal, D. 2000. Does modernization require westernization? *The Independent Review*, V, 5–24.

Maegawa, H., Nakamura, T. & Saito, K. 2014. Regulation of traditional herbal medicinal products in Japan. *Journal of Ethnopharmacology*, 158(Pt B), 511–515.

Miyoshi, A., Yachi, A., Masamune, O., Ishikawa, M., Fukutomi, H., Niwa, H., Matsuo, Y., Mori, H. & Harasawa, S. 1991. Clinical evaluation of rikkunshito (TJ-43 TSUMURA & Co.) for indeterminate digestive complaints including chronic gastritis, etc. Multi-institutional study with reference drug of Cisapride. *Progress in Medicine*, 11, 1605–1631.

Mori, K., Kondo, T., Kamiyama, Y., Kano, Y. & Tominaga, K. 2003. Preventive effect of Kampo medicine (Hangeshashin-to) against irinotecan-induced diarrhea in advanced non-small-cell lung cancer. *Cancer Chemotherapy and Pharmacology*, 51(5), 403–406.

Moschik, E. C., Mercado, C., Yoshino, T., Matsuura, K. & Watanabe, K. 2012. Usage and attitudes of physicians in Japan concerning traditional Japanese medicine (Kampo medicine): A descriptive evaluation of a representative questionnaire-based survey. *Evidence-Based Complementary and Alternative Medicine*, 2012, 13.

Nagai, T., Kataoka, E., Aoki, Y., Hokari, R., Kiyohara, H. & Yamada, H. 2014. Alleviative effects of a Kampo (a Japanese herbal) medicine maoto (Ma-Huang-Tang) on the early phase of influenza virus infection and its possible mode of action. *Evidence-Based Complementary and Alternative Medicine*, 2014, 12.

Nagata, T., Toume, K., Long, L. X., Hirano, K., Watanabe, T., Sekine, S., Okumura, T., Komatsu, K. & Tsukada, K. 2016. Anticancer effect of a Kampo preparation daikenchuto. *Journal of Natural Medicines*, 70(3), 627–633.

Nishino, T., Yoshida, T., Goto, M., Inoue, S., Minato, T., Fujiwara, S., Yamamoto, Y., Furukita, Y., Yuasa, Y., Yamai, H., Takechi, H., Toba, H., Takizawa, H., Yoshida, M., Seike, J., Miyoshi, T. & Tangoku, A. 2018. The effects of the herbal medicine Daikenchuto (TJ-100) after esophageal cancer resection, open-label, randomized controlled trial. *Esophagus*, 15(2), 75–82.

Pachuta, D.M. 1991. Chinese medicine: The law of five elements. *India International Centre Quarterly*, 18(2/3), 41–68.

Saita, M., Naito, T., Boku, S., Watanabe, Y., Suzuki, M., Oka, F., Takahashi, M., Sakurai, T., Sugihara, E., Haniu, T., Uehara, Y., Mitsuhashi, K. Fukuda, H., Isonuma, H., Lee, K. & Kobayashi, H. 2011. The efficacy of ma-huang-tang (maoto) against influenza. *Health*, 3(5), 300–303.

Shimada, K. 2007. Beginning of modern anatomy in Japan from the perspective of anatomical bibliographies of the Meiji era. *Kaibogaku Zasshi*, 82, 9–20.

Sunaga, T. & Watanabe, K. 2017. Kampo medicine for human homeostasis. In: Somasundaram Arumugam S. & Watanabe K. (eds.) *Japanese Kampo Medicines for the Treatment of Common Diseases: Focus on Inflammation*. London, UK, Academic Press, pp. 13–22.

Tatsuta, M. & Iishi, H. 1993. Effect of treatment with Liu-Jun-Zi-Tang (TJ-43) on gastric emptying and gastrointestinal symptoms in dyspeptic patients. *Alimentary Pharmacology & Therapeutics*, 7, 459–462.

Terasawa, K. 2004a. Evidence-based reconstruction of Kampo medicine: Part I-Is Kampo CAM? *Evidence-Based Complementary and Alternative Medicine*, 1, 11–16.

Terasawa, K. 2004b. Evidence-based reconstruction of Kampo medicine: Part II-the concept of Sho. *Evidence-Based Complementary and Alternative Medicine*, 1, 119–123.

Terasawa, K., Shimada, Y., Kita, T., Yamamoto, T., Tosa, H., Tanaka, N. Saito, Y., Kanaki, E., Goto, S., Mizushima, N., Fujioka, M., Takase, S., Seki, H., Kimura, I., Ogawa, T., Nakamura, S., Araki, G., Maruyama, I., Maruyama, Y. & Takaori, S. 1997. Choto-san in the treatment of vascular dementia: A double-blind, placebo-controlled study. *Phytomedicine*, 4(1), 15–22.

Tominaga, K. & Arakawa, T. 2013. Kampo medicines for gastrointestinal tract disorders: A review of basic science and clinical evidence and their future application. *Journal of Gastroenterology*, 48(4), 452–462.

Watanabe, K., Matsuura, K., Gao, P., Hottenbacher, L., Tokunaga, H., Nishimura, K., Imazu, Y., Reissenweber, H. & Witt, C. M. 2011. Traditional Japanese Kampo medicine: Clinical research between modernity and traditional medicine-the state of research and methodological suggestions for the future. *Evidence-Based Complementary and Alternative Medicine*, 2011, 513842.

CHAPTER 8

Back to the Future – The Prospects of African Indigenous Crops as Future Foods

Callistus Bvenura, Estonce T. Gwata, and Felix D. Dakora

CONTENTS

ABSTRACT

Background

Food and nutrition insecurity are among the major challenges the world faces today. Incessant droughts caused by a rise in global temperatures and agricultural crop pests and diseases are some of the leading causes. Sub-Saharan Africa has the worst prevalence of hunger, and yet the continent is endowed with numerous indigenous crops that can potentially ameliorate this situation. However, these crops are still neglected and their demand is very low. The cost of production may be too high to justify cultivation, given the low yields achieved per hectare. It is therefore not surprising that cultivation is limited to small plots or intercropped with staple crops. However, some non-food industrial applications have been reported but not yet fully explored.

Relevance

In this chapter, we explore both the nutraceutical aspects and the possible industrial uses of four of these crops. We posit that if value-added products are formulated, incentives offered for cultivation, or contract farming initiated, these crops could be revitalized. These efforts could create employment for small-scale farmers and local communities and also help to protect biodiversity. If these crops continue to be underutilized and neglected, they could eventually be forgotten even in communities where they were previously known.

INTRODUCTION

Drought resulting from the effects of global warming and ever-evolving pests and crop diseases are arguably some of the most important threats to food production – and thus to food and nutrition security. In fact, the impact of the El Niño weather phenomenon recently left over 38 million people at risk in 17 African countries alone (IRIN 2017). Global hunger, which at one point was thought to be declining, has actually increased. In 2015, 777 million people were affected by hunger, and this increased to 815 million in 2016 (FAO 2017). These figures paint a gloomy picture of failing food production systems and increasing threats to food and nutrition security.

The consumption of underutilized indigenous crops has been the subject of discussion for over a decade globally (Bvenura and Sivakumar 2017). Indigenous plants are well adapted to local climates and pests, and have the ability to survive drought periods (Eriksen 2006). The substitution of the staple cereal maize with more drought-, pest-, and disease-resistant crops is therefore an option. Additionally, indigenous plants may avoid perceived negative characteristics of global staple crops. For example, the demand for and consumption of gluten-free foods have been on the increase globally. The negative effects of consumption of gluten-containing foods are associated with celiac disease, a hereditary disorder that leads to damage of the small intestine and nutrient malabsorption resulting in vitamin deficiency and anaemia (Cawthorn et al. 2010). Although data for Africa are not available, reports indicate that celiac disease might affect fewer people globally (about 1% of people of European descent) than generally thought (Mearin et al. 2005). Regardless of the low incidence of this condition, consumption of gluten-free foods has become much more popular, with more than 30% of US adults reportedly reducing or eliminating gluten in their diets (Moore 2014). This has been driven by multiple factors, including social media, marketing targeted directly at consumers, medical reports, and mainstream media which tout the clinical benefits of consuming a gluten-free diet (Niland and Cash 2018). Kim et al. (2017) showed that a gluten-free diet led to weight loss over a year and reduced waist circumference but had no significant effect on the prevalence of metabolic syndrome and cardiovascular disease risk in gluten-tolerant participants. Thus, although the consumption of gluten-free foods is clinically beneficial to gluten-intolerant people, the effects on gluten-tolerant populations need to be verified by further studies.

This chapter will review the literature on the phytochemical, nutritional, and health benefits of some gluten-free crops indigenous to the global South, with a special focus on Bambara groundnut (*Vigna subterranea* (L.) Verdc.), finger millet (*Eleusine coracana* (L.) Gaertn.), cassava (*Manihot esculenta* Crantz), and quinoa (*Chenopodium quinoa* L.). Their supply chains, industrial uses, and potential for innovation are also explored. Further, their antinutritional aspects are commented on. Most global staple crops contain some compounds that have an anti-nutritive effect, and this does not mean that they are not safe or nutritious foods. Although adverse health effects of antinutrients have been documented, some beneficial ones have also been reported, suggesting that the so-called antinutrient effect is concentration-dependent (Shahidi 1997). For example, lectins, phytic acid, phenolic compounds, saponins, and some enzyme inhibitors reduce blood glucose and reduce the risk of cancer when consumed in low quantities (Yoon et al. 1983). However, when a crop is used as a staple, preparation methods should perhaps be chosen to decrease these compounds to preserve or increase its nutritive value. The mainstreaming of these underutilized crops into the diet could help to lower food and nutrition insecurity in the face of climate variability.

BAMBARA GROUNDNUT (*VIGNA SUBTERRANEA* (L.) VERDC.)

Current Status

Bambara groundnut is an underutilized African indigenous crop of which there is a surviving wild variety (Gulzar and Minnaar 2017; Aremu et al. 2006). Bambara groundnut grows easily on all soil types with pH ranging from 5.0 to 6.5 in areas receiving rainfall between 600 and 1200 mm annually, with ideal temperatures ranging from 20 to 28 °C (Baryeh 2001; Goudoum et al. 2016). The crop sprouts between 7 and 15 days after planting and needs about 110 to 150 days to mature (DAFF 2011). Bambara groundnut can be intercropped with other crops, although it does well as a sole crop. The nuts are subterranean, making them difficult to mechanically harvest commercially. Bambara groundnut is therefore ideal for production by small-scale farmers (Hillocks et al. 2012). This leguminous crop lives in symbiosis with rhizobia and plays a role in nitrogen fixation and therefore participates in preserving soil fertility (Koné et al. 2011). Seven diverse Bambara groundnut colors have been recorded, viz., black, red, cream/black eye, cream/brown eye, cream/no eye, brown, and speckled/flecked/spotted (DAFF 2011). On the African continent, Nigeria, Burkina Faso, Ghana, Mali, Cameroon, and Ivory Coast are the known leading producers of this crop (DAFF 2016). However, Karikari et al. (1997) reported cultivation of Bambara groundnut by 90% of farmers in Botswana, of whom 63% produce for personal consumption, 12% for sale, and 25% for both. In South Africa, although production statistics are scanty, reports indicate cultivation in KwaZulu-Natal, the Eastern Cape, Mpumalanga, Limpopo, and North-West Provinces (Gqaleni 2014).

Bambara groundnut is of importance due to its drought, disease, and pest resilience. Misinformation about the crop has led to neglect and/or underutilization (Mkandawire 2007). The crop can be consumed as a snack, fresh, boiled, or grilled, and eaten alone or mixed with maize. Flour and nut milk have been extracted from the nuts (Gulzar and Minnaar 2016). In Zimbabwe, the seed is roasted (Gqaleni 2014) or mixed with maize and peanuts to make an afternoon snack known as "mutakura" in the Shona language. In Zambia, the flour is used for bread (Brough and Azam-Ali 1992). In Kenya, dry seeds are pounded to cook in stew, then mixed with cowpea and potatoes or ugali (pap or cornmeal porridge) and served as a side dish. In Botswana, the seeds are boiled with maize and seasoned with salt to taste. When dry and ripe, the seeds can be ground into a flour to bake cakes. In South Africa, nuts are boiled and stirred to prepare "tshipupu", a type of porridge considered a delicacy by the Venda-speaking people. The Sotho-speaking people make a thick dough mixed with maize called "dithaku" (Bamishaiye et al. 2011). Puddings and low fat yogurt have also been developed from this crop (Oyeyinka and Oyeyinka 2017).

Nutraceutical Effects

According to Bull et al. (2000), a nutraceutical is a food that provides health and medical benefits, including disease prevention and treatment. It may also be called a functional food. It combines aspects of food and drug, when those are treated as distinct conceptual or regulatory categories.

A wide range of foods, such as fortified and processed foods, supplements, and herbal products, have been identified as nutraceutical. The nutraceutical potential of Bambara groundnut has been explored by many authors, including Okafor and

Table 8.1 Comparison of the Nutritional Composition of Maize and Four Indigenous Crops

	Maize[20]	Finger millet	Bambara groundnut	Quinoa	Ref
Macronutrients (%)					
Protein	9.42	3.49–13	11.05–25	13.00–14.12	1, 2, 10, 14, 18,19
Lipids	4.74	1.2–5.2	1.4–8.62	6.07–7.40	3, 4, 10, 11, 16, 18, 19
Fiber	7.3	3.6–11.5	2.05–10	7.00–11.70	5, 6, 10, 11, 13, 19
Carbohydrates	74.3	72.6	49–73.87	64.16–71.00	5, 10, 14, 18, 19
Vitamin B1/ Thiamine (mg/100g)	0.385	0.42–0.48	0.47	0.36–0.60	5, 7, 15, 18, 19
Minerals (mg/100g)					
Mg	127	130–188.83	20.9–335	207.00–502.00	6, 8, 12, 13, 17, 20
Fe	2.71	1.90–6.3	4.25–5.9	2.60–15.00	8, 6, 13, 14, 17, 20
Mn	0.485	3.5–51	1.8–2.3	3.50–4.44	6, 8, 14, 20
Ca	7	207–368	37–77.7	54.00–149.00	8, 12, 13, 17, 20
Na	35	11	7.4–25.2	6.00–31.00	9, 12, 14, 20
Cu	0.314	0.47	0.12	0.68–9.50	9, 14, 20
Zn	2.21	1.95–4.27	3.9–25.6	0.79–4.00	8, 13, 14, 20

1. Virupaksha et al. (1975); 2. Stabursvik and Heide (1974); 3. Sridhar and Lakshminarayana (1994); 4. Singh and Raghuvanshi (2012); 5. Kumar et al. (2016b); 6. Gull et al. (2014); 7. Devi et al. (2014); 8. Hiremath et al. (2018); 9. Chandra et al. (2016); 10. Murevanhema and Jideani (2013); 11. Yao et al. (2015); 12. Amarteifio et al. (2006); 13. Olaleye et al. (2013); 14. Aremu et al. (2006); 15. Fadahunsi (2009); 16. Goudoum et al. (2016); 17. Hayat et al. (2014); 18. Steduto et al. (2012); 19. Norwak et al. (2016); 20. USDA (2009).

Jideani (2016). Bambara groundnut is rich in micronutrients (Table 8.1), flavonoids, and phenolic acids, among other polyphenols (Table 8.2).

Considered a complete food, Bambara groundnut contains up to 25%, 74%, 10%, and 8% protein, carbohydrates, fiber and fat, respectively (Table 8.1). Cow milk, on the other hand, contains 3.15% protein, 4.80% carbohydrates, 3.25% fat, 0% fiber, and 10mg/ 100g cholesterol (USDA 2018). However, milk contains 113 mg/100g of calcium in comparison with 78 mg/100g in Bambara groundnut (USDA 2018; Olaleye et al. 2013). Overall, the nutritional composition of Bambara groundnut is superior in comparison to milk. Globally, Heyman (2006) estimated about 70% of the world's population to be lactose intolerant. Lactose-intolerant individuals are unable to digest lactose, a sugar found in milk and dairy products. Lactose intolerance varies with ethnic origins and dairy product usage in the diet, and is common in much of Africa. Milk and dairy products also contain a high polyunsaturated to saturated fatty acid ratio, hence consumption of large volumes would increase lower-density lipoprotein (LDL) cholesterol in the blood (Vijaya Kumar et al. 2015). Therefore, the use of milk derived from Bambara groundnuts would be a major advantage, especially to the lactose-intolerant community.

Table 8.2 Some Polyphenolic Compounds Found in *Vigna subterranea*

Flavonoids	Phenolic acids
Apigenin[3]	Caffeic acid[3]
Apigenin –6,8-di C-hexoside[3]	Caffeic acid hexoside[3]
Catechin glucoside[1,3]	Caffeic acid derivative[3]
Catechin[1,3]	Chlorogenic acid[3]
Daidzein[3]	Gallic acid[3]
Di-hydrokaempherol deoxyhexoside[3]	Isoferulic acid[3]
Epicatechin[1,3]	p-Coumaric acid[1]
Epigallocatechin[3]	Pyrogallic acid[3]
Eriodictyol-7-rutinoside	Quinic acid[1,3]
Gallocatechin[3]	Salicylic acid[1]
Gallocatechin-hexoside[3]	Syringic acid derivative[3]
Gallocatechin-hexoside derivative[3]	trans-Ferulic acid[1]
Kaempherol hexoside[3]	Vanillic acid hexoside[3]
Kaempherol[2,3]	
Medioresinol[1,3]	
Myrecetin hexoside[1,3]	
Myricetin[2]	
Quercetin[3]	
Quercetin deoxyhexoside[3]	
Quercetin di-hexoside[3]	
Quercetin hexoside[3]	
Quercetin-3-O-rutinoside[1]	
Rutin[2,3]	

1. Nyau et al. (2005); 2. Harris et al. (2018); 3. Tsamo et al. (2018).

Antinutritional Composition

Although Bambara groundnut is a nutritionally rich and promising food, its gas-forming ability after consumption has proven to be one of its major concerns (Oyeyinka and Oyeyinka 2017). This happens when fructans and galacto-oligosaccharides, some poorly absorbed short chain carbohydrates, are rapidly fermented in the large bowel by gut bacteria, resulting in gas (Iacovou et al. 2015). This leads to hard-to-ignore flatulence (Mubaiwa et al. 2018). However, mixing the Bambara groundnut flour with a solution of 60% alcohol and malting has been shown to reduce flatus sugars (Hillocks et al. 2012). This treatment can also decrease antinutrient factors present in the flour (Alain et al. 2007).

Phytates, which are present in considerable amounts in Bambara groundnut, can have detrimental effects on health because they block iron and zinc absorption (Nagel 2010). In addition, phytates can lead to calcium and phosphorus deficiencies and inhibit enzymes like pepsin and amylase. However, besides their negative effects, phytates have beneficial effects such as inhibiting cancer (colon, mammary), heart disease, diabetes, and renal calculi (Jenab and Thompson 2001). Soaking, blanching, addition of salts, boiling, cooking, sprouting, and roasting reduce phytates and other antinutrients such as tannins, oxalates, and trypsin in the nuts (Ndidi et al. 2014). Bambara groundnut is among many edible plants that contain oxalic acid. Oxalic acid causes the formation of kidney stones, which can block the urinary tract (Moe 2006).

Industrial and Other Applications

Based on its chemical composition, Bambara groundnut has various potential uses in the pharmaceutical industry. Different types of proteins from the dried seeds have been extracted and their functions are well known. For example, 7s globulins are markers for allergic reactions, while 11s globulins are used as emulsifiers and surfactants. Also, vicilin is used in micro-particle preparation (Okpuzor et al. 2010; Nehete et al. 2013; Adebowale et al. 2011).

Due to the expensive nature of animal protein, the seeds are used as animal feed. This is because the crop is high in nitrogen and phosphorous (Bioversity International 2013; Goudoum et al. 2016). Also, Bambara groundnut has previously shown some potential as a source of biofuel, although the yield can be as low as 19% (Eze 2012). Due to the structure and size of starch granules in the parenchymatous cells of Bambara groundnut, some industrial starch has been extracted for use (Ashogbon 2017; Oyeyinka and Oyeyinka 2017; Mubaiwa et al. 2018). However, these have not been fully explored, although there is potential for a significant role.

FINGER MILLET (*ELEUSINE CORACANA* GAERTN.)

Current Status

Finger millet is one of the minor millets belonging to the grass family (Poaceae). Pearl millet (*Pennisetum glaucum* (L.) R. Br.), proso millet (*Panicum miliaceum* L.), barnyard millet (*Echinochloa esculenta* (A. Braun) H. Scholz), foxtail millet (*Setaria italica* (L.) P. Beauv.), and little millet (*Panicum sumatrense* Roth) are some of the minor millets belonging to the same family. Unlike these others, which belong to the subfamily Panicoideae, finger millet is placed in subfamily Chloridoideae. Finger millet is a short-season crop, highly drought- and pest-resistant, requiring about 300 mm of rainfall in comparison with 500–600 mm required by maize (Taylor and Duodu 2010). This makes it an important famine crop, especially in much of Southern Africa, which often experiences erratic onset of rains, low average seasonal rainfall, and dry spells during the rainy season or early end of rains. This crop is also known as the African millet, koracan, bulo (Uganda), ragi (India), wimbi (Swahili), telebun (the Sudan), wimbi (Swahili) (FAO 1995), uphoko (IsiNdebele), and njera (Shona). It is the sixth most important cultivated cereal in the world and is widely consumed in Africa, Asia, and Europe (de Wet 2006). By the year 2002, about 3.7 million tons were being annually produced in the world, with an average yield of 0.6-4 t/ha depending on environmental factors, cultivar, geographical region, and management factors (Gomez and Gupta 2003). In Asia, India was the major producer of the crop followed by Nepal, China, and Afghanistan; in Africa, Uganda followed by Ethiopia, Tanzania, and Zimbabwe were the top producers (Gomez and Gupta 2003).

Multiple finger millet uses in food are known. A multigrain flour containing finger millet is used to prepare chapatti in India (FAO 1995). Puffed or popped grains are an afternoon snack while puffed grains are ground to prepare a powder consumed by children, pregnant women, or lactating mothers (Kakade and Hathan 2015). Finger millet is mixed with wheat flour or soy to make noodles or vermicelli, and malted grains are used to prepare baby weaning food (Patel and Verma 2015). In addition, the grains are used as an ingredient to ferment foods such as dosa and idli in India and sour porridge in parts of Southern Africa. Bakery products such as bread and biscuits are also prepared from finger millet (Kakade and Hathan 2015). In most parts of Africa, the crop is used to prepare traditional opaque beer, while in Ethiopia, concentrated distilled alcohol known as apare is prepared from the flour (NRC 1996).

Nutraceutical Effects

The nutritional and health benefits of finger millet are seen in the nutritional composition of millet from multiple geographical regions (Table 8.1). Finger millet is arguably nutritionally superior to maize. Maize is Southern Africa's most important staple cereal crop. Although geographical regions, agronomic practices, and cultivars, among other factors, exert an influence on nutritional composition, studies indicate the general superiority of finger millet. For example, finger millet contains about 28, 9, 37, and 105% more protein, fat, fiber, and thiamine, respectively, in comparison with maize (Table 8.1). Mineral compositions are even higher, with the content of Mg, Fe, Mn, Ca, Cu, and Zn, respectively, 33, 57, 99, 98, 33, and 48% higher in comparison with maize. On average, carbohydrates are slightly lower in finger millet, but the difference seems negligible.

Cereals normally provide the starch required in a diet and are consumed with a relish, normally meat and/or vegetables. Finger millet has the potential to supply both the macronutrients (protein, fat, fiber, carbohydrates) and vitamins and minerals, including thiamine, required on a daily basis for a balanced diet. The inclusion of this crop in the diet could therefore benefit efforts to lower micro-nutritional deficiencies, especially in children, and contribute to the United Nations' (2014) Sustainable Development Goal for "Zero hunger". This is partly because crop failure will be minimized by finger millet's ability to withstand severe water stress. "Good health and well-being" is potentially achieved due to the superior nutritional content and dependability of finger millet.

A current literature search revealed 21 flavonoids and two phenolic acids that have been identified in finger millet (Table 8.3). Space does not permit a review of the voluminous literature that has revealed the antioxidant, anticancer, anti-inflammatory, and cardioprotective roles of many of these polyphenols.

Table 8.3 Polyphenolic Compounds Found in Finger Millet

Flavonoids	Phenolic acids
Apigenin[2]	Caffeic acid[1,2]
Catechin[2]	Cinnamic acid[1,2,4]
Daidzein[2]	Ferulic acid[1,2,4]
Epicatechin[2]	Gallic acid[1,2,4]
Epigallocatechin[2]	Gentisic acid[3,4]
Gallocatechin[2]	p-Coumaric acid[2,4]
Isoorientin[4]	p-hydroxybenzoic acid[1,2,4]
Isovitexin[2]	Protocatechuic acid[2,4]
Kaempherol[2]	Sinapic acid[2,4]
Luteolin[2]	Syringic acid[2,4]
Myricetin[2]	trans-Ferulic acid[2]
N-(p-coumaroyl) serotonin[2]	Vanillic acid[1,2,4]
Naringenin[2]	
Orientin[4]	
Procyanidin B1[2]	
Procyanidin B2[2]	
Quercetin[1,4]	
Saponarin[4]	
Taxifolin[2]	
Tricin[2]	
Violanthin[4]	
Vitexin[2,4]	

1. Mehta ot al. (2015); 2. Shahidi and Chandrasekar (2013); 3. McDonough and Rooney (2000), 4. Chethan and Malleshi (2007).

Suffice it to say that there is reason to believe polyphenols are a major reason for the observed health benefits of a plant-rich diet. Quercetin, catechin, epicatechin, and epigallocatechin, flavonoids found in finger millet, are active against breast, colon, and prostate tumour cell lines (Ghasemzadeh and Ghasemzadeh 2011). Gallocatechin has been linked with the prevention of metabolic syndrome, which increases risk of cardiovascular diseases (Legeay et al. 2015). Vitexin and isovitexin (vitexin isomer) are antioxidant, anticancer, anti-inflamatory, anti-hyperalgesic, and neuroprotective (He et al. 2016). Anticancer activities have also been reported for caffeic, gallic, and ferulic acids (Ghasemzadeh and Ghasemzadeh 2011). Dok-Go et al. (2003) found taxifolin (also known as dihydroquercitin) to inhibit oxidative stress and inflammation. Myricetin has been reported to ameliorate insulin resistance in some animal studies (Li and Ding 2012). Tricin, luteolin, and apigenin play cardioprotective roles and have anticancer and anti-aging properties (Lan et al. 2016; Luo et al. 2017; Yan et al. 2017). High dietary fiber has been shown to reduce the risk of diabetes (Devi et al. 2014). A diet high in fiber lowers postprandial insulin and glucose response due to the formation of non-absorbable complexes with available carbohydrates (Lafiandra et al. 2014).

Antinutritional Composition

In finger millet, Thompson (1993) reported the presence of trypsin inhibitors, phytate (0.48%), and tannins (0.61%). Higher values of tannins have been reported in brown-colored cultivars in numerous studies in comparison with white varieties. For example, in India, Udayasekhara Rao and Deosthale (1988) found tannins in white finger millet to be below the limit of detection, while brown cultivars contained 351 to 2392 mg/100g. Some African cultivars contain between 3.42 and 3.47% tannins (Kumar et al. 2016a). However, sprouting/germination drastically lowers the tannin content in finger millet (Mbithi-Mwikya et al. 2000). In general, cultivar genetics, environmental factors, and agronomic practices play a role in the concentration of these compounds.

Phytate which was initially present at 851.4 mg/100 g in whole flower, was reduced to about 238.5 mg/100 g after sprouting and to 333.1 mg/100 g after popping finger millet grains (Chauhan and Sarita 2018). Sharma et al. (2015) found that germinating seeds of foxtail millet (*Setaria*) reduced the phytate content from 2.803 to 0.341 mg/100 g. The hydrolysis of phytate phosphorus into inositol monophosphate during germination is thought to decrease the concentration of phytate in finger millet. On the other hand, the heat-labile nature of phytic acid also leads to its decrease in popped finger millet.

An early study reported oxalic acid content in finger millet of up to 45.7 mg/100 g (Rachic and Peters 1977). A more recent study reported lower values, between 21 and 29 mg/100 g depending on cultivar (Ravindran 1991). In addition, finger millet contains higher total oxalic acids than other cereals but interestingly contains the lowest soluble oxalate (45.9%). The oxalic acid in cereals is responsible for 15-20% inhibition of calcium bioavailability. Although finger millet contains more total oxalate, this cereal still contains more bioavailable calcium than other cereals. The work of Akbar et al. (2015) suggests that there is no correlation between calcium accumulation and oxalate accumulation in finger millet. In another study, Chauhan and Sarita (2018) found germination and popping to decrease oxalic acid content in finger millet. The high levels of oxalates in the cereal can therefore be lowered by cooking, sprouting, or fermentation.

The study of Panwar et al. (2016) showed that saponins were lower in finger millet, ranging between 2.13 and 2.63 mg/g, in comparison with 8.38 and 10.26 mg/g in some barnyard millet cultivars. Recently, although saponins were detected in fermented finger millet, a negative reading was shown in the roasted grains, indicating that saponins were drastically reduced during cooking (Aisoni et al. 2018). In addition, Rao et al. (2011) detected the presence of saponins in finger millet, although these authors did not quantify them.

Industrial and Other Applications

Due to its high energy and better protein and mineral compositions, finger millet is used to replace maize in poultry feed (Cisse et al. 2017). The straw contains about 61% digestible nutrients, which is better than sorghum, wheat, or pearl millet (FAO 1995). This makes finger millet a good fodder crop for both small and large livestock.

About 70% of finger millet seed is starch, hence the crop's potential in the non-food industry (Zhu 2014). Biodegradable polymer mud has been produced from pre-gelatinized starch for drilling operations (Ukachukwu et al. 2010). In addition, oxidized or acetylated starch from this crop has been a component of pharmaceutical tablets and capsules for drugs such as sulphadimidine, chloroquine phosphate, and chloroquine (Afolabi et al. 2012). The utilization of finger millet starch is rather peripheral in both the food and non-food industries. This is due to the abundant supply of starch from the traditional sources of maize and potato (Zhu 2014).

CASSAVA (*MANIHOT ESCULENTA* CRANTZ)

Current Status

Cassava is the third most important source of calories in the tropics after rice and maize (IFAD/FAO 2005). Although cassava grows under conditions of high humidity in more than 100 tropical and subtropical countries, the cultivated species is thought to have originated in the Amazon (Howeler et al. 2013). It was introduced to Africa in the 16th century by Portuguese traders from Brazil (IFAD/FAO 2005). This crop grows to 30° North and South of the equator at up to 2000 m above sea level. It prefers annual mean temperatures between 25 and 29°C and rainfall of 1000–1500 mm, although in some regions it thrives on about 500 mm. This makes the crop also drought-resilient. Soils where cassava yield is excellent are acidic (pH 4.5–6.5) and deep with transitional fertility (Lim 2016). Worldwide, about 129.02 million tons of cassava tubers are produced annually on about 14.150 million hectares (Balagopalan 1988). About 800 million people depend on the tuber (Howeler et al. 2013). About 60% of the sub-Saharan African population relies on cassava as a staple food (Oti et al. 2011). In South Africa, cassava is a secondary crop grown by smallholder farmers for commercial and food-grade starch, with about 20,000 tons produced annually (DAFF 2010).

Some of the advantages of cassava are its calorie production efficiency, flexibility in terms of timing of planting and harvesting, and pest and disease resilience (Oti et al. 2011). Fresh cassava is consumed right after harvesting, as its shelf life is about three days (Oti et al. 2011). At least five products are derived from cassava: dried and fresh "roots" (botanically a tuber), cassava leaves, and granulated and pasty products (FAO 2005). Roots of sweet cultivars are consumed raw, deep fried, boiled in water, or roasted in fire. The fermented or unfermented dried roots are ground into flour which can be turned into fufu or gari in West Africa. More recently, conventional wheat bread has found a competitor in cassava bread in Trinidad and Tobago (FAO 2014). Uncooked and steamed pastes can be derived from cassava and turned into various products, such as chickwangu in the DRC (Tewe and Lutaladio 2004).

In addition, granulated products such as attieke, tapioca, and gari are also common in West Africa. The leaves, on the other hand, are popular vegetables in Sierra Leone, Congo (pondu), and Tanzania (mpiru) (Aro et al. 2010). The leaves are eaten with a starch such as chikwange, rice, fufu, or as part of a sauce (Li et al. 2017).

Nutraceutical Effects

The leaves of cassava are consumed in Sierra Leone, Congo, Tanzania, and other parts of the world due to their rich nutritional content (Table 8.4). Cassava can produce 42, 127, 25 and 119% more calories per hectare per day than rice, wheat, maize, and sorghum, respectively (Balagopalan et al. 1988). Compared with maize, the roots are lower in lipids but higher in fiber and the majority of essential minerals (Table 8.1).

The antioxidant potential of cassava flour has been documented (Eleazu et al. 2011). Yi et al. (2010) were able to extract balanophonin, isovanillin, 6-deoxyjacareubin, syringaldehyde, p-coumaric acid, coniferaldehyde, pinoresinol, ethamivan, ficusol, and scopoletin, some important phenolic compounds, from the stem. Organic fertilizers increase polyphenolic compounds, including flavonoids and phenolic acids, and thereby increase the antioxidant activities of cassava tubers (Omar et al. 2012). In vivo studies have also shown cassava flavonoid leaf extracts to ameliorate carbon tetrachloride-induced injury in the livers of mice (Tao et al. 2014). The anticancer activity of esculentoic acids A and B, isolated from leaf extracts, showed moderate in vitro cytotoxicity against the A2780 human ovarian cancer cell line (Lim 2016). Cassava stem oil showed significant cytotoxicity against human

Table 8.4 Nutritional Composition of Cassava Roots and Leaves

	Cassava	
	Root	**Leaf**
Proximate (%)		
Protein	0.3–3.5	1.0–10
Lipids	0.03–0.5	0.2–2.9
Fiber	0.1–3.7	0.5–10
Carbohydrates	25.3–35.7	7–18.3
Vitamin B/ Thiamine (mg/100g)	0.03–0.28	0.06–0.31
Minerals (mg/100g)		
Mg	30	120
Fe	0.3–14	0.4–8.3
Mn	3000	72000
Ca	19–176	34–708
Na	76000	51000
Cu	2000	3000
Zn	14	71

Source: Montagnac et al. (2009).

erythromyeloblastoid leukemia cell lines with IC_{50} value of 7 μg/ml. Other potential bioactivities such as anti-inflammatory, antimicrobial, hypotensive, antihyperglycemic, antioxidant, antipyretic, analgesic, hepatoprotective, anthelmintic, amoebicidal, and prebiotic properties have also been reported (Lim 2016).

Although lactic acid bacteria can rarely convert starch into lactic acid, a probiotic beverage has been developed using fermented cassava flour, indicating the potential of cassava as a probiotic base (Vasudha and Mishra 2013).

Antinutritional Composition

The cyanogenic glycoside found in cassava is known as linamarin. Cyanide is a highly toxic end product of the hydrolysis of cyanogenic glycosides either in the presence of an enzyme or spontaneously. Cassava is well known for its high cyanogenic glycoside concentration, which can vary from 10–500 mg/kg in the roots and 53–1300 mg/kg in the leaves (Bolarinwa et al. 2016). Levels depend on environmental factors, plant age, and cultivar. According to the European Food Safety Authority (EFSA 2004), the lethal cyanide dose in humans lies between 0.5 and 3.5 mg/kg. However, these high concentrations of cyanogenic glycosides are drastically reduced by pounding, grinding, soaking, and fermentation (Centre for Food Safety 2007).

Antinutritional compounds such as oxalate, phytates, tannins, and non-protein amino acids have been reported in varying amounts in cassava (Li et al. 2017). However, Montagnac et al. (2009) showed that oven drying and fermentation of the roots reduced phytate by 85.6%. These authors also noted that sun-drying the leaves, with or without prior treatment, reduced phytate by 60%. In another study, Wobeto et al. (2007) found that cyanide, saponin, trypsin inhibitor, and oxalate concentrations in cassava leaves were age-dependant. In a 17-month study, the contents were lowest in the 12th month.

Industrial and Other Applications

Besides its culinary use, cassava has a wide range of uses, including the production of biofuels such as methane and ethanol, starch, and products such as biopolymers, medicine, cosmetics, and feed (Balagopalan 1988). Large volumes of suspended solids and organic matter waste resulting from these industries can potentially be bio-refined or fermented into value-added products and biochemicals (Zhang et al. 2016). Cassava is also useful in the textile production, adhesives, plastic, and detergent industries (Li et al. 2017; Tonukari et al. 2015).

Cassava is also useful in the manufacture of corrugated cardboard, wallpaper, and remoistened gums, in foundries, in the paper industry, and in drilling of wells as a viscosity-boosting agent (FAO 2014). In addition, the discarded stalks are used to form particle boards, while cassava alcohol and dried yeast have been made from the roots (FAO 2014). Biomedical application in paracetamol tablet manufacturing is another important use of this plant's roots (Lim 2016). Also, according to the same author, about 1000 mg/L cassava leaf tannin extract is sprayed on cassava plants to repel the cassava mealy bug (*Pseudococcus jackbeardsleyi*).

QUINOA (*CHENOPODIUM QUINOA* WILLD.)

Current Status

Quinoa is an ancient pseudo-cereal, belonging to the family Amaranthaceae, which originated from the Andes in South America (Navruz-Varli and Sanlier 2016). Also known by the Inca people as the "mother plant", this crop dates back 7,000 years (Vega-Gálvez et al. 2010). According to the cited authors, quinoa can be cultivated in temperatures ranging between –1° and 35°C, with annual rainfall as low as 200-400 mm and as high as 3000 mm, and soil pH from 6.0 to 8.5. Furthermore, the cultivation cycle can be as short as four months in arid areas and as long as eight months in rainy and humid areas. The yield can range from 0.85 to at least 4.5 t/ha depending on agro-ecological conditions and the burden of diseases and pests, including birds and rodents (Steduto et al. 2012). The crop is generally frost tolerant but only if the frost occurs before flowering. The edible parts of the plant are the seed, young leaves, and ears.

Over 35 food preparations derived from quinoa have been documented. These include use in drinks, pastries, soups, dry snacks, desserts, and beverages, including beer (Lim 2016). Due to its gluten-free nature, alternative bread, cakes, biscuits, and other flour-based products have been produced to cater for celiac disease-prone individuals (Steduto et al. 2012). A beverage rich in proteins made using extracts of quinoa, mesquite (*Prosopis chilensis* (Molina) Stuntz), and lupine (*Lupinus albus* L.) has been produced to tackle nutritional deficiencies in children (Cerezal et al. 2012). Some semi-processed ready-to-eat products, including flaked, grated, extruded, puffed, and hot cereals have been produced (Steduto et al. 2012). Hot or cold water is added to these cereals just before consumption. Also, quinoa sprouts are added to fresh salads while young leaves replace vegetables such as spinach (Lim 2016).

Nutraceutical Effects

The nutraceutical benefits of quinoa are associated with this crop's nutritional and polyphenolic compositions (Tables 8.1 & 8.5). Although the protein content is lower than that of the common bean, the lipid and carbohydrate values are higher. In comparison with maize, quinoa is superior in nutritional value. Phenolic acids constitute the major part of the polyphenols associated with quinoa (Table 8.5).

The use of quinoa seeds and leaves in traditional medicine is well known and well documented. It is now understood that the medicinal potential of this plant is associated with its diverse polyphenols, especially the phenolic acids. Antioxidant, anti-inflammatory, antimicrobial, anticancer, antidiabetic, anti-obesity, skin protective, immunoregulatory, hypocholesterolemic, and gastroprotective effects have been reported (Vilcacundo and Hernández-Ledesma 2017). In an antioxidant activity study, Yao et al. (2014) concluded that water- and alkali-based polysaccharides extracted from quinoa possess immunoregulatory effects. The saponin extract of

Table 8.5 Polyphenolic Compounds Found in Quinoa

	Phenolic acids	Flavonoids
1	2,4-dihydroxybenzoic acid[3]	Catechin[3]
2	4-hydroxybenzoic acid[1]	Hesperidin[1]
3	8,5′-diferulic acid[3]	Isoquercetin[1]
4	Caffeic acid[2,3]	Isorhamnetin[2]
5	Chlorogenic acid[1]	Kaempferol[1,2,3]
6	Ferulic acid[1,2,3]	Myricetin[2]
7	Gallic acid[1]	Neohesperidin[1]
8	5-Hydroxymethylfurfural[3]	Quercetin[1,2,3]
9	o-coumaric acid[1,3]	Rutin[1]
10	p-coumaric acid[1,2,3]	
11	p-OH-benzoic acid[2]	
12	Protocatechuic acid[3]	
13	Rosmarinic acid[1,3]	
14	Sinapic acid[3]	
15	Syringic acid[1,3]	
16	Vanillic acid[1,2,3]	

1. Pellegrini et al. (2018); 2. Repo-Carrasco-Valencia et al. (2010); 3. Tang et al. (2016).

quinoa was shown to inhibit the growth of *Candida albicans*, the predominant cause of fungal infections in people (Coral and Cusimamani 2014). Tang et al. (2016) found some phenolic compounds in quinoa to inhibit pancreatic lipase and α-glucosidase, critical enzymes involved in lipid and sugar digestion respectively. The inhibition of these enzymes is important since they are linked to obesity and diabetes management. In Ecuador, it was found that feeding four-year-old children 100 g of food containing quinoa grains increased their plasma insulin-like growth factor, suggesting that the quinoa was able to boost the essential nutrients needed by children for growth and development (Ruales et al. 2002). Furthermore, Farinazzi-Machado et al. (2012) found that daily consumption of quinoa by individuals aged between 18 and 45 over 30 days reduced LDL cholesterol, total cholesterol, and triglyceride levels. The same study reported non-significant decreases in body weight, blood glucose, and blood pressure levels.

Antinutritional Composition

Quinoa contains tannins, saponins, phytate, and protease inhibitors, as well as phenol derivatives, alkaloids, ecdysteroids, and monoterpenoids (Vega-Gálvez et al. 2010). In quinoa, phytate is found in the external layers and the endosperm of the seed (Filho et al. 2017). Saponins are found in appreciable amounts in this crop, and these give the grains a bitter taste. Washing of quinoa before cooking is customarily done to remove much of the saponin. In the presence of trypsin inhibitor in the intestinal tract, there can be elevated pancreatic enzyme production resulting in hypertrophy of this organ and growth reduction (Filho et al. 2017). However, in quinoa trypsin inhibitor is less than 0.05 mg/g, concentrations that are too low to be a cause for concern (Vega-Gálvez et al. 2010). Oxalates are abundant in plants from the families Amaranthaceae and Polygonaceae, including quinoa. However, this antinutrient is most abundant in the stems and leaves. In humans, oxalic acid cannot be metabolized and is therefore excreted through urine. Gastrointestinal irritation, decreased ability of blood to clot, excretory organ injury, and decreased bioavailability of some minerals are some of the negative effects potentially posed by high oxalic acid foods (Lopes et al. 2009). However, some popular foods (e.g., spinach) are high in oxalic acid but cause no harm to people who are not unusually susceptible.

A few cases involving allergic reactions to quinoa have been reported. A 52-year-old man in France reportedly had an allergic reaction, while a 29-year-old woman developed a rash and urticaria with itchy palms and soles of feet shortly after consuming quinoa-containing food (Navruz-Varli and Sanlier 2016). A 38-year-old female diagnosed with eosinophilic esophagitis due to wheat consumption was found in lab tests to have cross-sensitivity to quinoa (El-Qutob Lopéz et al. 2014).

Industrial and Other Applications

Quinoa is considered to be a pseudo-cereal and a complete food due to its very good protein quality and its unique balance between oil, protein, and fat (James 2009; Vega-Gálvez et al. 2010). However, besides its food uses, quinoa has diverse non-food applications. The leaves and seed coat can be used to produce an insect repellent, anti-aging gel, anthelmintic, and anti-obesity agent (Vega-Gálvez et al. 2010).

Technological properties like solubility, water-holding capacity, gelation, emulsifying, and foaming have given rise to diversified applications of quinoa (James 2009). Due to its small grains and high viscosity, quinoa starch exhibits a high potential for specialized industrial applications (Galwey 1992). Small-sized starch granules would be beneficial as dusting starches in cosmetics and as rubber tire mould release agents (Bhargava et al. 2006). Quinoa would be beneficial in emulsion products due to its gelatinization process starting at about 56–58°C, its single-stage swelling occurring between 65–95°C, and finally its opaque starch nature (Bhargava et al. 2006). Quinoa starch could be used as well in low-density polyethylene films as biodegradable fillers (Ahamed et al. 1996). Carrier bags can also be manufactured from quinoa starch because of its good mechanical properties (Bhargava et al. 2006). Quinoa starch has potential in frozen- and emulsion-type food products due to its freeze–thaw stability which has been shown to withstand retrogradation (Ahamed et al. 1996).

Quinoa saponins can be used in the detergent industry, as an anti-inflammatory, and as a disinfectant (Lim 2016). Also, Miranda ct al. (2018) showed that 0.05 and 0.20 g of lyophilized quinoa extract inhibited lipid oxidation, hydrolysis, and microbial activity in Atlantic chub mackerel (*Scomber colias*) fish stored in an ice system, indicating the potential of quinoa in fish preservation technologies.

SUPPLY CHAINS AND SCALABILITY

According to the National Research Council (1996), finger millet has the potential to attain 2 t/ha under optimum conditions. In the early 1990s, Africa produced about 2 million tons of the 4.5 million tons produced annually at a global scale. Recent statistics on the production of Bambara groundnut are scanty. However, some reports estimate production at 330,000 tons in Africa, with Nigeria leading (100,000 tons), followed by Burkina Faso, Ghana, Mali, Cameroon, and Ivory Coast (DAFF 2016). Although estimates indicate that about 172 million tons of cassava were globally produced in the year 2000, this figure has gone up to 200 million tons (IFAD/FAO 2005; DAFF 2010). Nigeria produces about 34 million tons, while South Africa currently produces about 200 000 tons, but for the starch industry (DAFF 2010). Since quinoa shows great potential as a nutraceutical and for non-food applications, production has been reported in more than 70 countries, including the Andean region, more recently in Europe, India, and the United States, as well as Kenya in Africa (FAO 2019). In South Africa, although the Agricultural Research Council has been involved in trials with quinoa since 2013, production by farmers has not been established (van Rensburg 2014).

Although these indigenous crops show immense food and nutrition security potential, their production is very low, particularly in drought prone regions. Cultivation is often limited to small plots or intercropping with staple crops such as maize. In the case of finger millet, the grain is often reserved for use in religious ceremonies and as malting agents when brewing traditional beer. In fact, Matsuhira (2013) reported the use of finger millet in rain-making ceremonies in the Nyandoro village of Zimbabwe. The millet is used to prepare a porridge for the spirit medium while the beer prepared from the grains is poured on the ground during the ceremony.

The potential for scaling up production of underutilized crops, particularly in Southern Africa, depends on a number of factors, including the development of commercial markets for the products. The right market price as determined by demand is critical for upscaling production. In Zimbabwe, for example, the 2018 price of finger millet and maize was $390/t each; however, some private companies were charging as much as $1200/t. In comparison, the South African Grain Marketing Board was selling white maize at an equivalent of about $205/t (depending on the foreign currency exchange rate). However, finger millet is generally not produced at a commercial level and is imported from India at varying prices. Due to the low demand as well as the low supply of finger millet, the crop is often sold by vendors at about $0.67/100g at the Marabastad market in Pretoria, South Africa. The majority of the vendors import the crop from Zimbabwe. These prices show the potential high returns in finger millet production.

The use of sorghum in the brewing industry in Southern Africa often results in contract farming. In this case, farmers are contracted by brewing companies to produce the grain. This same approach could be applied to other underutilized crops for commercial production. However, a market needs to be established beforehand. Formulation of new products could positively impact the production of the crops since such products have the potential to expand the demand for the crops, hence cultivation.

CHALLENGES WITH INDIGENOUS CROPS

The decline in knowledge, cultivation, and consumption of indigenous crops in Southern Africa and many parts of the world has been widely reported (Cordeiro 2013). This decline has been attributed to a number of factors, including westernization of the African diet, the stigma that indigenous crops are only famine or poor people's food, and the viewing of some foods as sacred and only to be handled by a specific group of people (Bvenura and Sivakumar 2017). One of the major challenges facing indigenous crops is low market demand due to some of the above-mentioned perceptions. This makes indigenous crops uneconomical for farmers, reinforcing the peripheral role they play in the diet. Formulation of new products and promotion of indigenous crops and nutraceuticals may stimulate demand and make their cultivation commercially viable.

Also, due to the low demand for the crops, the cost of production may be too high to make the low yields achieved per hectare for some crops profitable. For example, maize can produce over 5 t/ha in a good season in comparison with about 3 t/ha for finger millet. Furthermore, there is no ready market for finger millet in comparison with maize. Farmers, therefore, would rather produce crops such as finger millet on a contractual basis to avoid both crop and monetary losses.

There is a general lack of diversity, particularly in Southern African diets that limits the use of indigenous crops. In addition, the preparation of indigenous crops foods is often lengthy and requires a lot of electricity. This may not be ideal for city people who are concerned with saving electricity. In addition, the elderly are usually the holders of knowledge regarding the modes of cooking, and hence there is a general association of the foods with the rural populace and the elderly.

The majority of indigenous crops are an acquired taste and therefore need more time and familiarity before they can be appreciated. Their scarcity does not aid revitalization and successful introduction to new areas. The reintegration of indigenous crops into the diet requires an aggressive approach which may include agronomists, nutritionists, plant scientists, and environmentalists, among others, working together to achieve the same goal.

FUTURE PROSPECTS

The threat to biodiversity posed by the neglect of indigenous crops cannot be overemphasized. For a long time, researchers have been concerned about the role of indigenous crops in food and nutrition security. Although value addition for consumption has been explored by numerous authors (e.g., Dandsena and Banik 2016), non-food applications have not been given full attention.

For example, about 80% of the world starch market is held by maize, as it is used as a thickener, stabilizer, colloidal gelling agent, water retention agent, and adhesive in food manufacturing (Santana and Meireles 2014). However, indigenous crops such as finger millet, cassava, Bambara groundnut, and quinoa, among others, contain appreciable amounts of starch. These crops often require less than half the moisture required by maize for cultivation, meaning that they may not need costly irrigation interventions. Pharmaceutical companies could therefore engage resource-poor, small-scale farmers to produce these crops on a contractual basis. For the companies, this could become part of their contribution to community development.

Indigenous crops can also be useful as food, feed, and biofuels at lower production costs than conventional crops (Li et al. 2017). Some of these crops have the potential to be developed into livestock feed. Some highly nutritious and healthy multi-crop, gluten-free products have the potential to be produced from indigenous crops in addition to some food fortification options.

CONCLUSION

The benefits of these and other indigenous crops as food and their potential as sources of feed or in industries such as pharmacy are beyond doubt. However, sheer neglect and eventual underutilization seriously undermine their potential. Some of the advantages of these crops lie in their ability to resist drought, pests, and diseases. Although cultivation of some requires less moisture, the yield is lower than that of major crops such as maize, making cultivation less attractive to commercial farmers. Indigenous crops need to be revitalized through value addition in food, feed, and industrial uses. Production can be up-scaled through contract farming, and this could generate employment for small-scale farmers. If indigenous crops continue to be neglected, they could soon be forgotten even in areas where they were previously known and used. Revitalization efforts will require multidisciplinary efforts from agriculturalists, nutritionists, and environmentalists, and should be undertaken as quickly as possible.

REFERENCES

Adebowale YA, Schwarzenbolz U, Henle T (2011) Protein isolates from Bambara groundnut (*Voandzeia Subterranean* L. [sic]): Chemical characterization and functional properties. *Int J Food Prop* 14(4):758–775.

Afolabi TA, Olu-Owolabi BI, Adebowale KO, Lawal OS, Akintayo CO (2012) Functional and tableting properties of acetylated and oxidised finger millet (*Eleusine coracana*) starch. *Starch* 64:326–337.

Ahamed NT, Singhal RS, Kulkarni PR, Kale DD, Pal M (1996) Studies on *Chenopodium quinoa* and *Amaranthus paniculatas* starch as biodegradable fillers in LDPE films. *Carbohydr Polym* 31(3):157–160.

Aisoni JE, Yusha'u M, Orole OO (2018) Processing effects on physicochemical and proximate composition of finger millet (*Eleusine coracana*). *Greener J Biol Sci* 8(2):14–20.

Akbar N, Tiwari A, Pathak RK, Singh BR, Singh KP, Kumar A (2015) Is high grain calcium resulted due to oxalate accumulation in finger millet (*Eleusine coracana*)? Bioinformatics approaches for characterization of metabolic networks of oxalic acid biosynthesis. Paper presented at the National Symposium on germplasm to genes: harnessing biotechnology for food security and health, ICAR-National Research Centre on Plant Biotechnology, New Delhi, August 9–11 2015.

Alain MMM, Israël ML, René MS (2007) Improving the nutritional quality of cowpea and Bambara bean flours for use in infant feeding. *Pak J Nutr* 6(6):660–664.

Amarteifio JO, Tibe O, Njogu RM (2006) The mineral composition of Bambara groundnut (*Vigna subterranea* (L) Verdc) grown in Southern Africa. *Afr J Biotechnol* 5(23):2408–2411.

Aremu MO, Olaofe O, Akintayo ET (2006) Chemical composition and physicochemical characteristics of two varieties of Bambara groundnut (*Vigna subterrenea* [sic]) flours. *J Appl Sci* 6(9):1900–1903.

Aro SO, Aletor VA, Tewe OO, Agbede JO (2010) Nutritional potentials of cassava tuber wastes: A case study of a cassava starch processing factory in south-western Nigeria. *Livest Res Rural Dev* 22(11):42–47.

Ashogbon AO (2017) Evaluation of compositional and some physicochemical properties of Bambara groundnut and cocoyam starch blends for potential industrial applications. *Am J Food Nutr* 5(2):62–68.

Balagopalan C, Padma G, Nanda SK, Moorthy S (1988) *Cassava in Food, Feed and Industry*. CRC Press, Boca Raton, FL.

Bamishaiye OM, Adegbola JA, Bamishaiye JA (2011) Bambara groundnut: An under-utilized nut in Africa. *Adv Agric Biotechnol* 1:60–72.

Baryeh EA (2001) Physical properties of Bambara groundnuts. *J Food Eng* 47(4):321–326.

Bhargava A, Shukla S, Ohri D (2006) *Chenopodium quinoa* – An Indian perspective. *Ind Crops Prod* 23(1):73–87.

Bioversity International (2013) Nutritious underutilized species. Bambara groundnut (*Vigna subterranea*). Bioversity International, Rome, Italy.

Bolarinwa IF, Oke MO, Olaniyan SA, Ajala AS (2016) A review of cyanogenic glycosides in edible plants. In: Soloneski S, Larramendy ML (eds), *Toxicology – New Aspects to This Scientific Conundrum*. IntechOpen, London, UK. doi: 10.5772/64886.

Brough SH, Azam-Ali SN (1992) The effect of soil moisture on the proximate composition of Bambara groundnut (*Vigna subterrannea* L. Verdc.). *J Sci Food Agric* 60:197–203.

Bull E, Rapport L, Lockwood B (2000) 1What is a nutraceutical? *Pharm J* https://www.pharmaceutical-journal.com/learning/learning-article/1-what-is-a-nutraceutical/20002095.article.

Bvenura C, Sivakumar D (2017) The role of wild fruits and vegetables in delivering a healthy and balanced diet. *Food Res Int* 99:15–30.

Cawthorn DM, Steinman HA, Witthuhn RC (2010) Wheat and gluten in South African food products. *Food Agric Immunol* 21(2):91–102.

Centre for Food Safety (2007) Natural toxins in food plants. Risk assessment studies. Report No. 27. Food and Environmental Hygiene Department Hong Kong.

Cerezal Mezquita P, Acosta Barrientos E, Rojas Valdivia G, Romero Palacios N, Arcos Zavala R (2012) [Development of a high content protein beverage from Chilean mesquite, lupine and quinoa for the diet of pre-schoolers]. *Nutr Hosp* 27(1):232–243.

Chandra D, Chandra S, Pallavi, Sharma AK (2016) Review of finger millet (*Eleusine coracana* (L.) Gaertn): A power house of health benefiting nutrients. *Food Sci Hum Wellness* 5:149–155.

Chauhan ES, Sarita (2018) Effects of processing (germination and popping) on the nutritional and anti-nutritional properties of finger millet (*Eleusine coracana*). *Curr Res Nutr Food Sci* 6:2. doi: 10.12944/CRNFSJ.6.2.30.

Chethan S, Malleshi NG (2007) Finger millet polyphenols: Characterization and their nutraceutical potential. *Am J Food Technol* 2:582–592.

Cisse RS, Hamburg JD, Freeman ME, Davis AJ (2017) Using locally produced millet as a feed ingredient for poultry production in Sub-Saharan Africa. *J Appl Poult Res* 26(1):9–22

Coral LLT, Cusimamani EF (2014) An Andean ancient crop, Chenopodium quinoa Willd.: A review. *Agric Trop Subtrop* 47(4):142–146.

Cordeiro LS (2013) The role of African indigenous plants in promoting food security and health. In: Juliani HR, Simon JE, Ho CT (eds) *African Natural Plant Products, Vol. II: Discoveries and Challenges in Chemistry, Health, and Nutrition.* ACS Symposium Series, vol. 1127. American Chemical Society, Washington, D.C., pp. 273–287.

DAFF [Department of Agriculture, Forestry and Fisheries] (2010) Cassava – production guidelines. DAFF, Pretoria.

DAFF [Department of Agriculture, Forestry and Fisheries] (2011) Production guidelines for Bambara groundnut. DAFF, Pretoria.

DAFF [Department of Agriculture, Forestry and Fisheries] (2016) Bambara groundnut- (*Vigna subterranea*). DAFF, Pretoria.

Dandsena N, Banik A (2016) Processing and value addition of the underutilized agriculture crops and indigenous fruits of Bastar region of Chhattisgarh. *Int J Multidisc Res Dev* 3(3):214–223.

Devi PB, Vijayabharathi R, Sathyabama S, Malleshi NG, Priyadarisini VB (2014) Health benefits of finger millet (*Eleusine coracana* L.) polyphenols and dietary fiber: A review. *J Food Sci Technol* 51(6):1021–1040.

De Wet JMJ (2006) *Eleusine coracana.* In: Brink M, Belay G (eds) *Plant Resources of Tropical Africa, Vol. 1. Cereals and Pulses.* PROTA Foundation, Wageningen; Backhuys Publishers, Leiden; CTA, Wageningen, pp. 60–65.

Dok-Go H, Lee KH, Kim HJ, et al. (2003) Neuroprotective effects of antioxidative flavonoids, quercetin, dihydroquercetin and quercetin 3-méthyl ether isolated from *Opuntia ficus-indica* var. *saboten. Brain Res* 965(1–2):130–136.

EFSA (2004) Opinion of the Scientific Panel on Food Additives, Flavourings, Processing Aids and Materials in Contact with Food (AFC) on hydrocyanic acid in flavourings and other food ingredients with flavouring properties. Question no EFSA-Q-2003-145. *EFSA J* 105:1–28.

El-Qutob López D, Bartolomé Zavala B, Ortiz I (2014) Cross-reactivity between buckwheat and quinoa in a patient with eosinophilic esophagitis caused by wheat. *J Investig Allergol Clin Immunol* 24(1):56–71.

Eleazu CO, Amajor JU, Ikpeama AI, Awa E (2011) Studies on the nutrient composition, antioxidant activities, functional properties and microbial load of the flours of 10 elite cassava (*Manihot esculenta*) varieties. *Asian J Clin Nutr* 3:33–39.

Eriksen S (2006) The role of indigenous plants in household adaptation to climate change: The Kenyan experience. In: Low PS (ed.) *Climate Change and Africa.* Cambridge University Press, Cambridge, UK, pp. 248–259.

Eze SOO (2012) Physico-chemical properties of oil from some selected underutilized oil seeds available for biodiesel preparation. *Afr J Biotechnol* 11(42):10003–10007.

Fadahunsi S (2009) The effect of soaking, boiling and fermentation with *Rhizopus oligosporus* on the water soluble vitamin content of Bambara groundnut. *Pak J Nutr* 8:835–840.

FAO (1995) *Sorghum and Millets in Human Nutrition.* FAO Food and Nutrition Series, vol 27. Food and Agriculture Organization, Rome, Italy.

FAO (2014) Contributing to the development of a cassava industry. Subregional Office for the Caribbean. Issue Brief #11. Food and Agriculture Organization, Christ Church.

FAO (2017) How close are we to zero hunger? The state of food security and nutrition in the world 2017. http://www.fao.org/state-of-food-security-nutrition/en/.

FAO (2019) Quinoa. http://www.fao.org/quinoa/en/.

Farinazzi-Machado FMV, Barbalho SM, Oshiiwa M, Goulart R, Junior PO (2012) Use of cereal bars with quinoa (Chenopodium quinoa W.) to reduce risk factors related to cardiovascular diseases. *Cienc Tecnol Alim* 32(3):239–244.

Filho AMM, Pirozi MR, Da Silva Borges JM, Sant'Ana HMP, Chaves JPB, Coimbra JSDR (2017) Quinoa: Nutritional, functional, and antinutritional aspects. *Crit Rev Food Sci Nutr* 57(8):1618–1630.

Galwey NW (1992) The potential of quinoa as a multi-purpose crop for agricultural diversification: A review. *Ind Crops Prod* 1(2–4):101–106.

Ghasemzadeh A, Ghasemzadeh N (2011) Flavonoids and phenolic acids: Role and biochemical activity in plants and human. *J Med Plants Res* 5(31):6697–6703.

Gomez MI, Gupta SC (2003) Millets. In: Trugo L, Finglas PM (eds) *Encyclopedia of Food Sciences and Nutrition*, 2nd edn. Academic Press, Amsterdam, the Netherlands, pp. 3974–3979.

Goudoum A, Ngamo Tinkeu LS, Madou C, Djakissam W, Mbofung CM (2016) Variation of some chemical and functional properties of Bambara groundnut (Voandzeia Subterranean L. Thouars) during sort time storage. *Food Sci Technol* 36(2):290–295.

Gqaleni P (2014) Nutritional value of Bambara groundnut (Vigna subterranea (L.) Verdc.): A human and animal perspective. Dissertation, University of KwaZulu-Natal.

Gull A, Jan R, Nayik GA, Prasad K, Kumar P (2014) Significance of finger millet in nutrition, health and value added products: A review. *J Environ Sci Comput Sci Eng Technol* 3(3):1601–1608.

Gulzar M, Minnaar A (2016) Underutilized protein resources from African legumes. In: Nadathur SR, Wanasundara JPD, Scanlin L (eds) *Sustainable Protein Sources*, 1st ed. Academic Press, London, UK, pp. 197–208.

Harris T, Jideani V, Le Roes-Hill M (2018) Flavonoids and tannin composition of Bambara groundnut (Vigna subterranea) of Mpumalanga, South Africa. *Heliyon* 4:e00833.

Hayat I, Ahmad A, Masud T, Ahmed A, Bashir S (2014) Nutritional and health perspectives of beans (Phaseolus vulgaris L.): An overview. *Crit Rev Food Sci Nutr* 54:580–592.

He M, Min JW, Kong WL, He XH, Li JX, Peng BW (2016) A review on the pharmacological effects of vitexin and isovitexin. *Fitoterapia* 115:74–85.

Heyman MB (2006) Committee on Nutrition. Lactose intolerance in infants, children, and adolescents. *Pediatrics* 118(3):1279–1286.

Hillocks RJ, Bennett C, Mponda OM (2012) Bambara nut: A review of utilisation, market potential and crop improvement. *Afr Crop Sci J* 20(1):1–16.

Hiremath N, Geetha K, Vikram SR, Nanja YA, Joshi N, Shivaleela HB (2018) Minerals content in finger millet [Eleusine coracana (L.) Gaertn]: A future grain for nutritional security. *Int J Curr Microbiol Appl Sci* 7:3448–3455.

Howeler R, Lutaladio N, Thomas G (2013) Save and grow: Cassava. A guide to sustainable production intensification. FAO, Rome, Italy.

Iacovou M, Tan V, Muir JG, Gibson PR (2015) The low FODMAP diet and its application in East and Southeast Asia. *J Neurogastroenterol Motil* 21(4):459–470.

IFAD/FAO (2005) A review of cassava in Africa with country case studies on Nigeria, Ghana, the United Republic of Tanzania, Uganda and Benin. Proceedings of the validation forum on the global cassava development strategy, vol 2. Food and Agriculture Organization, Rome, Italy.

IRIN (2017) Drought in Africa. IRIN, Nairobi. https://www.irinnews.org/feature/2017/03/17/drought-africa-2017.

James LEA (2009) Quinoa (*Chenopodium quinoa* Willd.): Composition, chemistry, nutritional, and functional properties. *Adv Food Nutr Res* 58:1–31.

Jenab M, Thompson LU (2001) Role of phytic acid in cancer and other diseases. In: Reddy NR, Sathe SK (eds) *Food Phytates*. CRC Press, Boca Raton, FL, pp. 225–248.

Kakade SB, Hathan BS (2015) Finger millet processing: Review. *Int J Agric Innov Res* 3(4):1003–1008.

Karikari SK, Wigglesworth DJ, Kwerepe BC, Balole TV, Sebolai B, Munthali DC (1997) Country reports: Botswana. In: Heller J, Begemann F, Mushonga J (eds), *Bambara groundnut (Vigna subterranea (L.) Verdc.)*. Proceedings of the Workshop on Conservation and Improvement of Bambara groundnut (*Vigna subterranea* (L.) Verdc.), 14–16 November 1995, Harare, Zimbabwe. IPGRI, Rome, Italy, pp. 11–18.

Kim HS, Demyen MF, Mathew J, Kothari N, Feurdean M, Ahlawat SK (2017) Obesity, metabolic syndrome, and cardiovascular risk in gluten-free followers without celiac disease in the United States: Results from the National Health and Nutrition Examination Survey 2009–2014. *Dig Dis Sci* 62(9):2440–2448.

Koné M, Paice AG, Touré Y (2011) Bambara groundnut [*Vigna subterranea* (L.) Verdc. (Fabaceae)] usage in human health. In: Preedy VR, Watson RR, Patel VB (eds) *Nuts and Seeds in Health and Disease Prevention*. Academic Press, London, UK, pp. 189–196.

Kumar A, Metwal M, Kaur S, et al. (2016a) Nutraceutical value of finger millet [*Eleusine coracana* (L.) Gaertn.], and their improvement using omics approaches. *Front Plant Sci* 7:934.

Kumar SI, Babu CG, Reddy VC, Swathi B (2016b) Anti-nutritional factors in finger millet. *J Nutr Food Sci* 6(3):491.

Lafiandra D, Riccardi G, Shewry PR (2014) Review – Improving cereal grain carbohydrates for diet and health. *J Cereal Sci* 59:312–326.

Lan W, Rencoret J, Lu F, et al. (2016) Tricin-lignins: Occurrence and quantitation of tricin in relation to phylogeny. *Plant J* 88:1046–1057.

Legeay S, Rodier M, Fillon L, Faure S, Clere N (2015) Epigallocatechin gallate: A review of its beneficial properties to prevent metabolic syndrome. *Nutrients* 7:5443–5468.

Li S, Cui Y, Zhou Y, Luo Z, Liu J, Zhao M (2017) The industrial applications of cassava: Current status, opportunities and prospects. *J Sci Food Agric* 97(8):2282–2290.

Li Y, Ding Y (2012) Minireview: Therapeutic potential of myricetin in diabetes mellitus. *Food Sci Hum Wellness* 1(1):19–25.

Lim TK (2016) *Edible Medicinal and Non-Medicinal Plants, Vol 5, Fruits*. Springer, Dordrecht, the Netherlands.

Lopes CO, Dessimoni GV, Da Silva MC, Vieira G, Pinto NAVD (2009) Aproveitamento, composição nutricional e antinutricional da farinha de quinoa (*Chenoipodium quinoa*). *Alim Nutr* 20(4):669–675.

Luo Y, Shang P, Li D (2017) Luteolin: A flavonoid that has multiple cardio-protective effects and its molecular mechanisms. *Front Pharmacol* 8:692.

Matsuhira Y (2013) Rain making ceremony in the Nyandoro region, Zimbabwe. *Afr Relig Dyn* 1:165–182.

Mbithi-Mwikya S, van Camp J, Yiru Y, Huyghebaert A (2000) Nutrient and antinutrient changes in finger millet (*Eleusine coracan*) during sprouting. *Lebensm Wiss Technol* 33:9–14.

McDonough CM, Rooney LW (2000) The millets. In: Kulp K, Ponte Jr JG (eds) *Handbook of Cereal Science and Technology*. Marcel Dekker, Inc., New York, pp. 177–201.

Mearin ML, Ivarsson A, Dickey W 2005. Coeliac disease: Is it time for mass screening? *Best Pract Res Clin Gastroenterol* 19:441–452.

Mehta JP, Fultariya CR, Parmar PH, Vadia SH, Golakiya BA (2015) Determination of phenolic acids and a flavonoid in *Eleusine coracana* L. by semi-preparative HPLC photo diode array detector. *J Anal Chem* 70(3):369–373.

Miranda JM, Carrera M, Pastén A, Vega-Gálvez A, Barros-Velázquez J, Aubourg SP (2018) The impact of quinoa (*Chenopodium quinoa* Willd.) ethanolic extracts in the icing medium on quality loss of Atlantic chub mackerel (*Scomber colias*) under chilling storage. *Eur J Lipid Sci Technol* 120:1800280.

Mkandawire CH (2007) Review of Bambara groundnut (*Vigna subterranea* (L.) Verdc.) production in sub-Saharan Africa. *Agric J* 2:464–470.

Moe OW (2006) Kidney stones: Pathophysiology and medical management. *Lancet* 367(9507):333–344.

Montagnac JA, Christopher DR, Tanumihardjo SA (2009) Processing techniques to reduce toxicity and antinutrients of cassava for use as a staple food. *Compr Rev Food Sci Food Saf* 8(1):17–27.

Moore LR (2014) "But we're not hypochondriacs": The changing shape of gluten-free dieting and the contested illness experience. *Soc Sci Med* 105:76–83.

Mubaiwa J, Fogliano V, Chidewe C, Linnemann AR (2018) Bambara groundnut (*Vigna subterranea* (L.) Verdc.) flour: A functional ingredient to favour the use of an unexploited sustainable protein source. *PLoS ONE* 13(10):e0205776.

Murevanhema YY, Jideani VA (2013) Potential of Bambara groundnut (*Vigna subterranea* (L.) Verdc) milk as a probiotic beverage: A Review. *Crit Rev Food Sci Nutr* 53(9): 954–967.

Nagel R (2010) Living with phytic acid. The Weston A. Price Foundation, Washington, D.C. https://www.westonaprice.org/health-topics/vegetarianism-and-plant-foods/living-with-phytic-acid/.

National Research Council (1996) *Lost Crops of Africa. Volume 1: Grains. Finger Millet.* National Academy Press, Washington, D.C.

Navruz-Varli S, Sanlier N (2016) Nutritional and health benefits of quinoa (*Chenopodium quinoa* Willd.). *J Cereal Sci* 69:371–376.

Ndidi US, Ndidi CU, Aimola IA, Bassa OY, Mankilik M, Adamu Z (2014) Effects of processing (boiling and roasting) on the nutritional and antinutritional properties of Bambara groundnuts (*Vigna subterranea* [L.] Verdc.) from Southern Kaduna, Nigeria. *J Food Process* 2014:9.

Nehete JY, Bhambar RS, Narkhede MR, Gawali SR (2013) Natural proteins: Sources, isolation, characterization and applications. *Pharmacogn Rev* 7(14):107–116.

Niland B, Cash BD (2018) Health benefits and adverse effects of a gluten-free diet in non-celiac disease patients. *Gastroenterol Hepatol* 14(2):82–91.

Nowak V, Du J, Charrondière RU (2016) Assessment of the nutritional composition of quinoa (*Chenopodium quinoa* Willd.). *Food Chem* 193:47–54.

Nyau V, Prakash S, Rodrigues J, Farrant J (2005) Identification of nutraceutical phenolic compounds in Bambara groundnuts (*Vigna subterranea* L. Verdc) by HPLC-PDA-ESI-MS. *Br J Appl Sci Technol* 6(1):77–85.

Okafor JN, Jideani VA (2016) Bambara groundnut (*Vigna subterranea* L. verdc) nutraceutical: An unexplored resource fobolar functional and health foods. Presented at the 10th Annual Global Summit on Food Safety, Processing & Technology, San Antonio, TX, December 5–7, 201.

Okpuzor J, Ogbunugafor HA, Okafor U, Sofidiya MO (2010) Identification of protein types in Bambara nut seeds: Perspectives for dietary protein supply in developing countries. *EXCLI J* 9:17–28.

Olaleye AA, Adeyeye EI, Adesina AJ (2013) Chemical composition of bambara groundnut (*V. subterranea* L. Verdc) seed parts. *Bangladesh J Sci Ind Res* 48(3):167–178.

Omar NF, Hassan SA, Yusoff UK, Abdullah NA, Wahab PE, Sinniah U (2012) Phenolics, flavonoids, antioxidant activity and cyanogenic glycosides of organic and mineral-base fertilized cassava tubers. *Molecules* 17(3):2378–2387.

Oti E, Olapeju O, Dohou S, et al. (2011) *Training Manual. Processing of Cassava into Gari and High-Quality Cassava Flour in West Africa.* CORAF/WECARD, Dakar.

Oyeyinka SA, Oyeyinka AT (2017) A review on isolation, composition, physicochemical properties and modification of Bambara groundnut starch. *Food Hydrocolloids* 75:62–71.

Panwar P, Dubey A, Verma AK (2016) Evaluation of nutraceutical and antinutritional properties in barnyard and finger millet varieties grown in Himalayan region. *J Food Sci Technol* 53(6):2779–2787.

Patel S, Verma V (2015) Ways for better utilization of finger millet through processing and value addition and enhance nutritional security among tribals. *Global J Med Res (L) Nutr Food Sci* 15(1):23–29.

Pellegrini M, Lucas-Gonzales R, Ricci A, et al. (2018) Chemical, fatty acid, polyphenolic profile, techno-functional and antioxidant properties of flours obtained from quinoa (*Chenopodium quinoa* Willd) seeds. *Ind Crops Prod* 111:38–46.

Rachic KO, Peters LV (1977) *The Eleusines: A Review of the World Literature.* International Crops Research Institute for the Semi-Arid Tropics, Hyderabad.

Rao BR, Nagasampige MH, Ravikiran M (2011) Evaluation of nutraceutical properties of selected small millets. *J Pharm Bioall Sci* 3:277–279.

Ravindran G (1991) Studies on millets: Proximate composition, mineral composition and phytate and oxalate contents. *Food Chem* 39:99–107.

Repo-Carrasco-Valencia R, Hellström JK, Pihlava JM, Mattila PH (2010) Flavonoids and other phenolic compounds in Andean indigenous grains: Quinoa (*Chenopodium quinoa*), kañiwa (*Chenopodium pallidicaule*) and kiwicha (*Amaranthus caudatus*). *Food Chem* 120:128–133.

Ruales J, De Grijalva Y, Lopez-Jaramillo P, Nair BM (2002) The nutritional quality of infant food from quinoa and its effect on the plasma level of insulin-like growth factor-1 (IGF-1) in undernourished children. *Int J Food Sci Nutr* 53:143–154.

Santana ÁL, Meireles MAA (2014) New starches are the trend for industry applications: A review. *Food Public Health* 4(5):229–241.

Shahidi F (1997) Beneficial health effects and drawbacks of antinutrients and phytochemicals in foods. In: Shahidi F (ed.) *Antinutrients and Phytochemicals in Food.* ACS Symposium Series 662. American Chemical Society, Washington, D.C., pp. 1–9.

Shahidi F, Chandrasekar A (2013) Millet grain phenolics and their role in disease risk reduction and health promotion: A review. *J Funct Foods* 5:570–581.

Sharma S, Saxena DC, Riar CS (2015) Antioxidant activity, total phenolics, flavonoids and antinutritional characteristics of germinated foxtail millet (*Setaria italica*). *Cogent Food Agric* 1:1081728.

Singh P, Raghuvanshi RS (2012) Finger millet for food and nutritional security. *Afr J Food Sci* 6(4):77–84.

Sridhar R, Lakshminarayana G (1994) Content of total lipids and lipid classes and composition of fatty acids in small millets: Foxtail (*Setaria italica*), proso (*Panicum miliaceum*), and finger (*Eleusine coracana*). *Cereal Chem* 71(4):355–358.

Stabursvik A, Heide OM (1974) Protein content and amino acid spectrum of finger millet [*Eleusine coracana* (L.) Gaertn.] as influenced by nitrogen and sulphur fertilizers. *Plant Soil* 41(3):549–571.

Steduto P, Hsiao TC, Fereres E, Raes D (eds) (2012) Crop yield response to water. FAO irrigation and drainage paper 66. Food and Agriculture Organization, Rome, Italy.

Tang Y, Zhang B, Li X, et al. (2016) Bound phenolics of quinoa seeds released by acid, alkaline, and enzymatic treatments and their antioxidant and a-glucosidase and pancreatic lipase inhibitory effects. *J Agric Food Chem* 64:1712–1719.

Tao HT, Qiu B, Du FL, et al. (2014) The protective effects of cassava (*Manihot esculenta* Crantz.) leaf flavonoid extracts on liver damage of carbon tetrachloride injured mice. *Afr J Tradit Complement Alternat Med* 12(1):52–56.

Taylor JRN, Duodu KG (2010) Sorghum and millets: Characteristics and quality requirements. In: Wrigley CW, Batey IL (eds) *Cereal Grains: Assessing and Managing Quality*. Woodhead Publishing, Cambridge, UK, pp. 215–262.

Tewe OO, Lutaladio N (2004) *Cassava for Livestock Feed in Sub-Saharan Africa*. Food and Agriculture Organization, Rome, Italy.

Thompson LU (1993) Potential health benefits and problems associated with antinutrients with foods. *Food Res Int* 26:131–149.

Tonukari NJ, Ezedom T, Enuma CC, et al. (2015) White gold: Cassava as an industrial base. *Am J Plant Sci* 6(7):972–979.

Tsamo AT, Papoh Ndibewu PP, Dakora FD (2018) Phytochemical profile of seeds from 21 Bambara groundnut landraces via UPLC-qTOF-MS. *Food Res Int* 112:160–168.

Udayasekhara Rao P, Deosthale YG (1988) In vitro availability of iron and zinc in white and coloured ragi (Eleusine coracana): role of tannin and phytate. *Plant Foods Hum Nutr* 38(1):35–41.

Ukachukwu CO, Ogbobe O, Umoren SA (2010) Preparation and characterization of biodegradable polymer mud based on millet starch. *Chem Eng Commun* 197:1126–1139.

UN (2014) Press release – UN General Assembly's Open Working Group proposes sustainable development goals. https://sustainabledevelopment.un.org/content/documents/4538pressowg13.pdf. Retrieved 2018-07-05.

USDA (2009) National nutrient database for standard reference. http://www.nal.usda.gov/fnic/foodcomp/search/.

USDA (2018) National nutrient database for standard reference. Legacy release, April 2018. https://ndb.nal.usda.gov/ndb/foods/show/70.

Van Rensburg WJ (2014) Cultivating pseudo-grains in South Africa. *Farmers Weekly*. https://www.farmersweekly.co.za/crops/field-crops/cultivating-pseudo-grains-in-south-africa/.

Vasudha S, Mishra HN (2013) Non-dairy probiotic beverages. *Int Food Res J* 20(1):7–15.

Vega-Gálvez A, Miranda M, Vergara J, Uribe E, Puente L, Martínez EA (2010) Nutrition facts and functional potential of quinoa (*Chenopodium quinoa* Willd.), an ancient Andean grain: A review. *J Sci Food Agric* 90:2541–2547.

Vijaya Kumar B, Vijayendra SVN, Reddy OVS (2015) Trends in dairy and non-dairy probiotic products – A review. *J Food Sci Technol* 52:6112.

Vilcacundo R, Hernández-Ledesma B (2017) Nutritional and biological value of quinoa (*Chenopodium quinoa* Willd.). *Curr Opin Food Sci* 14:1–6.

Virupaksha TK, Ramachandra G, Nagaraju D (1975) Seed proteins of finger millet and their amino acid composition. *J Sci Food Agric* 26:1237–1246.

Wobeto C, Corrêa AD, Abreu CM, Santos CD, Pereira H (2007) Antinutrients in the cassava (*Manihot esculenta* Crantz) leaf powder at three ages of the plant. *Cienc Tecnol Alim* 27(1):108–112.

Yan X, Qi M, Li P, Zhan Y, Shao H (2017) Apigenin in cancer therapy: Anti-cancer effects and mechanisms of action. *Cell Biosci* 7:50.

Yao DN, Kouassi KN, Erba D, Scazzina F, Pellegrini N, Casiraghi MC (2015) Nutritive evaluation of the Bambara groundnut ci12 landrace [*Vigna subterranea* (L.) Verdc. (Fabaceae)] produced in Côte d'Ivoire. *Int J Mol Sci* 16:21428–21441.

Yao Y, Shi Z, Ren G (2014) Antioxidant and immunoregulatory activity of polysaccharides from quinoa (*Chenopodium quinoa* Willd.). *Int J Mol Sci* 15(10):19307–19318.

Yi B, Hu L, Mei W, et al. (2010) Antioxidant phenolic compounds of cassava (*Manihot esculenta*) from Hainan. *Molecules* 16:10157–10167.

Yoon JH, Thompson LU, Jenkins DJ (1983) The effect of phytic acid on in vitro rate of starch digestibility and blood glucose response. *Am J Clin Nutr* 38(6):835–842.

Zhang M, Xie L, Yin Z, Khanal KS, Zhou Q (2016) Biorefinery approach for cassava-based industrial wastes: Current status and opportunities. *Bioresour Technol* 215:50–56.

Zhu F (2014) Structure, physicochemical properties, and uses of millet starch. *Food Res Int* 64:200–211.

CHAPTER 9

The Role of Traditional Health Practitioners in Modern Health Care Systems

Norman Z. Nyazema

CONTENTS

ABSTRACT

Background: In Africa and much of the developing world, 80% of the population uses traditional medicines to meet their primary health care (PHC) needs. These needs, particularly in the rural areas, are met by services provided by traditional health practitioners (THPs), who are members of the community. Their services are the only available, accessible, and affordable sources of health care. Traditional health care, in Africa, in spite of its deep rootedness, has unresolved issues of efficacy and safety which hinder its potential benefits from being safely and effectively harnessed. The United Nations' Sustainable Development Goals (SDG) call for universal health coverage (UHC) as the center of the overall health goal. Certainly professionalized THPs whose practices are fully documented can play a very important role in this goal. Applying principles of evidence-based medicine (EBM) to African traditional medical practice (TMP), and indeed to any culture's TMP, is vital for the practice to gain acceptance by Western-trained healthcare providers. EBM is a necessary step when exploring African TMP as a cost-effective treatment and management option to help tackle healthcare issues such as functional diseases, chronic illnesses in aging populations, and rising healthcare costs.

***Relevance*:** It is now 40 years since the Alma-Ata Declaration on Primary Health Care (PHC) officially recognized the important role that traditional medicine and its practitioners can make in the healthcare delivery system. In 2000, the WHO also recognized THPs as an important resource in achieving healthcare for all. Policy makers still need to recognize the importance of THPs when re-engineering PHC, in spite of what has been achieved so far. By adapting WHO tools for institutionalization of traditional medicine, national policies can, among other things, facilitate:

a. the strengthening of multidisciplinary and intersectoral mechanisms to support national and regulatory frameworks
b. the collaboration between traditional health practitioners and Western-trained healthcare providers, most of whom work in the current healthcare systems
c. the inclusion of traditional medicine research and development on the national health research agenda in order to produce scientific evidence on the quality, safety, and efficacy of traditional medicines and practices that policy makers can utilize
d. the intensification of integration of certain aspects of traditional health/medical practices into training programs of relevant institutions

Policy makers should not just pay lip service to SDGs. SDGs recognize that eradicating poverty and inequality, creating inclusive economic growth, and preserving the planet are inextricably linked – not only to each other – but also to population health, and that the relationships between each of these elements are dynamic and reciprocal.

CONTEXT

The World Health Organization report "Health in 2015: From MDGs to SDGs" (WHO 2015) aimed to describe global health in 2015. It looked back 15 years at the trends and positive forces during the Millennium Development Goals (MDG) era and assessed the main challenges for the coming 15 years as addressed by the Sustainable Development Goals (SDGs; United Nations 2015a). The 2030 Sustainable Development Agenda is of unprecedented scope and ambition, applicable to UN member states, and obviously goes well beyond the MDGs. Doubts were expressed about the purpose of the MDGs. For example, the MDGs have been applied as one-size-fits-all development planning instruments with targets that every country should meet, even though the MDGs were never meant as targets for individual countries. Global targets are less useful for countries with a low starting point or in conflict situations.

Many countries in the developing world continue to face several serious health challenges. Health indicators in these countries are among the worst in the world, and diseases such as HIV and AIDS, tuberculosis, diarrheal diseases, malaria, and measles, to name but a few, continue to be dominant causes of morbidity and mortality.

For instance, the HIV/AIDS epidemic has emerged as one of the most serious in Africa and its impact has significantly worsened the health status of the region. The epidemic, together with other health problems such as nutritional deficiencies, is already having major effects on African societies and prospects for the development of generations to come. Africa needs to take seriously the broader sense of the 1948 WHO Constitution preamble definition of health, which states that "health is a state of complete physical, mental, and social well-being and not merely the absence of disease or infirmity" (WHO 2006). Others have redefined health as "a dynamic state of well-being characterized by a physical and mental potential, which satisfies the demands of life commensurate with age, culture, and personal responsibility" (Bircher 2005) or as "a condition of well-being, free of disease or infirmity, and a basic and universal human right" (Saracci 1997) . However, questions continue to be raised as to the acceptable definition of health (Huber et al. 2011). Whatever definition is adopted, it is important to situate it within the context of SDGs.

SUSTAINABLE DEVELOPMENT GOALS

On 25 September 2015, the UN General Assembly adopted the new development agenda "Transforming our world: the 2030 agenda for sustainable development" (United Nations 2015a). The agenda builds upon the outcome document of the UN Conference on Sustainable Development (Rio+20 conference), which took place in June 2012 and led to the establishment of the Open Working Group on Sustainable Development Goals. This was a group of member states tasked with preparing a proposal on the SDGs. The Open Working Group proposal was welcomed by the UN General Assembly in September 2014 and became the principal guideline for integrating SDGs into the post-2015 development agenda. Further intergovernmental negotiation processes resulted in the final document for the 69th UN General Assembly in 2014, which also included the outcomes of major global meetings such as the Sendai Framework for Disaster Risk Reduction 2015–2030 (UNISDR 2015) and the Addis Ababa Action Agenda (United Nations 2015b), as well as inputs such as the synthesis report of the Secretary-General on the post-2015 agenda, "The road to dignity: ending poverty, transforming all lives and protecting the planet", published in December 2014. The 17 goals for the new development agenda adopted by the General Assembly are enumerated in Table 9.1.

The SDGs integrate all three dimensions of sustainable development (economic, social, and environmental) around the themes of people, planet, prosperity, peace, and partnership. The SDGs continue to prioritize the fight against poverty and hunger while also focusing on human rights for all and the empowerment of women and girls as part of the push to achieve gender equality. They also hope to build upon and extend the MDGs in order to tackle the "unfinished business" of the MDG era. The SDGs recognize that eradicating poverty and inequality, creating inclusive economic growth, and preserving the planet are inextricably linked, not only to each other but also to population health, and that the relationships between each of these elements are dynamic and reciprocal.

Table 9.1 The 17 Sustainable Development Goals (SDGs) to Transform Our World (United Nations 2015a)

GOAL 1: No Poverty
GOAL 2: Zero Hunger
GOAL 3: Good Health and Well-Being
GOAL 4: Quality Education
GOAL 5: Gender Equality
GOAL 6: Clean Water and Sanitation
GOAL 7: Affordable and Clean Energy
GOAL 8: Decent Work and Economic Growth
GOAL 9: Industry, Innovation, and Infrastructure
GOAL 10: Reduced Inequality
GOAL 11: Sustainable Cities and Communities
GOAL 12: Responsible Consumption and Production
GOAL 13: Climate Action
GOAL 14: Life below Water
GOAL 15: Life on Land
GOAL 16: Peace, Justice, and Strong Institutions
GOAL 17: Partnerships to Achieve the Goal

While poverty eradication, health, education, food security, and nutrition remain priorities, the SDGs comprise a broad range of economic, social, and environmental objectives and offer the prospect of more peaceful and inclusive societies. In a few countries in Africa, progress towards the MDGs, on the whole, was remarkable in poverty reduction, education improvements, and increased access to safe drinking water. Progress on the three health goals and targets was considerable. Globally, the HIV, tuberculosis (TB), and malaria epidemics were "turned around", and child mortality and maternal mortality decreased considerably (53% and 44%, respectively, since 1990), despite falling short of the MDG targets. It is fair to say that the MDGs went a long way to changing the way Africans think and talk about the world, shaping the international discourse and debate on development, and contributing to major increases in development assistance. Unfortunately, several limitations of the MDGs became apparent, including a limited focus resulting in verticalization of health and disease programs in countries, a lack of attention to strengthening health systems, the emphasis on a "one-size-fits-all" development planning approach, and a focus on aggregate targets rather than equity. This "one-size-fits-all" approach appears to have been abandoned with the SDGs.

SDG goal 3 is broad: "Ensure healthy lives and promote well-being for all at all ages." The health goal has 13 targets, and their achievement will be expected to integrate the three dimensions mentioned earlier. Health has a central place as a major contributor to and beneficiary of sustainable development policies. There are many linkages between the health goal and other goals and targets, reflecting the integrated approach that underpins the SDGs. Universal health coverage (UHC), one of the 13 health goal targets, provides an overall framework for the implementation of a broad and ambitious health agenda in all countries. Monitoring and review of progress will be a critical element of the SDGs. An indicator framework is still being developed and is scheduled to be adopted soon.

UNIVERSAL HEALTH COVERAGE

The main text of the SDG declaration endorsed by heads of government in February 2015 puts UHC at the center of the overall health goal, and makes progress towards the UHC target a prerequisite for the achievement of all the others. Under SDG 3, UHC is assigned the specific Target 3.8: "Achieve universal health coverage, including financial risk protection, access to quality essential health care services, and access to safe, effective, quality and affordable essential medicines and vaccines for all" (United Nations 2015a). The goal of UHC (all people and communities receiving the needed quality services, including health protection, health promotion, preventive services, treatment, rehabilitation and palliation without financial hardship) is relevant to all countries and offers an unprecedented opportunity to increase coherence in health-related actions and initiatives. Accountability – defined as a cyclical process of monitoring, review, and remedial action – will be critically important in ensuring progress towards UHC. WHO and the World Bank have developed a UHC monitoring framework based on a series of country case studies and technical expertise (WHO/World Bank 2014)

The framework focuses on the two core components of UHC: coverage of the population with *quality, essential health services* (some of which to a large extent are provided by traditional health practitioners [THPs, also called traditional medical practitioners] in the communities in which they reside) and coverage of the population with financial protection, the key to which is reducing dependence on payment for health services out-of-pocket (OOP) at the time of use. The proposed indicators are a "coverage index" of essential services, disaggregated by key stratifiers where possible, and a measure of the lack of financial protection against the costs of health services. These two indicators need to be interpreted together to assess the state of UHC in Africa, and indeed globally.

HEALTH INEQUITY IN AFRICA

In spite of the MDGs, SDGs, and a host of other international conventions, there are clear indications of growing inequities in health and healthcare, both within and among countries, especially in Africa. A recent first-ever global study conducted by The Institute of Health Metrics and Evaluation found massive health care inequity in, for example, South Africa and Botswana despite their relative levels of wealth and development (GBD 2015 Healthcare Access and Quality Collaborators 2017). This is hardly surprising, as these indications often seem to be ignored and are persistently downplayed as countries on the African continent implement policy changes that affect healthcare provision, including the role that can be played by THPs. To this day, the lives of the African people, to a large extent, lie in the hands of health systems inherited from their colonial masters, where more money is invested. At the time of the evolution of these systems, healthcare was implemented primarily to cater for colonial administrators and expatriates, with separate or second-class provision made – if at all – for Africans. The systems so evolved, however, now have a vital and continuing responsibility to people throughout their life span and are crucial to the healthy development of individuals, families, and communities in Africa. This inevitably makes healthcare provision in Africa a practical as well as a political issue.

This was clearly illustrated by the report of the Organization of African Union (now African Union, [AU]) secretary general to the 6th Conference of the African Ministries of Health (CAMH 6) held in Cairo 20 years ago, in 1999 (OAU 1999). It stated that access to basic healthcare is a *fundamental human right* and a foundation for socio-economic development. In his address to the meeting, at the time, the AU secretary general also asked the following questions about globalization: Is globalization an answer to the African health dilemma? Or is it just one more challenge facing African health?

In the report, it was stated that obligations and commitments for implementation of the agreements under the World Trade Organization Global Agreement would definitely have a negative impact on the health sector in Africa. The cost of health services in general would be increased beyond the capacity of most African countries, and the gap between the developing and developed countries would widen more and more. Surely, back then this created an opportunity for African countries to seriously interrogate the role that could be played by the traditional medical system.

The report pointed out that the critical challenges that African health systems faced at that time included:

- The implementation of an effective public health care program based on Primary Health Care (PHC), a concept first elaborated in 1978 at the Alma-Ata WHO Conference (WHO 1978). It stressed the principles of equity, community participation, intersectoral action, use of appropriate technology, and the central role played by the health system. It became a core concept for WHO's goal of Health for All by the year 2000 (WHO 1978). Different approaches have been used to apply it. In Africa, the approaches have often, unfortunately, excluded THPs, who are healthcare professionals working at the community level with close community ties. Since the early 1970s, WHO has repeatedly advocated for the recognition of THPs as PHC providers and for the integration of TM (traditional medicine) in the national health systems.
- Access to medicines based on the essential drug concept. Essential medicines in many African countries are not always available (Cameron et al. 2009) and as a result more than 75% of the population rely on traditional remedies for their PHC needs (Mhame et al. 2010). Almost 25% of modern medicines on the market are derived from plants that have been used by THPs, and more are being developed (WHA 1978; Farnsworth et al. 1985; Gurib-Fakim 2010).
- The achievement of strong national, regional, and continental partnerships in the field of health development.
- The building up of an efficient human resource system and health sector infrastructure that lead to the strengthening of the contemporary health system.

HEALTH SYSTEMS AND SERVICES

In today's complex world, it can be difficult to say exactly what a health system is, what it consists of, and where it begins and ends (Murray and Frenk 2000). The 2000 World Health Report (WHO 2000) gives the example of the baby boy designated the six billionth person on the planet, who was born in Sarajevo on 13 October 1999. He was born in a hospital staffed with trained midwives, doctors, nurses, and technicians with access to a variety of equipment and medicines. That hospital was part of a nationwide health service that would seek to measure, maintain, and improve his health throughout his life. All of these services contribute to that baby's life expectancy of 73 years, the Bosnian average at the time. Health systems therefore include all the people and institutions who provide services, finance health care, or set policies to administer it (WHO 2000).

Thus, health systems include all the activities whose primary purpose is to promote, restore, or maintain health and are defined as comprising all the organization, institutions, and resources that are devoted to producing or maintaining health. Surely, in Africa, some of those resources include indigenous knowledge systems, whose custodians are THPs.

Health action is defined as any action, whether in personal health care, public health services, or through intersectoral initiatives, whose primary purpose is to improve health. This, however, is not the only objective of a health system. The objective of a good public health system is really twofold, namely:

- *Goodness* – the best attainable average level of health and health services. A good health system responds well to people's needs and expectations. Many communities in Africa have high expectations of traditional medical practice because of its apparent holistic approach to health, and hence they have a great reliance on it.
- *Fairness* – the smallest possible disparities among individuals and groups in health status and access to services. A fair health system responds to everyone's needs and makes services available to all, without discrimination. Traditional health practice does not discriminate within a community; access to services is not so constrained by ability to pay as is often the case in Western medicine.

The system can respond by:

- Allocating equal resources or expenditure for equal need/demand (i.e. allocating resources to a particular group or geographical area in proportion to its health needs/demand);
- Making sure that there is equal access for equal need/demand (i.e. ensuring that for all individuals with the same need/demand, they will have the same opportunity to use health services);
- Promoting equal utilization for equal need/demand (which would involve devising a system whereby the use of health services would be allocated *pro rata* with need/demand);
- Adjusting in certain ways and standardizing at least for age and sex to ensure equality in health.

Demand is normally seen as involving the preferences (through willingness and ability to pay) of the patient and perhaps his family. Need, on the other hand, is based on the value judgments of healthcare professionals on behalf of the patient and his family and/or society at large. A health system will have to adopt either a "demand" or a "need" stance in order to ensure equity in health, at least at a primary healthcare level, involving THPs.

One of the criticisms of PHC as a route to achieving affordable UHC – the goal of healthcare for all – is that it gave little attention to people's *demand* for healthcare. It instead, concentrated almost exclusively on their perceived *needs*. Health systems in Africa have failed because these two concepts have not been made to match, and the supply of services offered cannot possibly align with both. What we now see happening is that only the simplest and most basic care for the poor, rather than all possible care for everyone, is being provided. Politically, we are seeing governments limiting promises and expectations about what they should do and, unfortunately at the same time, people's expectations of health systems are greater than ever before. There is an increase in demands and pressures on health systems. Governments in Africa are facing constraints in the face of new developments in health technology, new drugs or treatments, and further advances in medicine. What this means is that if services are to be provided for all, then not all services can be provided.

Nonetheless, there is the potential to improve each component of the health systems and to achieve the objectives of the current health systems on the continent, in order to ensure that people are treated promptly and with respect for their dignity and their demands and needs. This is possible to a large extent if THPs are integrated into the health systems to provide certain health services of good quality. Attempts to accomplish this integration of THPs have been made during the scaling up of comprehensive HIV and AIDS care and prevention strategies in some African countries where health resources are scarce (Diallo et al. 2003; Homsy et al. 2003; PROMETRA 1995). In spite of all these efforts, unfortunately, THPs appear to be ignored in the re-engineering of PHC in some countries.

The intensifying struggle around scarce health resources, globally and in many African countries in particular, requires the recognition that equity (whichever way it is defined) needs to define and build a more active role for important stakeholders in health, including THPs. Health systems around the world are reported to be experiencing increased levels of chronic illness and escalating healthcare costs. Today, patients and healthcare providers alike are demanding that healthcare services be revitalized, with a stronger emphasis on individualized, person-centered care (Roberti di Sarsina et al. 2012). With the involvement of THPs, genuinely people-centered initiatives can be strengthened, leading to a reduction in ever-increasing health care costs. Worldwide there is now demand for traditional and complementary medicine that is expanding beyond products to focus on practices and practitioners (WHO 2013). One of the reasons for this demand is the increased burden of non-communicable diseases, which are often inadequately addressed by conventional medicine.

TRADITIONAL MEDICINE AND COMPLEMENTARY MEDICINE

Traditional medicine (TM) is a broad, comprehensive term generally used to refer to various forms of indigenous healing systems and medicine that are found in every culture. It is sometimes used so broadly as to include all medical practices with a significant history of use other than conventional Western medicine, ranging from commonplace folk medical practices to complex codified systems practiced by a professional class of physicians or the equivalent, such as Traditional Chinese Medicine. WHO defines TM as:

> "the sum total of the knowledge, skill, and practices based on theories, beliefs, and experiences indigenous to different cultures, *whether explicable or not* [emphasis added], used in the maintenance of health as well as in the prevention, diagnosis, improvement or treatment of physical and mental illness" (WHO 2000).

Complementary medicine (CM), on the other hand, is defined as:

> "a broad set of health care practices that are not part of that country's own tradition or conventional medicine and are not fully integrated into the dominant health-care system. They are used interchangeably with traditional medicine in some countries" (WHO 2013).

In Africa during colonial days, indigenous medicine was greatly suppressed by European colonizers, and yet Europe had and still has a very long history of TM. TM was the fundamental method used by Africans, before the advent of the Europeans, to preserve their health and avoid disease since the dawn of time. African TM is not alternative or complementary for many rural people who live far from medical facilities, in places where there are no modern healthcare providers and drugs are too expensive. WHO estimated that up to 80% of people in Africa make use of TM (WHO 2011). The ratio of TM practitioners to the population is approximately 1:500, while that of medical doctors to the population is 1:40,000 (Abdullahi 2011). Some people argue that there is little or no evidence to support the figures; however, anecdotal and observational evidence exists in the popular press of the central role of THPs to an average African's life regardless of his or her socio-economic status. There are many terms in African cultures used to describe THPs and recognize them as people of high standing in a community, someone who is open and available to serve others when they need healthcare services. However, the great diversity of practices and the unfamiliarity of most to Western-trained practitioners means that, even when it is acknowledged that there is a role for THPs in the healthcare system, integrating them into an evidence-based system poses challenges (Ncayiyana 2016).

Further, even where there are good physicians and therapies are available, people take the lead in their own healthcare, mainly because they have their own interpretation of illness. This is the case in many African communities, where people prefer to use alternative medicine and therapies. In this environment traditional beliefs,

African and Western religious beliefs and myths, and even quackery, intermingle and mix.

In traditional medical practice (TMP), therapies consist of complex interactions. Treatment sometimes cannot be reduced to a single specific therapeutic ingredient but may also encompass a specific setting as well as diagnostic and interaction processes (Hewson 1998). Thus people's beliefs about illness and health may contribute to the benefits of a practice.

Many people in both industrialized and developing countries rely on and trust in TM for resolution of many minor and sometimes severe diseases (Firenzuoli and Gori 2007). In Europe, the dominant healthcare system is based on allopathic medicine, and also TM in places in which it has been partially or fully incorporated into the public healthcare system, such as in Tuscany (Italy). TM is often termed and categorized as CM, "alternative medicine" or "non-conventional" medicine. There are different models of CM being integrated into conventional medicine throughout Europe, each with its own strengths and limitations. Studies are ongoing to understand why CM is so popular in Europe, in order to consider employing it as part of the solution to healthcare, health creation, and self-care challenges (Fischer et al. 2014). In Europe, the pharmacopoeias of folk practitioners as well as professional medical healthcare providers contain thousands of medicines made from leaves, herbs, roots, bark, animals, insects, mineral energetic substances, and other materials found in nature. Switzerland, the home of big pharmaceutical companies Hoffmann-La Roche, Novartis, and Sandoz, has become the first country in Europe to integrate TM and CM into its health system (MacKenzie 2011; Nikolai 2013)

Traditional Chinese medicine (TCM) has spread to almost all the countries of the world and has grown into an international industry. There are about 100,000 Chinese medicine clinics, about 300,000 Chinese medicine practitioners, and no less than 1,000 Chinese medicine education institutes worldwide. In China and in both North and South Korea, TCM is practiced alongside conventional medicine, and health insurance is available for users of that healing modality. The same status prevails in Vietnam, where TM practitioners are able to practice in both public and private hospitals and clinics. The government insurance in Vietnam fully covers acupuncture, herbal medicine, and TM treatment (WHO 2013).

The difference in Africa is that most traditional practices have not been fully documented and professionalized. This situation is the result of colonial rule, which sought to suppress African cultural and religious practices. In many African cultures, illness is thought to be caused by psychological and spiritual conflicts or disturbed social relationships that create a disequilibrium expressed in the form of physical or mental problems. Disequilibrium may be caused by psychological or spiritual factors, or both, that relate to African cosmology and threaten the intactness of the person. Thus, in African traditional cultures, healing emphasizes righting this disequilibrium, which makes African TM holistic, i.e. providing treatments for physical illness as well as psycho-spiritual ones (Tabuti et al. 2003).

The role of TM in addressing gaps in the health care delivery system has been discussed by many national and international political and health bodies (WHO 2013).

A result has been the adoption of a number of resolutions and declarations on TM. Notable among these are (1) the resolution "Promoting the role of traditional medicine in health systems: A strategy for the Africa Region" which was adopted by the WHO Regional Committee for Africa in Ouagadougou, Burkina Faso, in 2000 (WHO 2001), and (2) the declaration on the Decade of African Traditional Medicine (2001–2010) by the Heads of State and Government in Lusaka in 2001 (Kasilo et al. 2010). These efforts were all harnessed into the WHO TM Strategy 2002–2005, whose goals were the following:

- Harness the potential contribution of TM to health, wellness, and people-centered health
- Promote the safe and effective use of TM by regulating, researching, and integrating TM products, practitioners, and practice into health systems (WHO 2002)

As noted earlier, great progress has been achieved in many countries around the globe, and there is continuing demand for TM services. These services are not only in demand for treatment and management of diseases, especially chronic diseases that are becoming prevalent in Africa as people get urbanized and eat Western types of foods; they are also widely used in disease prevention, health promotion, and health maintenance. TM services have proved to be cost-effective (Korthals-de Bos et al. 2003; Kooreman and Baars 2012). Unfortunately, in many African countries there are still many challenges which can hamper the role of THPs in the health systems, particularly in the era of evidence-based medicine (EBM). The fact that many of the Ministries of Health in African countries are staffed by individuals trained in conventional medicine makes the situation even more challenging. EBM, by integrating individual clinical expertise with the best available clinical evidence from systematic research, has in recent years been established as the standard of modern medical practice for greater treatment efficacy and safety. African traditional medical practice, on the other hand, evolved as a system of medical practice many years ago based on empirical knowledge as well as theories and concepts which have yet to be mapped by scientific equivalents. Whether the practices are explicable or not, they equally fall under the WHO definition of TM (WHO 2000)

TM in Africa, in spite of its deep rootedness, is surrounded by issues of efficacy and safety which remain unresolved and hinder its potential benefits from being safely and effectively harnessed. The WHO definition of TM does not make the situation easy, either. It goes against the philosophy of scientism, which purports to define what the world really is like. Scientism adopts what the philosopher Thomas Nagel called "an epistemological criterion of reality" (Nagel 1986), defining what is real as that which can be discovered by certain methods of scientific investigation. In this belief system, as Loughlin et al. (2013) explain, all human experiences not amenable to study (or chosen for study) "by those methods are deemed 'subjective' in a way that suggests they are either not real, or lie beyond the scope of meaningful rational inquiry." This philosophy is quite distinct from the African view of reality and, as Loughlin et al. (2013) argue, devalues human capacities that "are in fact essential components of good reasoning and virtuous practice." Adopting scientism as a belief system therefore limits reasoning and decision-making abilities in clinical contexts, not to mention integration of Western and TM practitioners.

THE WAY FORWARD

Applying EBM to African TM is still vital for the practice to gain acceptance by those who have adopted the philosophy of scientism, particularly Western-trained healthcare providers. In turn, the curriculum of these providers should include training in some aspects of African TMP. Otherwise as things stand, given the challenges still to be faced, integration of TM into modern healthcare systems is going to take a while, in spite of what the World Health Report 2006, "Working together for health" recommended (WHO 2006b).

Nevertheless, EBM is a necessary step when exploring African TM as a cost-effective treatment and management option to help tackle healthcare issues such as functional diseases, chronic illnesses in aging populations, and the rising healthcare costs. There is also a need for research into the comparative effectiveness of TMP, utilizing mixed methods, in order to obtain data to understand its effects and applicability in healthcare. Everybody needs to know in what situations TMP is a reasonable choice. A clear emphasis on concurrent evaluation of the overall effectiveness of TMP as an additional or alternative treatment strategy in real-world settings is recommended. High-quality evidence via randomized clinical trials (RCTs) assessing its efficacy or effectiveness and safety are still needed to substantiate the use of TMP in suitable conditions, including psychiatric and psychological conditions. (The commonly used term "efficacy" refers to "the extent to which a specific intervention is beneficial under ideal conditions"; by contrast, "effectiveness" is a measure of the extent to which an intervention, when deployed in routine circumstances, does what it is intended to do for a specific population [Last et al. 2001].) It would appear that TMP safety is usually not a great concern, because of its cultural rootedness in many African communities and its long history of use (though safety becomes a challenge where herbs are used by untrained personnel). However, for it to be acceptable to and integrated with allopathic medicine, the following domains in which safety plays an important role need to be taken into consideration:

(i) TM in the legislative and regulatory area – even though some attempts at regulation have been made in some African countries, there appears to be no seriousness of purpose when it comes to implementation.
(ii) The competency (difficult to determine, given how one becomes a THP, particularly in Africa) and safety of the THPs and their therapies and products.
(iii) Interactions between TM and conventional medicine.

Regarding RCTs, it would be important to take a leaf from the European CAMbrella project's recommendations about how RCTs should be conducted (Weidenhammer et al. 2011). The recommendation is that RCTs should be pragmatic by taking into consideration patients' preferences regarding their therapy of choice, using patient preference designs or documentation of patient preferences. Pragmatic RCTs (pRCTs) are intended to maintain internal validity while testing interventions in the full spectrum of everyday clinical settings in order to maximize applicability and generalizability. The trials also measure a wide spectrum of outcomes, mostly patient-centered (Patsopoulos 2011) and the order of research should really be in reverse of that used in developing novel conventional interventions, with the aim of establishing comparative effectiveness and safety before assessing the efficacy of individual components. Cluster-randomized trials could also be used when appropriate (Zwarenstein 2008; Mullins et al. 2010; Fischer et al. 2014).

There are major healthcare challenges in all African countries as a result of increased urbanization and attendant nutritional and demographic transition. African TM is going to be employed in the management of chronic long-term conditions, health promotion, and illness prevention. EBM practice is going to be vital to future health planning. At the same time, many communities are forever going to favor TM utilization, while incorporating their long-held cultural beliefs based on indigenous knowledge. Their needs and views will have to be the key priorities, and their interests shall need to be investigated if efforts are going to be made towards integrative healthcare provision.

Collaboration between THP and conventional health practitioners is needed. However, before all that is done, there is an urgent need to obtain communities' views on the prerequisites for the collaboration, as advocated by Kaboru et al. (2006) during their studies on STI/HIV/AIDS care. The following goals, proposed almost two decades ago at a WHO-Afro regional meeting (WHO 2003), are still applicable today because since then nothing has really changed:

- Political leadership that defines the social goals for the health system. This involves ensuring policy frameworks exist and are combined with oversight, coalition building, attention to system design, and accountability (e.g. health professions councils closely working together).
- A range of interventions for health promotion, prevention, and rehabilitation as well as treatment. Good health services are those which deliver *effective, safe, quality* personal and non-personal health interventions to those that need them, when and where needed, with minimum waste of resources (e.g. male circumcision campaigns to reduce HIV transmission).
- The right number and mix of health workers with appropriate skills. A well-performing health workforce is one that works in ways that are *responsive, fair, and efficient* in order to achieve the best health outcomes possible, given available resources and circumstances (e.g. PHC re-engineering that includes THPs while recognizing their skills and knowledge).
- The required medicines, technologies, and facilities. A well-functioning health system ensures equitable access to essential medical and medicinal products, vaccines, and technologies of assured *quality, safety, efficacy, cost-effectiveness, and scientific soundness* (e.g. encouragement of the production and use of efficacious traditional herbal and animal medicinal products).
- Timely and reliable information, research evidence, and capabilities in knowledge management. The acquisition, generation, sharing, and use of information, research evidence, and knowledge is critical so that the system can be adapted to changing

circumstances and can improve and develop (e.g. joint congresses, conferences, symposia etc.).

- Robust and equitable mechanisms and institutions for long-term financing. A good health financing system raises adequate funds for health in ways that ensure people can use needed services and are protected from financial catastrophe or impoverishment associated with having to pay for them. It provides incentives for providers and users to be efficient (e.g. establishment of a well-managed National Social Security System or National Health Insurance).

REFERENCES

Abdullahi AA. Trends and challenges of traditional medicine in Africa. *African Journal of Traditional, Complementary and Alternative Medicine* 2011, *8* (Suppl):115–123.

Bircher J. Towards a dynamic definition of health and disease. *Medicine, Health Care and Philosophy* 2005; 8:335–341.

Cameron A, Ewen M, Ross-Degnan, Ball D and Laing R. Medicines prices availability, and affordability in 36 developing countries and middle-income countries: a secondary analysis. *Lancet* 2009, *373* (9659):240–249. DOI:10.1016/S0140-6736(08)61762-6.

Diallo D, Koumare M, Traore AK, Sanogo R and Coulibaly D. Collaboration between traditional health practitioners and conventional health practitioners: The Malian experience. In: *Traditional Medicine: Our Culture, Our Future, African Health Monitor*, Volume 4(1). Jan–June 2003. WHO Afro, Brazzaville, 35–36.

Farnsworth N, Akerele AO, Bingel AS, Soejarto DD and Guo Z. Medicinal plants in therapy. *Bulletin of the World Health Organization* 1985, *63*:965–981.

Firenzuoli F and Gori L. Herbal medicine today: clinical and research issues. *Evidence-Based Complementary and Alternative Medicine* 2007, *4* (S1):37–40.

Fischer F, Lewith G, Witt CM, et al. A research roadmap for complementary and alternative medicine – what we need to know by 2020. *Forschende Komplementärmedizin* 2014, *21*:e1–e16. DOI:10.1159/000360744.

GBD 2015 Healthcare Access and Quality Collaborators. Health care Access and Quality Index based on mortality from causes amenable to personal health care in 195 countries and territories 1990–2015: a novel analysis from the Global Burden of Disease study 2015. *Lancet* 2017, *390* (10091):231–266.

Gurib-Fakim A and Kasilo OSMJ. Promoting African medicinal plants through an African herbal pharmacopoeia. In: *The African Health Monitor*. Special Issue 14. August 2010. WHO Afro, 65–67.

Hewson M. Traditional healers in Southern Africa. *Annals of Internal Medicine* 1998, *128* (12 Part 1):1029–1034.

Homsy J, King R and Tenywa J. Building a regional initiative for traditional medicine an AIDS in Eastern and Southern Africa. In: *Traditional Medicine: Our Culture, Our Future, African Health Monitor*, Volume 4 (1). Jan–June 2003. WHO Afro, Brazzaville, 24–26.

Huber M, Knottnerus JA, Green L, et al. How should we define health? *BMJ* 2011, *343:*d4163.

Kaboru BB, Falkenberg T, Ndulo J, et al. Communities' views on prerequisites for collaboration between modern and traditional health sectors in relation to STI/HIV/AIDS care in Zambia. *Health Policy* 2006, *78* (2–3):330–339.

Kasilo OMJ, Trapsida J-M, Mwisika CN and Lusamba-Dikassa PS. An overview of the traditional medicine situation in the African region. In: *The African Health Monitor*. Special Issue 14. August 2010. WHO Afro, 7–15.

Kooreman P and Baars EW. Patients whose GP know complementary medicine tend to have lower costs and live longer. *European Journal of Health Economics* 2012, *13* (6):769–776.

Korthals-de Bos IBC, Hoving JL, van Tulder MW, et al. Cost effectiveness of physiotherapy, manual therapy and general practitioner care of neck pain economic evaluation alongside a randomized controlled trial. *BMJ* 2003, *326*:911–916.

Last J, Spasoff R and Harris S. *Dictionary of Epidemiology*. Oxford, UK: Oxford University Press, 2001.

Loughlin M, Lewth G and Falkenberg T. Science, practice and mythology: a definition and examination of the implications of scientism in medicine. *Health Care Analysis* 2013, *21* (2):130–145.

MacKenzie D. Swiss recognise 'alternative' medicine – for now. *New Scientist*, 24 January, 2011.

Mhame PP, Busia K and Kasilo OMJ. Collaboration between traditional health practitioners and conventional health practitioners: Some country experiences. In: *The African Health Monitor*. Special Issue 14. August 2010. WHO Afro, 40–46.

Mullins CD, Whicker D, Reese ES and Tunis S. Generation evidence for comparative effectiveness research using more pragmatic randomized controlled trials. *Pharmacoeconomics* 2010, *28*:969–976.

Murray CJL and Frenk J. A framework for assessing the performance of health systems. *Bulletin of the World Health Organization* 2000, *28* (6):717–731.

Nagel T. *The View from Nowhere*. 1986. New York, NY: Oxford University Press.

Ncayiyana D. Above all, do no harm. Health Professions Council of South Africa, HPCSA. *The Bulletin* 2016, 39–40.

Nikolai T. Report on Swiss Report on the Complementary Medicine Evaluation Programme (PEK). European Committee for Homeopathy (ECH), 2005 (Available at http://www.portaldehomeopatia.com.br/documentos/Report%20on%20PEK%20study.pdf).

OAU. Report to the 6th Conference of the African Ministries of Health (CAMH 6), 18–26 October, 1999, Cairo, Egypt.

Patsopolous N. A pragmatic view on pragmatic trials. *Dialogues in Clinical Neuroscience* 2011, *13* (2):217–227.

PROMETRA. FAPEG Method – Traditional healer's self proficiency training for their involvement in the health care challenge and its links to development. PROMETRA, 1995.

Roberti di Sarsina P, Alivia M and Guadagni P. Widening the paradigm in medicine and health: person centred medicine as the common ground of traditional, complementary, alternative and non-complimentary medicine. In: *Health Care Overview: New Perspectives, Advances in Predictive, Preventive and Personalized Medicine*. 2012. Dordrecht, the Netherlands: Springer, *1*:335–353.

Saracci R. The World Health Organisation needs to reconsider its definition of health. *BMJ* 1997, *314*:1409–1410.

Tabuti JRS, Dhillion SS and Lye KA. Traditional medicine in Bulamogi county, Uganda: its practitioners, users and viability. *Journal of Ethnopharmacology* 2003, *85* (1):119–129.

United Nations. Transforming our world: the 2030 Agenda for Sustainable Development. Resolution adopted by the General Assembly on 25 September 2015. A/RES/70/1. United Nations General Assembly, Seventieth session, agenda items 15 and 116; paragraph 26. United Nations, 2015a.

United Nations. Third International Conference on Finance for Development. 13–16 July, Addis Ababa, Ethiopia: United Nations, 2015b.

UNISDR. The Sendai Framework for Disaster Risk Reduction 2015–2030. Geneva, Switzerland: United Nations, 2015.

Weidenhammer W, Lewith G, Falkenberg T, et al. EU FP7 project 'CAMbrella' to build European research network for complementary and alternative medicine. *Forschende Komplentärmedizin* 2011, *18*:69–76.

WHA. Resolution WHA31,33 on Medicinal plants. *The Thirty-First World Assembly*, 1978.

WHO. Declaration of Alma-Ata. *International Conference on Primary Health Care*. Alma-Ata: USSR, 6–12 September, 1978.

WHO. Why do health systems matter? In: *The World Health Report 2000 – Health Systems: Improving Performance*. 2000. Geneva, 3–19.

WHO. Promoting the role of Traditional Medicine in health systems: A strategy for the African Region. WHO-Afro. Document AFR/RC50/9 and Resolution AFR/RC50/R3, 2001.

WHO. The Regional Meeting on Integration of Traditional Medicine in Health Systems: Strengthening Collaboration between Traditional and Conventional Health Practitioners, Harare, Zimbabwe, 26–29 November 2001. *Final Report*. WHO Regional Office for Africa. Brazzaville (AFR/TRM/03.2), 2002.

WHO. *Constitution of the World Health Organization – Basic Documents*. Forty fifth edition, October 2006a.

WHO. The World Health Report 2006 – Working Together for Health. Geneva, 2006b.

WHO. Traditional medicine strategy 2014–2023. Geneva, Switzerland: World Health Organization, 2013.

WHO. *Health in 2015 from MDGs to SDGs*. Geneva, Switzerland: World Health Organization, 2015.

WHO/World Bank. Monitoring progress towards UHC at country and global level: framework, measure and targets. *Joint WHO/World Bank Group Paper*, May 2014.

Zwarenstein M, Treweek S, Gagnier JJ, et al. Improving the reporting of pragmatic trials: an extension of the CONSORT statement. *BMJ* 2008, *337*:a2390.

CHAPTER 10

Interrogating the Framework for the Regulation of Complementary Medicines in South Africa

David R. Katerere, Kaizer Thembo, and Renée A. Street

CONTENTS

ABSTRACT

Background Research and development of traditional medicines is one of the areas identified in South Africa's National Drug Policy (NDP). Worldwide, the safety, quality, efficacy, availability, and regulation of traditional, complementary, and alternative medicines (T/CAMS) are debated by policy makers, healthcare professionals, and the public. In 2013, regulations for the registration of Complementary Medicines (CM) sold in South Africa were gazetted by the South African Minister of Health.

Guidelines and associated documentation for prospective and retrospective product registration have now been finalised to facilitate the process. However, there appears to be continued work needed with respect to achieving exact and appropriate definitions of what constitute complementary medicines, including the place of African Traditional Medicines (ATM), the disciplines currently included (and excluded), and the implementation plan by the Medicines Control Council (MCC) (now known as the South African Health Products Regulatory Authority [SAHPRA]). The framework has been hailed by many as a progressive step which will bring rationality to an industry that has been perceived as dangerously unregulated, to the detriment of public health.

Relevance This chapter provides insight into the evolution of the complementary medicine regulatory framework in South Africa and highlights important issues around Quality, Safety, and Efficacy (QSE) requirements of the registration regime currently unfolding. Overall, this chapter creates awareness among the public, researchers, policy makers, and industry about the scientific approach to medicine regulation and the potential opportunities and challenges that lie ahead. It also identifies gaps which need to be addressed.

INTRODUCTION

In South Africa, various forms of medicine appear to exist in the healthcare space, namely, conventional medicine, indigenous African traditional medicine (ATM), and complementary and alternative medicine (CAM). Up until a decade ago, there was much focus on trying to mainstream ATM by the then-Minister of Health, but the promotion of untested traditional products for HIV and AIDS and an antipathy to conventional anti-retroviral therapy (ART) (1, 2) may have inflicted damage to the cause of traditional medicine. It need not have, because as Wrenford (1) argued, it should have been an opportunity for collaboration between the two systems of healthcare. Since the introduction of the mass rollout of ART therapy in 2009, and the change of administration at the National Department of Health, traditional medicines seem to have taken a back seat. Nonetheless, the more formalised, *albeit* related CAM industry has seen regulatory advances in the last few years, and CAM is now subject to regulatory oversight. We give an overview of CAM and describe what steps South Africa has taken to regulate these medicines and what challenges lie ahead. We also propose that the capacity being built by the regulatory authority in this sector can be used to extend regulation to pharmaceutically packaged ATM but that lessons should also be drawn from other African countries.

COMPLEMENTARY AND ALTERNATIVE MEDICINE: THE NEED FOR REGULATION

Health and healing practices that are neither inherent to a country nor well established within the country's conventional healthcare system are frequently referred to as CAM (2). The terms 'complementary' and 'alternative' practices are commonly used interchangeably; however, they refer to different concepts. 'Complementary' implies that healing modalities are being used together with mainstream conventional medical interventions, while 'alternative' implies that the particular intervention is used alone and to the exclusion of conventional medicine (3).

In the 1980s, there was a rapid acceptance of complementary medicine in key markets in the West, leading to its mainstreaming and commercialization (4, 5). This came about because of the belief that these medicines, particularly herbal, were 'natural' in contrast to synthetic pharmaceuticals and hence innately safe and associated with fewer side effects (6, 7). However, problems soon emerged in terms of the quality of raw materials and finished products (4, 8, 9). In addition, clinically significant herb-drug interactions (10) have also been noted that point to the fact that use of CAMs is not without danger.

Owing to public health concerns, regulatory regimes started to emerge to mitigate against contamination and adulteration of products as well as variability in manufacturing and storage. This was the first introduction of quality management systems (QMS) across the CAM value chain. Germany was probably the first country to institute the regulation of herbal drugs going back to 1978 when the so-called Commission E (phytotherapy and herbal substances) was established as an independent division of the Federal Institute for Drugs and Medicinal Devices (Bundestinstitut für Arzneimittel und Medizinprodukte). It evaluates herbal medicine for safety and efficacy and has published more than 300 monographs to date (11). In the early 2000s, most of Europe followed the regulatory approach largely spurred on by the Belgium slimming clinic incident (12). A simplified registration process for what are considered as traditional herbal medicinal products (THMP) (i.e. products with a history of use of over 30 years (a generation) and at least 15 years in Europe) was introduced by the Committee on Herbal Medicinal Products (HMPC), established within the European Medicines Agency (EMA) (13).

ALLIED HEALTH PROFESSIONS IN SOUTH AFRICA

In South Africa, health professionals, be they conventional, complementary, or traditional, are envisaged to be regulated by their own statutory regulatory body, which falls under the ambit of the National Department of Health, and to advise and make recommendation to the Minister of Health on any matter falling within the scope of their legislation. While for the conventional and complementary modalities these regulatory bodies are long-established councils, professional regulation of ATM is a new initiative and appears to be drawn out (and still incomplete) in the establishment and operation of the Traditional Health Practitioner Council of South Africa (THPCSA). In South Africa, CAM practitioners are governed by a statutory health body, the Allied Health Professions Council of South Africa (AHPCSA), established in terms of the Allied Health Professions Act, 1982 (Act 63 of 1982). Its mandate is to regulate and control 12 CAM health professions (see Table 10.1).

It is a legal requirement for CAM practitioners to be registered with the AHPCSA. Chiropractics make up the biggest group, with 803 registered practitioners, followed

Table 10.1 Number of registered allied health practitioners in South Africa

Allied Health Profession	Number of practitioners
Ayurveda	13
Chinese medicine	158
Unani tibb	73
Acupuncture	58
Osteopathy	38
Chiropractics	803
Homeopathy	568
Naturopathy	89
Phytotherapy	51
Therapeutic aromatherapy	140
Therapeutic massage treatment	104
Therapeutic reflexology	508
Total	2603

Source: AHPCSA website (14).

by homeopathy (568 practitioners). The AHPCSA is a professional council and is therefore confined to regulating practices and not products. The medicinal articles of their trade are regulated by the MCC under the new regime of regulations promulgated in 2013.

COMPLEMENTARY MEDICINE PRODUCT REGULATION IN SOUTH AFRICA

In South Africa, the MCC is the statutory council set up by the Medicines and Related Substances Act (Act 101 of 1965 as amended), which through this act ensures medicine governance in the country. This entails the regulation of the 'manufacture, distribution, sale, and marketing of medicines'. There are only six allied health professions that dispense medicinal products, namely, Ayurveda, Traditional Chinese medicine, homeopathy, phytotherapy, therapeutic aromatherapy, and Unani-tibb. The first attempt at regulating the CAM sector started with the 2002 call-up notice that was meant to be a market survey to understand the size and nature of the CAM market. Up to 150,000 products were submitted at that stage (15). The problem with that call up, however, was that some of the industry took the call up as a registration process and subsequently claimed that their products were registered based on the call-up notice number which the MCC allocated to the products for administrative purposes. There was no legal framework for registration at that stage, and the establishment of such a framework took the next 11 years to institute and implement.

In 2013, the MCC gazetted regulations to provide for the registration of medicinal products used by selected CAM disciplines. The General Regulations in terms of the Medicines and Related Substances Act, 1965 (Act 101 of 1965), (Gazette Number

37032 promulgated 15 November 2013) (16), established Category D in which medicines defined as complementary medicines (CM) now reside (17). Conventional pharmaceuticals for humans are classified as Category A, while veterinary products are Category B. Complementary medicine is defined in the Regulations in terms of the Act as 'any substance or mixture originating from plants, fungi, algae, seaweeds, lichens, minerals, animals, or other substance as determined by the MCC, which is used [....] in maintaining, complementing or assisting healing of the physical or mental state. Such a product can also be used to diagnose, treat, mitigate, modify, alleviate, or prevent disease in human beings or animals and includes health supplements and probiotics' (17). The definition itself was subject to much consultation and discussion, and the final definition reflects a compromise and also underwrites the inclusivity of the public consultative process required, as intended by the Constitution and Promotion of Administrative Justice act, 2000 (Act 3 of 2000) (PAJA) (18). With specific reference to CM only, the Act's definition and associated regulations have no provision for *alternative* medicines. Such a provision would encourage the possible use of unproven treatments, evince an aversion to conventional therapies, and expose members of the public to potentially unsafe and inefficacious medicines (15). Thus, under South African law, Complementary Medicines (CMs), rather than the Complementary and Alternative Medicines (CAMs) are provided for.

While major steps have been taken to regulate ATM products, they remain outside the current medicine regulatory framework for various political and technical reasons. Politically ATMs cannot be rightly considered as complementary because they are not of foreign origin; technically, most of them are used in crude and/or extemporaneous form (i.e. not generally packaged into pharmaceutical packaging nor distributed across a wide geographic footprint). However, those products which are sold in pharmaceutical packaging and make medicinal claims fall under the purview of the Medicines and Related Substances Act 101 of 1965. Those that are formulated with popular herbal products which have found markets in the West may, however, be classified as 'complementary', as they may closely identify with the allied health professions of phytotherapy and aromatherapy. This is a point of contention, and accusations of cultural appropriation have been levelled against regulators who take this view.

DEVELOPMENT OF THE SA REGULATIONS AND RELATED CHALLENGES

When the regulations were promulgated in November 2013, a number of steps were to be undertaken, many of which the public did not fully or clearly understand. The road map (Figure 10.1) published a few weeks later provided some guidance as interpretation of the amended regulations (19). The first step was the immediate withdrawal from the market of all CM products which might contain banned substances or scheduled substances.

Thus, all products containing substances such as yohimbine, ephedrine, kava kava, cannabis (or derivatives), or cocaine alkaloids (which are Schedule 6 and 7

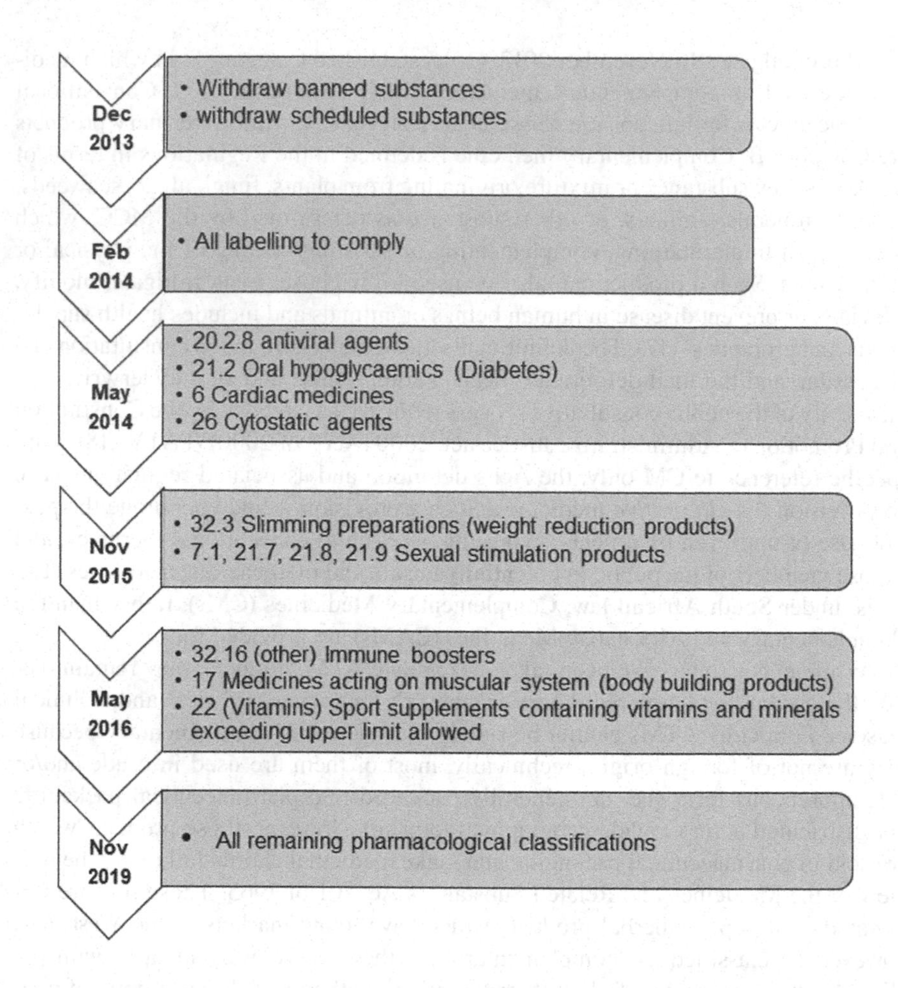

Figure 10.1 Roadmap to complementary medicine compliance (19).

substances, respectively) should have been removed from open sale by December 2013. Simultaneously, all medicines being sold as CMs but not fitting the definition were to be withdrawn immediately. This created problems with combinations of category A medicines with herbal products. These include, for instance, combinations of multivitamin products and ginseng or *Gingko biloba* extracts. While there might be scientific rationale for such a formulation, the challenge is finding the appropriate regulatory landscape for such products. No new products which fall under the definition of CM could be introduced onto the market without first being registered. *All products already on the market could only remain on the market as long as registration dossiers were submitted* (according to a schedule for application submission deadlines) and *subject to immediate labelling amendment with the words:* 'This medicine has not been evaluated by the Medicines Control Council. This medicine is not intended to diagnose, treat, cure or prevent any disease'. This was then deemed to be the pathway to registration or removal of the product, depending on the ultimate outcome from the MCC processes which can either be registration or rejection.

As for the registration deadlines related to products under this new framework, this was done according to the risk of the claims and purported therapeutic use/ claims of the products and was envisaged to commence in May 2014 for five-and-a-half years (ending in November 2019). By May 2014, all products claiming to treat or manage viral infections (HIV and AIDS being the most popular), diabetes mellitus, heart disease, and cancer were supposed to either amend their claims and/or have been submitted for registration in order to remain on the market. November 2015 marked the submission deadline for slimming preparations and sexual stimulants. Thereafter, in May 2016 sports supplements and body building products were expected to have been submitted for registration. All other classes of unregistered medicines would follow, culminating in a situation where all CM products on the market should be registered by November 2019. While there has been major progress on the road map, it is thought that the number of products which have been submitted for registration represents only a minuscule number of products in the marketplace. This is due to various reasons, including the delay in finalizing the guidelines on the registration processes of 'health supplements', i.e. all those products which are not aligned to a particular complementary discipline, such as probiotics, vitamins, and minerals. Furthermore, there has been poor information flow to manufacturers, especially those not aligned with any of the major trade associations, or quasi-formal entrepreneurs who are generally small businesses and have little understanding of government regulations. There is certainly a concerted need to inform and train such businesses around regulatory compliance. Another cause of delay appears to be initial resistance to regulations by some of the CM industry and in some cases a politicisation of regulatory affairs as exemplified by popular press articles; see for instance (15, 20). There are lobby groups which have not been particularly helpful, perhaps due to a misunderstanding or fear of a regulated future and some initially threatened court action in order to stop the process (21). Increasingly the SAHPRA seems to have become proactive in meeting stakeholders and sharing information which has increased awareness of the regulations. Whether awareness has led to increased compliance is subject to research.

More recently, as the product registration process has been unfolding, it has become apparent that the products recommended for registration can, in fact, not be registered because the manufacturers and/or importers (warehousing) are not licenced by SAHPRA. Therefore, the need for a parallel licensing process has emerged. In the past, MCC has not been very effective at timely inspecting and issuing manufacturing/warehousing licences.

The current challenges facing the new regulations are summarized below.

Delays in the Publishing of Final Guidelines

A problem that arose was that while products were being called in, the guidelines appeared not to have been finalised. These guidelines were subsequently published for comment in mid-2016.

With the finalization of this process, it is envisaged that both industry and regulators will be on the same page regarding registration requirements. While the listing system which industry was advocating for is not entirely reflected in the regulations and guidelines, the philosophy of a risk-based approach is central to the South African regulatory framework. Thus, using standard monographs and low-risk claims will facilitate registration as outlined in the guidelines. In essence, the riskier the therapeutic claim, the higher the burden of evidence. For example, in terms of efficacy a clear medical (high-risk) claim associated with the treatment of disease will require evidence of traditional use *plus* clinical studies or randomised clinical trial, while a low-risk claim for a self-limiting condition is registrable on literature evidence detailing its traditional use. The problem with this approach is that it might stifle innovation of new products and/or new applications of old products within South Africa where the resources for a clinical trial are limited.

Requirements for Quality Management Systems

A contentious issue for South African-based manufacturers appears to be the stance that quality and safety (which are interdependent) are non-negotiable. According to the Medicines and Related Substances Act (Section 22C), 'on application in the prescribed manner and on payment of the prescribed fee, issue to a [...], manufacturer, wholesaler or distributor of a medicine, Scheduled substance, [...] a licence to manufacture, import, export, act as a wholesaler of or distribute, as the case may be...'. Products will fail registration if they do not meet certain minimum quality management standards (in essence, current Good Manufacturing Practice [cGMP]), which local industry appear to consider too high and costly to implement. Thus, the issues around Good Manufacturing Practice (GMP) requirements for all manufacturers of CMs have been at the heart of the fightback strategy from industry. GMP compliance is generally expensive. The nascent nature of the CM industry in South Africa means that GMP compliance is an added cost that may hitherto not have been contemplated to be necessary. The MCC has suggested a phased approach to GMP compliance. In addition, because of the under-resourcing and low capacity in pharmacovigilance and market surveillance, it makes sense for low-resourced medicine regulatory authorities (MRAs) to concentrate resources on making sure that product quality is high by capacitating and prioritizing inspectorate rather than policing functions.

Food–Medicine Distinction

With many products which are marketed as health supplements, the distinction between their use as food substances and medicines can be difficult. This might explain why the US government tried to simplify the situation outside the mandate of the FDA. However, this can then compromise patient safety in some cases. For example, *Camellia sinensis* is a beverage normally packed in teabags, but in concentrated form it has been known to cause hepatitis (22). It might be that it is presented as a tea but is marketed with a medicinal claim or marketed as a normal tea with health benefits claims. Increasingly, aromatic African herbs are presented in food format, packed as tea bags or infusions, as they are popular 'non-medicalised' (and therefore 'natural') dosage forms. Thus, the South African market has anything from sutherlandia (*Sutherlandia frutescens*) (kankerbos) tea to hoodia (*Hoodia gordonii*) and lippia (*Lippia javanica*) teas and some combinations with more conventional herbal teas such as rooibos (red bush) or honeybush tea. Subtle claims are then inserted, which potentially move them from the food territory into medicine territory.

Multivitamins and minerals also present similar problems in straddling the food–medicine divide. The regulations have tried to provide for this by allowing them to be considered as food if sold in levels normally found in food, or to be registered as category D medicines if used for supplementation purposes (as envisaged in the definition of CMs), or to be registered as category A medicines if formulated in high doses with clear therapeutic intent, or in injectable form.

Labelling, Marketing, and Advertising

Another issue which is addressed in the new legal regime is that of labelling and to some extent marketing and advertising. The guidelines related to health supplements have a list of phrases which cannot be used on the label and/or marketing material. Words such as miraculous, 100% safe, world's best, anti-ageing, etc. are not allowable unless backed by compelling evidence, i.e. clinical evidence.

While the Medicines and Related Substances Act and regulations allude to advertising and marketing, enforcement is already seen as a problem, as evidenced by advertisements broadcast on any given radio station, television, or multimedia platform. For its part, the MCC website does not appear to have regulations or clear guidelines on the advertising of medicines, apart from provisions in the Medicines Act (Section 18C and 20) and Regulation 45. A Code of Marketing for Medicines, which is part of the Advertising Standards Authority (ASA) Code of Advertising Practice, was originally approved by the MCC in 1985 (23). The code envisaged a close working relationship between the MCC and the ASA in the 'registration and substantiation' of medicines. This would entail that all advertisements of medicines must be approved by the Registrar of Medicines and that such a product meets quality safety and efficacy requirements and thus 'the claims made are not false and misleading'. However, such a code could not be located on the Advertising Standards Association in the present review. A Code of Marketing for Complementary Medicines as part of the Marketing Code Authority was finalised in 2013 (24). This authority is less well known than the Advertising Standards Authority (ASA) (25) and brings together pharmaceutical industry players to promote ethical marketing of pharmaceutical products. Both bodies are voluntary bodies with no statutory powers and so will not be effective in protecting members of the public from unscrupulous marketing. Thus, while the MCC is concerned with sale and production of products, advertising appears to be outside its ambit and is currently regulated only by the National Consumer Commission (NCC) which might step in should there be complaints from consumers about the products at the point of use. The NCC is retrospective in its remedy and addresses individual consumer complaints. There is, therefore, an urgent need to create an advertising framework for CMs, as they are prone to being 'oversold' to the possible detriment of public safety.

Training

There is no institution currently focusing on training regulators for both industry and the MCC. There are some private providers, but their focus tends to be on specific items such as dossier compilation, as a study commissioned by the Southern African Generic Manufacturers Association (SAGMA) revealed (26). Attempts to start the so-called Institute of Regulatory Science (IRS), which was meant to be a virtual centre, have not yet taken off. Meanwhile, there are some in-house courses being initiated around the country. The University of Pretoria (UP) now offers a short course, 'Manufacturing and Regulation of Herbal Medicines in South Africa'. The evaluation of CM dossiers is potentially more complex because of the multicomponent nature of some of the products, the risk of adulteration and misidentification, and the risk of natural contaminants. The fact that these products have not previously been regulated also means that companies and regulators have to create a new path together.

RECOMMENDATIONS

From this review, it is evident that a lot of work has been done in developing and implementing a regulatory framework for CMs in South Africa. The next step should be enforcing the regulations to make sure that those companies that do not comply are prosecuted and their products removed from the market. The role of the SAHPRA should continue to be facilitatory to those stakeholders who genuinely want to comply. It is recommended that support be given to the training programmes currently in place so that they can grow into formal qualifications, such as diplomas and possibly degrees. While the idea of an institute of regulatory affairs is noble, it is unlikely to be achieved in the short term due to the territoriality of universities in South Africa.

The regulation of ATM, while necessary, should be approached with due sensitivity for the underlying traditional principles and with a sound understanding of the market and its origins. Those medicines which are being formalised and traded widely as finished pharmaceutical products (FPP) can certainly be gradually included in the current framework, but it will also mean that experts and practitioners in the discipline should be co-opted and trained in the science of regulation. This was the approach with allied health professionals who support the current CMs regulatory framework, particularly on the evaluation of clinical use.

The monetary value of the trade of CMs in South Africa is unknown. The main reason is because the sector is segmented into different associations. A 1996 survey of the Health Products Association (HPA) found that turnover was about R900 million, and by 2003 another survey showed an increase to R1.35 billion (27). There is a need to better understand the economic contribution of this sector from a pharmacoeconomic point of view. Such data is important for the right kind of support interventions, for sector development, and for understanding the impact assessment of legislative changes.

As submissions came into the regulator and products were ready to be approved for registration, it became apparent that the applicants did not have manufacturing or warehousing licences. These licences are required before any product can be registered and approved for sale in South Africa. This seemed to throw a wrench in the works and required a different approach, which is to license the applicants first, prior to calling up products for registration. This appears to be the two-pronged approach which SAHPRA will now adopt. This is an important lesson to other regulators – that this sector is different in its evolution and therefore different approaches should be attempted.

CONCLUSION

The thinking around regulating CMs in South Africa is informed by global trends, and the regulatory framework appears to have borrowed heavily from the Health Canada approach while learning lessons from other territories, particularly Australia and the EU. However, lessons from other African countries and from WHO will also be useful because of the unique position of the dichotomy of the existence of ATM alongside CM disciplines. Licensing of the facilities/applicants prior to calling up the products for registration has also emerged as an important lesson learnt by SAHPRA, of which other regulators should take note.

The risk-based approach is in keeping with current regulatory science thinking, and it appears to be the best use of resources, as it allows the regulator to focus on potentially risky manufacturers and also risky therapeutic areas. However, such an approach will need to be accompanied by resources and capacity building in both the regulatory and industry sectors, particularly around quality management systems compliance (at the manufacturer level) and inspectorate and compliance monitoring (at the SAHPRA level). This calls for collaborative public–private partnerships to develop this important economic and health sector to the benefit of all South Africans.

ACKNOWLEDGEMENTS

We wish to thank Dr Neil Gower the chairperson of the Complementary Medicines Advisory Committee for useful inputs into the draft chapter.

REFERENCES

1. Wreford J. Missing each other: Problems and potential for collaborative efforts between biomedicine and traditional healers in South Africa in the time of AIDS. *Social Dynamics.* 2005;31(2):34.

2. Management Sciences for Health (MSH). *MDS-3: Managing Access to Medicines and Health Technologies*. Arlington, VA: Management Sciences for Health; 2012.
3. National Institute of Health. Complementary, Alternative, or Integrative Health: What's In a Name? 2017 [Available from: https://nccih.nih.gov/health/integrative-health].
4. McRae C, Agarwal K, Mutimer D, Bassendine M. Hepatitis associated with Chinese herbs. *European Journal of Gastroenterology & Hepatology*. 2002; 14(5): 559–562.
5. Plaeger SF. Clinical immunology and traditional herbal medicines. *Clinical and Vaccine Immunology*. 2003; 10(3): 337–338.
6. Bent S. Herbal medicine in the United States: Review of efficacy, safety, and regulation. *Journal of General Internal Medicine*. 2008; 23(6): 854–859.
7. Stickel F, Patsenker E, Schuppan D. Herbal hepatotoxicity. *Journal of Hepatology*. 2005; 43(5): 901–910.
8. Ko RJ. Adulterants in Asian patent medicines. *New England Journal of Medicine*. 1998; 339(12): 847.
9. Nortier JL, Martinez M-CM, Schmeiser HH, Arlt VM, Bieler CA, Petein M, et al. Urothelial carcinoma associated with the use of a Chinese herb (*Aristolochia fangchi*). *New England Journal of Medicine*. 2000; 342(23): 1686–1692.
10. Magee K. Herbal therapy: A review of potential health risks and medicinal interactions. *Orthodontics & Craniofacial Research*. 2005; 8(2): 60–74.
11. Calixto J. Efficacy, safety, quality control, marketing and regulatory guidelines for herbal medicines (phytotherapeutic agents). *Brazilian Journal of Medical and Biological Research*. 2000; 33(2): 179–189.
12. Vanherweghem JL, Tielemans C, Abramowicz D, Depierreux M, Vanhaelen-Fastre R, Vanhaelen M, Dratwa M, Richard C, Vandervelde D, Verbeelen D, Jadoul M, Rapidly progressive interstitial renal fibrosis in young women: association with slimming regimen including Chinese herbs. *The Lancet*. 1993; 341(8842): 387–391.
13. Sahoo N, Manchikanti P, Dey S. Herbal drugs: Standards and regulation. *Fitoterapia*. 2010; 81(6): 462–471.
14. AHPCSA. The Allied Health Professions Council of South Africa Practitioners, 2017 [Available from http://ahpcsa.co.za/practitioners/].
15. Viall J. Leave complementary medicines alone. *Cape Times*. 2014. [Available from: https://mg.co.za/article/2014-07-29-bitter-pill-for-complementary-medicine-sector/], (accessed 10 January 2017).
16. Government of South Africa. General Regulations in terms of the Medicines and Related Substances Act, 1965. Pretoria: Government Gazette; 2013. p. 13.
17. Government of South Africa. Medicine and Related Substances Act, 101 (1965).
18. Government of South Africa. Promotion of Administrative Justice 3 (2000).
19. Medicines Control Council (MCC). Roadmap for registration of complementary medicines, 2013. [Available from: http://www.mccza.com/documents/66d8cd937.02_Roadmap_for_CAMs_Dec13_v1.pdf], (accessed 3 January 2017).
20. THNA. The CAM Crossroads – where are we heading? 2016. [Available from: http://www.tnha.co.za/the-cam-crossroads/], (accessed 10 January 2017).
21. Cronje M. Bitter pill for complementary medicine sector. *Mail and Guardian* [Internet]. 2014, (accessed 10 January 2017).
22. Javaid A, Bonkovsky HL. Hepatotoxicity due to extracts of Chinese green tea (*Camellia sinensis*): A growing concern. *Journal of Hepatology*. 2006; 45(2): 334–335.
23. Jobson R. Pet food advertising better regulated complementary medicines, 2013. [Available from: http://www.quackdown.info/article/pet-food-advertising-better-regulated-complementary-medicines/], (accessed 10 January 2017).

24. Marketing Code Authority (MCA). Marketing Code Authority's website. [Available from: http://www.marketingcode.co.za].
25. Advertising Standards Authority. The Advertising Standards Authority of South Africa website. [Available from: http://www.asasa.org.za] (accessed 17 January 2017).
26. Katerere D. Analysis of Training Needs in the Generic Medicines Sector in the South African Development Community. Vienna, Austria: UNIDO; 2014.
27. Gqaleni N, Moodley I, Kruger H, Ntuli A, McLeod H. Traditional and complementary medicine: Health care delivery. *South African Health Review.* 2007; (1): 175–188.

CHAPTER 11

Animal Health and Indigenous Knowledge Systems

Lillian Mukandiwa and Donald R. Sibanda

CONTENTS

ABSTRACT

Background The ethno-veterinary knowledge system plays a major role in veterinary health practices globally – and especially in Africa. Disregard for this knowledge system has heralded an epidemic of animal diseases across the globe.

Relevance This chapter reviews indigenous veterinary practices and their role in combating animal diseases.

INTRODUCTION

Veterinary medicine as practiced today has its origin in traditional medicine as practised in prehistoric times in China, India, the Middle East, and elsewhere (Schillhorn van Veen, 1997). Today various indigenous societies across the world each have a wealth of knowledge and immense experiential practice in animal healthcare and production. This knowledge is derived mainly from indigenous knowledge on livestock and their environment which has been collected and refined over centuries of living in specific physical environments. As such, the knowledge varies from region to region and from community to community. Some of this rich and time-proven indigenous knowledge can be and has been harnessed to improve animal health and production for food security and to develop new therapeutics in the present era. The systematic approach to the study of the indigenous bodies of knowledge on animal healthcare has given rise to a subject now commonly known as ethnoveterinary medicine (EVM), or sometimes, veterinary anthropology.

Research into EVM has been stimulated by the alarming observation that the promotion of Western veterinary medicine, to the exclusion of indigenous animal health knowledge, has led to the disuse or loss of a cost-effective, easily accessible, sustainable, environmentally benign, and socioculturally acceptable and often site-specific animal health management resource (Wanzala et al., 2005). Ultimately, this loss of indigenous knowledge has left pastoralists vulnerable and unable to cope with livestock disease and pests, more so in areas with limited access to modern veterinary medicine (Davis, 2016). In addition, the research in EVM is driven by the resurgent interest in the screening and application of plant and animal extracts for medicinal use as alternatives to current therapeutics and pesticides which are continually rendered ineffective by the development of resistance.

Even though indigenous knowledge does not have the techniques and quantitative tools that are at the disposal of Western science, some systems of indigenous knowledge seem to have developed ways to deal with animal health-related complexities. The mere fact of their survival over hundreds of years is a strong indication that these practices have been tested enough times to be significant. Any insights from indigenous wisdom in regard to animal health are of huge potential interest in the face of challenges that have come with Western veterinary medicine, such as the development of resistance in disease-causing microorganisms and parasites, and environmental contamination. In addition, climate change, the recent expansion of organic farming, and the increasing importance of consumer concerns drive the need to explore indigenous knowledge systems (IKS) for sustainable practices for management of animal health and welfare. In the developing world, the restrictions on the use of modern medicine and the failure of both government and private veterinary services (especially in developing countries) to meet the needs of pastoralists create a further need for research into IKS. This chapter will give a critical appraisal of some of the currently recorded indigenous animal health knowledge systems and explore how the application of this indigenous knowledge in the modern era can contribute to sustainable management of animal health in the face of the challenges outlined above.

ANIMAL HEALTH INDIGENOUS KNOWLEDGE

Pastoralists worldwide have extensive knowledge of animal diseases and their management (Mathias-Mundy and McCorkle, 1989; Niamir-Fuller, 1994). Their traditional knowledge encompasses the areas of what in modern science would be epidemiology, gross pathology, diagnostics, treatment, disease prevention and control, pharmacology, toxicology, and herd management. Many of their herding practices (McCorckle and Mathias-Mundy, 1992; Niamir-Fuller, 1994) are designed to prevent disease. These practices include:

- avoiding certain pastures at particular times of the year
- not staying long in one place, to avoid parasite build-up
- lighting smoke fires to repel insects, especially tsetse flies
- burning the grass just before the rains to kill ticks in various stages of development
- mixing species in the herd to avoid the spread of disease
- avoiding infected areas or moving upwind of them
- spreading livestock among different herds to minimize risks
- quarantining sick animals
- selecting only healthy animals for breeding

Details of these practices are given in subsequent sections in this chapter. This body of knowledge has evolved within the communities over time and is orally communicated from one generation to the next with the ultimate aim of ensuring survival and progress of livestock production. Examples of these indigenous knowledge systems include that of the Masaai of Kenya and Tanzania (Ole-Miaron et al., 2004), the Fulani of Nigeria (Leeflang, 1993), the Konda Reddi of India (Misra and Kumar, 2004), and the Koochi of Afghanistan (Davis et al., 1995), among others. It can be argued that the mere fact of the survival of such systems over hundreds of years should be a strong indication that these systems were tested enough times to be of significant importance and that it is thus unwise to dismiss all treatments and practices that have not been scientifically developed and proven.

Indigenous Epidemiology

Most of the communities using Animal Health IKS have been reported to have, for the majority of diseases, a basic understanding of the cause of the disease, how it is transmitted, and the conditions under which it is most likely to occur. It has been reported that the Masaai and the Koochi perceptions of diseases are similar to the current understanding of their aetiologies, transmission and/or epidemiology by the Western veterinary community (Davis et al., 1995; Ole-Miaron et al., 2004; Catley, 2006). Most stockmen worldwide understand the importance of ticks and, quite often, the relationship between ticks and tick-borne disease. West African pastoralists correctly associate *Amblyomma* ticks with heartwater disease and with the bacterial skin disease streptothricosis.

In a study on indigenous animal health care among the Koochi nomads of Afghanistan (Davis et al., 1995), the majority of study respondents knew that diarrhoea is often caused by gastrointestinal worms and that gastrointestinal worms come from the animals grazing on contaminated or wet pastures, and that liver fluke infection is much more likely to occur when animals graze at the banks of irrigation ditches or other areas bordering slow-moving or stagnant water. The snails necessary for the liver fluke's life cycle are much more likely to be found under these conditions. In the same study, the majority of Koochi also explained that foot-and-mouth disease (FMD) can pass from one animal to another by one animal walking in the footsteps of another animal that has the disease, under crowded conditions. FMD is a highly contagious virus that is spread primarily by aerosols and is transmitted very easily among animals kept in close contact. While the Koochi did not appear to understand the exact aetiology or mechanism of transmission of FMD, they understood very well the patterns of its occurrence, that is, the epidemiology of the disease. The Maasai correctly associate ticks with East Coast fever (ECF), dust with anthrax, tsetse fly with trypanosomosis. Liver flukes, malignant catarrhal fever (MCF) and FMD are considered to be water-borne diseases (Ole-Miaron et al., 2004). The inference that MCF and FMD are water-borne can be attributed to how the diseases are actually transmitted. In Africa, most wildebeest-associated MCF is seen around the time of wildebeest calving. Transmission is by transfer of virus-laden nasal secretions by direct contact or poorly defined airborne routes. Sharing of water bodies by cattle and wildebeest results in transmission to cattle. The same scenario occurs with FMD transmission where cattle infect each other at water bodies or by sharing with buffaloes, which are carriers of the FMD virus in Africa.

A study by Luseba and Van der Merwe (2006) shows that the Tsonga of South Africa also exhibited high epidemiological knowledge. According to the Tsonga farmers, heartwater is suspected when an over-excited animal runs and suddenly falls dead with froth from the mouth. This inference is not very far off from the modern veterinary medicine signs of heartwater. These include strange behaviour, such as twisting the head towards the body, a high-stepping walk, convulsions, or very hard kicking. Animals that die from heartwater often have foam and fluid coming out of the nose. It was also thought that tick-borne diseases such as heartwater and redwater are propagated through tick saliva. Indeed, both diseases are transmitted through tick bites. Tick bites were recognised by farmers as the biggest causes of wounds. Blackquarter was suspected when the animal failed to wake up and was limping and sometimes showed swellings on the legs. These clinical signs were reported to occur in spring or after frost in winter. Similarly, in modern veterinary medicine, blackquarter cases are also reported mostly in spring and late winter when animals graze short grasses that expose them to clostridia infections. Internal parasite infection was strongly linked to diarrhoea and lack of appetite but infections by nematodes from tapeworms or flukes were often non-specific and difficult to categorize. Newcastle disease in chickens was matched with signs of greenish diarrhoea and paralysis occurring when seasons change. While the Tsonga have some relatively accurate perceptions of disease when compared to modern veterinary medicine, not all the documented practices and inferences in this study are accurate, which is an observation common in other indigenous knowledge systems.

Indigenous Diagnostics

Indigenous animal health systems make use of a number of ethnodiagnostic techniques such as:

- smelling a sick animal's breath, urine, or dung
- tasting its milk
- listening to its respiration and vocalizations
- observing any subtle changes in behaviour, such as feed and water intake, stance, gait, excitement or lethargy, and social interaction (Bizimana, 1994)

Curasson (1947) mentioned a famous camel healer in Niger who was able to diagnose most diseases by the scent of their expired air; a phenomenon that makes sense today because of the fact that analysis of expired air as a diagnostic tool has become useful in modern medical science to diagnose various diseases (Amann et al., 2004; Sethi et al., 2013). Analysis of exhaled air is a promising tool for diagnosis of various diseases in veterinary medicine, too. Within the last two decades, a number of in vivo studies were carried out in domestic animals and large animal models to identify volatile biomarkers attributed to different diseases (Spinhirne et al., 2004; Fend et al., 2005; Knobloch et al., 2009; Purkhart et al., 2011; Peled et al., 2012; Bayn et al., 2013; Cho et al., 2015). Another example of how most traditional diagnosticians use their senses, either as such or after some manipulations, is in how desert camel owners in Punjab (Pakistan), Sudan, and Somalia developed traditional methods of diagnosis of surra, a *Trypanosoma evansi* infection. The method, known by some as the "Sand-ball test", entails mixing camel urine with mud and assessing the smell of the dried mud. The method was reported to be fairly accurate (Leese, 1927). In camels with high parasitemia, a characteristic sweet, pungent and "sickly" smell attributable to the presence of urochromes (Hunter, 1986) can be detected. In India, another traditional method of surra diagnosis involves pulling hairs out of the suspected animal's tail and applying the hair roots to the downward facing hand. The animals are considered healthy if the hairs stick to the hand; otherwise they are infected. The basis of this test ("Hair-stick test"), considered by Leese (1927) to be of some value, is that in healthy camels some meat is also pulled out with the hairs, which then sticks to the hand. The pastoralists of Eastern Tyrol (Austria) believe that to maintain and/or increase the health and overall performance of the animals, it is essential that the animals demonstrate "appropriate" digestion. Digestive processes are therefore evaluated by inspecting the dung. The consistency of a fresh cowpat has the look of a traditional flat loaf of bread: not too liquid, but not too dry and firm. To enhance and improve digestion in general, plants such as alpine meadow hay are used as fodder (Vogl et al., 2016).

The Maasai ability to diagnose livestock diseases and knowledge of the species affected by a particular disease condition has been found to compare favourably with that of modern veterinarians and does not greatly deviate from the published information in veterinary books (Ole-Miaron, 2003). The Maasai traditional disease diagnostic procedures are based on various techniques/information such as symptoms of disease, knowledge of known vectors of livestock diseases, knowledge of seasonality of disease outbreak, and species affected by specific diseases. The seasonal outbreak of diseases is an important diagnostic tool for the Maasai. They associate anthrax with the dry season. FMD and MCF are associated with the wet season. Incidences of these disease conditions are known to increase during the wet season.

Indigenous Preventative and Vaccination Practices

The Fulani of Nigeria, the Moors of Mali, and the Baggara of the Central African Republic all have a unique indigenous vaccination practice against contagious bovine pleuropneumonia (CBPP) (Leeflang, 1993; Bizimana, 1994; Alhaji and Babalobi, 2015). They slice lung tissue from a diseased animal and implant it under the skin on the foreheads of their cattle, sealing the incisions with mud. Alternatively, the lung tissues from infected dead cattle with CBPP are soaked in fresh milk and briefly placed on the nasal area of the healthy ones or wrapped in a rag and hung on a tree very close to the herd site. Some pastoralists also dry the lung and grind and spread the granules in the herd. This is related to modern-day veterinary science where vaccines contain either weakened or denatured forms of disease-causing viruses or bacteria, or parts of these that will stimulate a response by the immune system of the animal. This process then protects against the development of disease conditions. Transmission of CBPP usually happens as a result of direct contact between infected and susceptible animals, with fomites apparently playing no role in the spread of the disease. The infected lung tissue that is planted under the skin will induce an immune response but not disease. It is also thought that healthy animals that are introduced into environments that have previously been occupied by CBPP-infected animals do not become infected. Ingestion of contaminated feed or direct exposure to the organs of animals that have died of CBPP does not lead to transmission (http://www.afrivip.org).

The Koochi of Afghanistan, the Kikuyu of Kenya, and the Somalis also utilize similar indigenous vaccination techniques for contagious caprine pleuropneumonia (CCPP) and sheep and goat pox (SGP) (Baumann, 1990; Bizimana, 1994; Davis et al., 1995). For both of these diseases, an animal which has died from either CCPP or SGP is cut open and the lungs are retrieved. Small pieces of the diseased lung are crushed and placed in a slit made between the eyes or in the ear of a living animal. The entire herd is treated this way when a few are showing signs of the illness or have died. Another technique used is to dab a needle in the diseased tissue and then the ears of healthy animals are stabbed with the needle; sometimes the diseased lung is rubbed into the punctures made by the needle. For SGP, the fluid from a pox blister on the skin of the animal is sometimes used instead of the lung for making the vaccine. The technique has also been tried with anthrax without much success. This highlights the fact that indigenous animal healthcare practices will work depending on a specific disease transmission or depending on the virulence of a disease-causing agent. One of the dangers of a live vaccine such as this is the possible infection and production of serious disease in the vaccinated animals. This would be especially true in the case of anthrax. Anthrax is a particularly virulent disease, in part because it is highly and equally infective for any route of exposure (ingestion, inhalation, or percutaneous). CCPP, on the other hand, is most infective when inhaled by livestock and less virulent via percutaneous exposure. Thus, the ear vaccination is more successful in producing an immune response and in preventing disease with CCPP than with anthrax.

The Fulani management response to FMD has been recognised as an illustrations of how indigenous knowledge sometimes outstrips contemporary Western science (Leeflang, 1993). They sometimes move their cattle upwind of infected herds to prevent the disease from spreading, and sometimes they move them downwind to expose the animals to FMD, knowing that a mild case of the disease will not be fatal and will confer immunity. Western scientists learned that the FMD virus could be aerially transmitted over long distances only after outbreaks of the disease in Europe, decades after the Fulani had been practicing their indigenous vaccination practice. These pastoralists also know that the fluid in the tongue blisters of animals infected with FMD is infectious to other animals. To control the spread of the disease, they collect this fluid, dip a tree thorn in it, and scratch the tongue epithelium of apparently healthy animals in order to vaccinate them.

As a prophylactic measure against bovine brucellosis, the Fulani have been reported to mix the fluid of aborted foetuses with urine and rub it on the genital and udder areas of their herd (Alhaji and Babalobi, 2015).

Indigenous knowledge systems also comprise precautionary measures based on the knowledge of the cause and the vectors. For example, for centuries the Maasai were convinced that the wildebeest is a silent carrier of the MCF causative agent. They have no known traditional cure for MCF, but they are strict in keeping their cattle away from the wildebeest during their calving season (Ole-Miaron et al., 2004). Modern veterinary medicine also discourages grazing cattle in areas close to farms with wildebeest.

Preventative measures also include environmental controls which comprise steps taken to modify the animal's immediate environment to make it safer and healthier. These include poisoned baits and planting of pest-repellent plants or other preparations sprinkled, strewn, or burned in and around the animal's environment. Andean stock raisers spread quicklime or creosote in their corral to cleanse them of disease bearing pests and pathogens. Lime is commonly used to disinfect animal houses in modern veterinary medicine. It is also used for disinfecting environments and carcasses after epidemical outbreaks such as anthrax, foot and mouth disease, and African swine fever (EuLA, 2009). It destroys most bacteria and some viruses by saponification of the lipid components of cell membranes, which results in structural disruption of the microorganisms, since it has properties of an alkali. Furthermore, if the concentration of an alkali disinfectant can rise to a local pH of at least 10, it may have the additional microbiocidal effect where disruption of the structure of bacterial peptidoglycans occurs and causes hydrolysis of viral genome nucleotides (Himsworth, 2008). Nigerians plant tobacco around their farmyards and poultry coops to repel snakes. Likewise, Zimbabweans sprinkle a cold infusion of *Annona senegalensis* root in their hen runs to keep snakes away. In Trinidad and Tobago, chicken nests and litter are dusted with neem and other pesticidal leaves to combat ectoparasites (Martin and McCorkle, 2001)

Control of Contagious Diseases

The ethnoveterinary practices of most communities show that they have a clear understanding of contagious diseases and know about animal-to-animal transmission. For example, when a contagious disease strikes, the Fulani separate their sick from their healthy animals (Leeflang, 1993). They warn neighbouring herders and make arrangements to use separate rangeland and watering places. On the basis of this tradition, the veterinary service of the former Northern Nigerian government in the 1950s designated special routes for cattle being traded or sent to slaughter. The aim was to prevent contact between the national herd and cattle with an unknown history. Similarly, in modern veterinary medicine, quarantine of infected farms and regions and restricted movement of livestock are common practices as part of disease control measures.

Any time an outbreak of rinderpest or CBPP occurred, the Fulani would not graze their cattle in the affected areas for two months. Pastures infected with endemic diseases such as blackquarter and anthrax were not used for grazing for two years. Places where animals had died from these diseases would be covered with thorn bushes in order to prevent healthy cattle from grazing. The Fulani had no effective cure for rinderpest. Knowing that they might lose a number of diseased animals every year, their herds included a surplus to compensate for losses.

Selective Breeding

The combination of indigenous communities' environmental knowledge and livestock management techniques helps maintain domestic animal diversity via pastoralists' recognition and evaluation of genetic characteristics, types, and breeds of livestock and their understanding of how plant and animal genetic resources interact. In most indigenous systems, the animals were selected for breeding based on their health (Davis et al., 1995; Rajan and Sethuraman, 1997). Breeding objectives of traditional herders vary and may relate to short-term interests, such as for special hide-colour patterns, or long-term benefits, such as for the ability to survive in a hostile environment, for the agility to enable migration throughout rocky hillsides, and for drought resistance. These selection parameters also extend to breeding for disease resistance (whether on purpose or by default), especially in areas with a strong disease pressure (Schillhorn van Veen, 1997). This is especially the case in Africa, where disease pressure has been particularly high, both from imported diseases such as CBPP and rinderpest, and through diseases mainly occurring on the continent, such as trypanosomiasis, heartwater (cowdriosis), and ECF. The most commonly quoted example of natural disease resistance is the trypanotolerance of the indigenous taurine breeds of cattle in West and Central Africa, namely, the N'Dama and West African Shorthorn (D'Ieteren et al., 1988). Herders on the fringe of the West African forest zone showed their understanding of genetic disease resistance by cross-breeding with the above breeds and creating hybrids such as N'Gabou and Bambara. Another example is Africa's Akole cattle, which have been bred to resist ECF (Moran et al., 1996). Certain sheep breeds are also known to be resistant to gastrointestinal nematodes (Hohenhaus and Outteridge, 1995; Stear and Murray, 1994).

The breeding techniques included obtaining new genetic material through the borrowing of stud bulls or exchange of animals with merchants, family, and friends, and grazing one's herd near another herd with the desired traits in the hope that the animals will mingle and mate naturally (Boutrais, 1998). Controlled breeding practices included tying the penis to the scrotum and inserting a stone into the uterus (Martin and McCorkle, 2001). The BaSotho cattle sterilisation consists of thrusting a heated stone into the vagina of the cow, cauterizing the cervical os and thus preventing the discharge of oestrus fluid when the cow is in heat (Martin and McCorkle, 2001). The resilience of the indigenous livestock breeds is based on traditional veterinary, herding, and husbandry skills; hence, lessons can be learned by modern-era veterinary medicine.

Indigenous Management Practices

In indigenous animal health systems, herd health and composition are maintained through the strategic implementation and timing of several herd management practices and surgical treatments. They primarily involve mixing, separating, or moving stock to reduce their exposure to danger and disease. Arthropod-borne diseases are numerous, and most pastoralists are well aware of the relationship between certain diseases and the flies, ticks, mites, and mosquitoes that transmit them. Thus, strategies to control flies and ticks typically constitute an important part of pastoralists' response to these diseases.

Nomadic and transhumant pastoralists in Africa, the Middle East, South America, and Central Asia use a common strategy of disease prevention by moving animals through time and space in such a way as to avert infection (Eckert and Hertzberg, 1994; Schillhorn van Veen, 1997). The best-known example is the seasonal migration of livestock in the savannas of West and Central Africa. Pastoralists move their animals north during the rainy season to avoid the risk of tick- and fly-borne disease (especially trypanosomiasis) in the more humid and forested south. During the dry season, they take them south in search of better pasture. Detailed accounts of such migrations (e.g. Ford, 1971; Leeflang, 1993) indicate that herders are well aware that the risk of exposure to tsetse flies and other disease-bearing arthropods is much greater in the wet season and that these pests concentrate in certain areas that are often wet and/or wooded. Similar adaptive patterns of migration in time and space have been reported in the Middle East to prevent tick paralysis (Hadani and Shimshony, 1994). The biannual migrations that most Koochi make assist in lowering the incidence of internal parasites and certain diseases such as anthrax (Davis et al., 1995). This grazing pattern also helps to lower the incidence of most gastrointestinal worms, because they pose the worst threat under crowded conditions where animals are kept in the same area for prolonged periods of time.

In addition, pastoralists' efforts to avoid disease-transmitting biting insects and infested wetlands can have the added bonus of breaking the lifecycle of other parasitic infections such as liver fluke. Many herders correctly link liver fluke infection to grazing in swamps and floodplains. Unless drought intervenes, herders, whether in Africa or in the Andes (Balazar and McCorkle, 1989), try to avoid such areas or at least to minimize the time spent there, thereby reducing exposure to contaminated pastures. African savanna pastoralists have also used tactical (versus seasonal) movements to prevent disease outbreaks (Schillhorn van Veen, 1997). For example, if it is impossible for them to avoid moving their stock through known fly-belts, especially during the rainy season, they traverse these areas by night when the flies are inactive. Watering schedules and the length of time spent at watering places, too, are adjusted to take account of fly activity.

African herders also recognize that insects often concentrate in woody areas, and may deliberately overstock pastures so as to keep vegetation down, thus destroying potential tsetse habitat. For the same reason, pastoralists may purposely set fire to rangelands and/or cut down vegetation in order to control fly and tick populations, although other reasons (hunting, obtaining good regrowth, etc.) may complement or dominate the decision to burn. Pasture spelling has been used in Australia, controlling *Boophilus microplus*, but is difficult to apply in more traditional settings in Africa and Asia (Sutherst, 1983), where most ticks have more than one food source and longer survival times.

In addition, common preventative strategies against insect-borne diseases also include lighting of fires to drive away insects. To repel biting and sucking flies, the Fulani burn dried grasses or dried wood amongst the herds in the morning before embarking on grazing and immediately on return from grazing in the evening (Alhaji and Babalobi, 2015). The Fulani also have special ways of controlling ticks. They feed host animals salty plants so that ticks fall off; they pick off ticks and burn them; and they burn off infested rangeland.

Isolation of the young and sick animals is also one of the animal management practices used by many communities to reduce the incidence of disease among their livestock. For example, the Koochi were reported by Davis et al., (1995) to often keep the newborn and young animals at the tent with the women and children until they were old enough to go out to pasture with the herd. Sick animals were likewise kept at the family tent, effectively isolated from the herd, decreasing the chances of infection spreading throughout the flock.

Finally, and perhaps most importantly, more or less typical for most traditional animal husbandry systems, is the mixed herds used to help decrease the risk of epidemic disease and high mortality. A typical West African Fulani herd consists of cattle, sheep, and some goats that are herded together. In parts of Russia, floodplain pasture may be shared by ruminants and geese, and Kirghiz herders have a mixed herd of sheep and horses (Schillhorn van Veen, 1997). Only in more extreme natural conditions (and where disease pressure is probably lower), such as semi-desert, tundra, and high mountain pasture, is it more common to see single-species herds or herds sharing pasture with grazing wildlife. This pattern of spreading the family's resources widely to include many different species of livestock maximizes the chance of at least some of them surviving any kind of disaster which might decimate a single species. The advantage of mixed and alternate grazing (especially with non-related species such as sheep, cattle, horses, or in some African steppe systems, wildlife) on reducing parasite loads with gastrointestinal helminths has been reported (Schillhorn van Veen, 1997; Fraser et al., 2014).

Medicinal Plants

The use of plants to treat animal diseases, especially parasitic diseases, is also commonly described for many traditional societies. The use of medicinal plants is an integral part of all indigenous animal health knowledge systems. An extensive literature is available on the medicinal plants used in ethnoveterinary medicine by various communities. A search on ScienceDirect using "medicinal plants used in ethnoveterinary medicine" as keywords yields 279 articles, while on Google Scholar it yields over 5,000 results. A glance at this literature shows that most of the plants used in ethnoveterinary medicine are currently subjects of bioprospecting. These activities are focused on discovering new prevention and treatment agents of plant origin to combat the widespread resistance, to provide environmentally friendly products, and to cater for the expanding organic farming industry. There is a very high probability of discovering new medicines from these bioprospecting activities, because the ethnoveterinary practices of most traditional societies are well developed and compare favourably with modern veterinary practice (Mathias-Mundy and McCorkle, 1989), and such plants have often been tested by generations of indigenous people (Cox, 2000; Makhubu, 1998).

Sometimes animal based treatments are also used, and these make sense, too. For instance, the Fulani of Niger place fresh goat liver in the eyes of cattle with night blindness. This is an insightful treatment for this vitamin A deficiency disease, since the liver is the main body depot for vitamin A (Ross and Harrison, 2007).

Mechanical/Physical Techniques

In addition to the administration of the different ethnoveterinary medicines, disease is also treated through various mechanical techniques in indigenous animal health systems. Natural dips are utilized to remove parasites. These entail animals taking wallows, river runs, sea or dust baths, and day-long sun baths. Animals with sprains and lameness are made to stand for long periods in the cold or cold water. American Indians and Mongolians emphasize exercising their horses in order to work out stiffness and pain, build resistance to stress, and increase the cleansing flow of blood after snakebites (Bizimana, 1994). There are also various mechanical techniques to deal with ruminant bloat. In countries like Germany, Mexico, and Mongolia, the animal is strapped, bound, gagged, flogged, sat upon, or suspended over a cliff (Martin and McCorkle, 2001). In Germany, farmers fill the animal's mouth with its own dung and gag it with a rope while working a rubber tube around its anus. The Lozi of Zambia manually extract the dung of bloated cattle (Beerling, 1986). The Japanese shove papaya stalks up the anus while the Mexicans place bitter twigs in the animal's mouth in order to stimulate salivation, mastication, and gas release. In Bolivia, a stone is tied to a retained placenta so that gravity slowly pulls the placenta free. Another mechanical practice common in Africa is manual tick removal and burning.

Incorrect Epidemiological Information and Inferences and Harmful Practices

While the relevance of most of the practices in indigenous knowledge animal health systems can be explained by modern veterinary medicine, incorrect epidemiological information and inferences, and harmful practices, have also been identified. For example, the Tsonga farmers considered drinking dirty water as a cause of redwater (bovine babesiosis) and excessive accumulation of blood in the head as the cause of heartwater. For the latter, the condition was treated traditionally by cutting a small edge of the animal's ear to let the blood flow, a practice considered harmful in modern veterinary medicine.

The method of working a rubber tube through the anus in ruminal bloat is also controversial. In modern veterinary medicine, the method is used only when animals present with constipation. The stone method in dealing with retained placenta is also common in Southern Africa, but veterinarians discourage this method, as it is potentially harmful.

LIMITATIONS OF INDIGENOUS ANIMAL HEALTH KNOWLEDGE SYSTEMS

The limitations of most animal health indigenous knowledge systems have been identified as a lack of regulation and being prone to be affected by abuse and quackery. Other limitations include poor knowledge on differential diagnoses – for example, hemoparasitic infections cowdriosis and anaplasmosis are lumped together with ECF – and there is a lack of conventional knowledge to classify disease causative agents into viruses, bacteria, or fungi. These limitations did not however stop pastoralists from developing relatively effective herbal remedies to cure livestock diseases.

Indigenous knowledge systems largely make use of ethnoveterinary medicines, but these come with a number of disadvantages:

- They are site- or region-specific, and their wider dissemination may be limited.
- They vary in effectiveness according to seasons and preparation methods.
- Relatively few have been scientifically validated.
- They are much less standardized.
- They appear to have little to offer against acute viral diseases (Lans et al., 2007; Wanzala, 2017).

In addition, indigenous animal health knowledge systems are limited by incorrect epidemiological information; not all ethnoveterinary treatments or practices are effective (see section "Incorrect Epidemiological Information and Inferences and Harmful Practices"), and some may even be harmful, such as withholding water from diarrhoeal animals and colostrums from newborns, indiscriminate bloodletting, burning around and inside the anus to halt diarrhoea, and treating dermatological and ectoparasitic conditions with harsh/poisonous substances like engine oil and battery acid.

IMPACT OF MODERN/ORTHODOX VETERINARY MEDICINE ON INDIGENOUS ANIMAL HEALTH KNOWLEDGE SYSTEMS

The enforcement of orthodox veterinary practices as a result of a number of livestock epidemics (including rinderpest), the belief that orthodox medicine is superior to indigenous practices, and growing human population pressures have been detrimental to the application of traditional animal healthcare practices in many communities (Toyang et al., 1995; Ole-Miaron et al., 2004). Before the introduction of orthodox veterinary medicine in most African countries, pastoralists depended solely on indigenous health practices outlined above. With the advent of orthodox veterinary medicine, many orthodox veterinarians did not promote indigenous practices and in some places, it was even illegal to treat animals using local herbs without the permission of a veterinarian (Toyang et al., 1995; Wanzala, 2005). As a result, many livestock owners ceased to use local practices, while those who continued to rely on them did so in secret where these practices were banned. This meant that the knowledge and use of ethnoveterinary medicine declined. To illustrate, some of the Fulani practices in Northern Nigeria that were recorded by Leeflang in 1993 were unknown to the Fulani in the same region that were interviewed by Alhaji and Babalobi (2015).

The most significant impact of modern veterinary medicine could be illustrated by the increasing dependence on antibiotics, anthelmintics and acaricides. In Kenya, the Maasai were positively receptive and appreciative of the government effort in the eradication of CBPP/CCPP, FMD, and rinderpest, and they welcomed with open arms the inputs associated with modern veterinary medicine such as the antibiotic-based preparation “Terramycine”. The widespread use of “Terramycine” in response to any manifested symptoms, including non-infectious conditions and/or diseases that do not need antibiotic treatment such as piloerection, diarrhoea, and bloat, led to the erosion and regrettably, the atrophy of the Maasai ethnoveterinary medicine which was particularly effective in the treatment of many non-contagious diseases (Ole-Miaron et al., 2004). In addition, the acceptance and adoption of terramycine by the Masaai, who are known to treat their own animals, opened doors for opportunists who made a fortune by selling the injectable antibiotics without providing information on how to use them and the proper dosage. This led to repeated exposure of pathogenic bacteria to ineffective dosages of this antibiotic and ultimately led to the development of new terramycine-resistant strains which became prevalent in the Maasai pastoral region livestock (Ole-Miaron, 1997).

The terramycine era and modern animal husbandry are directly responsible for the disappearance of the ancient pastoral tradition of decentralizing stocks. As a cautionary measure against disease and starvation, the Maasai often sent some of their herds to relatives and friends who lived further away from their settlement

(Ole-Miaron, 1997). As a result of modern veterinary medicine, this practice stopped and centralisation of stocks led to overstocking and a greater disease risk. It is proposed that the introduction of centralised modern veterinary services has contributed to the decrease in pastoral movement/migration (Ole-Miaron, 2004), a well-tested strategy to escape foci of animal disease and from biting insects such as the tsetse fly.

APPLICATION OF INDIGENOUS ANIMAL HEALTH KNOWLEDGE IN THE MODERN ERA

Orthodox animal health care, while having largely brought an improvement in the control and management of animal diseases and production of livestock worldwide, is plagued by several challenges. Included are:

- The development of drug and pesticide resistance. Examples include the widespread resistance of nematode parasites to anthelmintic drugs; bacterial resistance to antibiotics; resistance to antiprotozoal drugs such as those used in the treatment of trypanosomiasis; the evolution of virus resistance to vaccines for diseases such as Marek's disease; and acaricide resistance in ticks. In the case of antibiotics, there are also concerns regarding residues in the food chain and the implications for human health of the emergence of antibiotic-resistant micro-organisms (BOA, 1999).
- Climate change, which has resulted in the expansion of the distribution zones of some pests and parasites.
- Environmental contamination.
- Increasing importance of consumer concerns and the need for control agents compatible with organic farming.
- In the developing world, the question of affordability and accessibility of treatments to poorer livestock keepers and the failure of both government and private veterinary services to meet the needs of pastoralists.

On the other hand, indigenous livestock disease management practices that are largely not dependent on modern medicines have been shown to be mostly effective, sustainable, environmentally friendly, cost-effective, and practical. It is indisputable that the development of drug and pesticide resistance is as a result of exposure to the chemical or related chemicals concerned. Reviewing the indigenous knowledge systems, it is also inarguable that there was no use of these synthetic chemicals and no issues of resistance, yet diseases and pests were controlled for decades before the introduction of the chemicals. In addition, there were also no issues of environmental contamination, the control agents would be organic farming friendly in the modern era, and most importantly, the control agents were easily accessible in terms of both cost and availability. This situation drives the impetus to look for insights from indigenous knowledge systems for sustainable practices for the management of animal health and welfare that can be applied in the modern era.

The first strategy from indigenous animal health systems that has been borrowed into the modern era is selective breeding for disease management. There is much evidence pointing to the greater disease resistance of livestock breeds indigenous to environments where they face a heavy disease challenge (Agyemang, 2005). Thus, the indigenous practice of selecting breeding animals based on health is well justified and is the basis of selective breeding as we know it today. Well-adapted local disease-resistant breeds are now increasingly recognized as the foundation for sustainable disease control and livestock production. This perspective is a departure from earlier approaches, which routinely sought to upgrade or even substitute indigenous farm animals with high-performance exotic breeds. Research in indigenous breeds, specifically, the identification of genes that confer disease and pest resistance and/or tolerance is currently being prioritized. A number of breeds have been identified for a number of diseases:

1. Trypanosomiasis: The N'Dama and West African Shorthorn cattle, as well as Djallonke sheep and goats, have been identified as the most trypano-tolerant breeds of trypanosomiasis, which is transmitted by tsetse flies and is one of the most important animal health problems in Africa (Agyemang, 2005).
2. Ticks and tick-borne diseases: Resistance or tolerance to ticks, and to a lesser extent to tick-borne diseases, is well documented. For example, a number of studies indicate that N'Dama cattle show a higher resistance than Zebu animals to ticks (Claxton and Leperre, 1991; Mattioli et al., 1993; Mattioli et al., 1995). Another example is provided by a study in Australia which found that pure-bred *Bos indicus* cattle were less susceptible to babesiosis than were cross-bred *Bos indicus* × *Bos taurus* animals (Bock et al., 1999). In the case of theileriosis caused by *Theileria annulata*, Sahiwal calves, a breed indigenous to India, were found to be less adversely affected than Holstein-Friesian calves when infected with the disease (Glass et al., 2005).
3. Internal parasites: Resistance or tolerance to *Haemonchus contortus*, an ubiquitous nematode worm that infests the stomachs of ruminant animals, has been subject to many studies. The Red Maasai sheep breed, for example, is noted for its resistance to gastrointestinal worms (Baker, 1998). Similarly, greater resistance and higher productivity was found in Small East African goats as compared to goats of the Galla breed under the same conditions (ibid.) The Indonesian Thin Tailed sheep have been found to show greater resistance to the liverfluke *Fasciola gigantica* than sheep of the St. Croix and Merino breeds (Roberts et al., 1997).

Secondly, there has been an increasing interest in the application of plant-derived chemotherapeutics in the animal health field, an approach borrowed from indigenous animal health systems. Issues of resistance to orthodox drugs used to treat or control pathogenic organisms, such as bacteria, chlamydia, fungi, helminths, protozoa, rickettsia, and viruses, which cause disease in animals and expanding organic farming systems (Mayer et al., 2014) have stimulated research into alternative products such as plant extract-based remedies. Plants comprise complex mixtures of chemicals, many of which possess efficacious and distinctive pharmacologic activities. Numerous chemical compounds from plants have unique structures that are impossible for human chemists to design or manufacture. It has been postulated that these remedies contain more than a single pharmacologically active compound, and the risk of resistance developing to a combination of phytochemicals is lower than the risk of resistance against a single active compound. In addition, the extensive variety of constituents in plants present a possibility that, in addition to those compounds with significant biologic activity, there may be other chemicals that enhance the activity of the bioactive compounds.

The mechanism of action of herbal medicine differs from that of conventional pharmacologic drugs (Wynn and Fougère, 2007). Many traditional healing systems employ combinations of plant species and plant parts in efforts to enhance efficacy, and possibly also to reduce or ameliorate toxicity or adverse side effects (McGaw, 2013). Plant medicines combine a range of chemicals that may have additive, antagonistic, or synergistic effects; the mechanism of action of synergy is of considerable importance and is receiving increasing scientific attention. Reports on the antagonistic interactions of different plant species are relatively few in comparison with those supplying evidence of synergistic effects. Synergy between compounds in plant extracts may have a pharmacokinetic or pharmacodynamic basis (Wynn and Fougère, 2007). Where one component enhances intestinal absorption or utilization of another constituent, pharmacokinetic synergy is apparent, but where two compounds interact with a single target or system, pharmacodynamic synergy comes into play (Wynn and Fougère, 2007).

REFERENCES

Alhaji, N.B. & Babalobi, O.O., 2015. Participatory epidemiology of ethnoveterinary practices Fulani pastoralists used to manage contagious bovine pleuropneumonia and other cattle ailments in Niger state, Nigeria. *Journal of Veterinary Medicine*, 2015: 10. http://dx.doi.org/10.1155/2015/460408.

Agyemang, K., 2005. *Trypanotolerant Livestock in the Context of Trypanosomiasis Intervention Strategies*. PAAT Technical and Scientific Series No. 7. Rome, Italy: FAO.

Amann, A., Poupart, G., Telser, S., Ledochowski, M., Schmid, A. & Mechtcheriakov, S., 2004. Applications of breath gas analysis in medicine. *International Journal of Mass Spectrometry*, 239: 227–233.

Baker, R.L., 1998. Genetic resistance to endoparasites in sheep and goats. A review of genetic resistance to gastrointestinal nematode parasites in sheep and goats in the tropics and evidence for resistance in some sheep and goat breeds in sub-humid coastal Kenya. *Animal Genetic Resources Information*, 24: 13–30.

Baumann, M.P.O., 1990. The Nomadic Animal Health System (NAHA-System) in Pastoral Areas of Central Somalia and its Usefulness in Epidemiological Surveillance. Master's thesis. University of California-Davis School of Veterinary Medicine.

Bayn, A., Nol, P., Tisch, U., Rhyan, J., Ellis, C.K. & Haick, H., 2013. Detection of volatile organic compounds in *Brucella abortus* seropositive bison. *Analytical Chemistry*, 85: 11146–11152.

Bazalar, H. & McCorkle, C.M. (eds), 1989. *Estudios Etnoveterinarios en Comunidades Altoandinas del Perú*. Lima, Perú: Lluvia Editores.

Beerling, M.-L.E.J., 1986. *Acquisition and Alienation of Cattle in the Traditional Rural Economy of Western Province, Zambia*. Mongu, Zambia: Department of Veterinary and Tsetse Control Service, Ministry of Agriculture and Water Development, Western Province.

Bizimana, N., 1994. *Traditional Veterinary Practice in Africa*. Eschborn, Germany: Deutsche Gesellschaft für Technische Zusammenarbeit (GTZ) GmbH.

BOA, 1999. *The Use of Drugs in Food Animals: Benefits and Risks*. Washington, D.C.: National Academy Press.

Bock, R.E., Kingston, T.G. & de Vos, A.J., 1999. Effect of breed of cattle on transmission rate and innate resistance to infection with *Babesia bovis* and *B. bigemina* transmitted by *Boophilus microplus*. *Australian Veterinary Journal*, 77(7): 461–464.

Boutrais, J., 1998. Les taurins de l'Ouest du Cameroun. In: Seignobos C. & Thys E. (eds), *Des taurins et des hommes: Cameroun, Nigéria*. Paris, France: ORSTOM, pp. 313–326.

Catley, A., 2006. Use of participatory epidemiology to compare the clinical veterinary knowledge of pastoralists and veterinarians in East Africa. *Tropical Animal Health and Production*, 38(3): 171–184.

Cho, Y.S., Jung, S.C. & Oh, S., 2015. Diagnosis of bovine tuberculosis using a metal oxide-based electronic nose. *Letters in Applied Microbiology*, 60: 513–516.

Claxton, J. & Leperre, P., 1991. Parasite burdens and host susceptibility of Zebu and N'Dama cattle in village herds in the Gambia. *Veterinary Parasitology*, 40: 293–304.

Cox, P.A., 2000. Will tribal knowledge survive the millennium? *Science*, 287: 44–45.

Curasson, G., 1947. *Le Chameau et ses Maladies*. Paris, France: Vigot Frères.

Davis, D.K., 2016. Political economy, power, and the erasure of pastoralist indigenous knowledge in the Maghreb and Afghanistan. In: Meusburger, P., Freytag, T. & Suarsana, L. (eds), *Ethnic and Cultural Dimensions of Knowledge. Knowledge and Space (Klaus Tschira Symposia), vol 8*. Cham, Switzerland: Springer, pp. 211–228.

Davis, D.K., Quraishi, K., Sherman, D., Sollod, A. & Stem, C., 1995. Ethnoveterinary medicine in Afghanistan: an overview of indigenous animal health care among Pashtun Koochi nomads. *Journal of Arid Environments*, 31: 483–500.

d'Ieteren, G.D.M., Authié, E., Wissocq, N. & Murray, M., 1998. Trypanotolerance, an option for sustainable livestock production in areas at risk from trypanosomosis. *Revue scientifique et technique (International Office of Epizootics)*, 17(1): 154–175.

Eckert, J. & Hertzberg H., 1994. Parasite control in transhumant situations. *Veterinary Parasitology*, 54: 103–125.

EuLA, 2009. Practical Guidelines on the Use of Lime for Prevention and Control of Avian Influenza Foot and Mouth Disease and Other Infectious Diseases. Version 3. EuLA aisbl (European Lime Association), Brussels, Belgium.

Fend, R., Geddes, R., Lesellier, S. et al., 2005. Use of an electronic nose to diagnose mycobacterium bovis infection in badgers and cattle. *Journal of Clinical Microbiology*, 43: 1745–1751.

Ford, J., 1971. *The Role of Trypanosomiases in African Ecology: A Study of the Tsetse Fly Problem*. Oxford, UK: Clarendon Press.

Fraser, M.D., Moorby, J.M., Vale, J.E. & Evans D.M., 2014. Mixed grazing systems benefit both upland biodiversity and livestock production. *PLoS ONE*, 9(2): e89054. https://doi.org/10.1371/journal.pone.0089054.

Glass, E.J., Preston, P.M., Springbett, A., Craigmile, S., Kirvar, E., Wilkie, G. & Brown, C.G.D., 2005. *Bos taurus* and *Bos indicus* (Sahiwal) calves respond differently to infection with *Theileria annulata* and produce markedly different levels of acute phase proteins. *International Journal for Parasitology*, 35: 337–347.

Hadani, A. & Shimshony, A., 1994. Traditional veterinary medicine in the Near-East: Jews, Arabs, Bedouins and felahs. *Scientific and Technical Review of the Office International des Epizooties*, 13: 581–597.

Himsworth, C.G., 2008. The danger of lime use in agricultural anthrax disinfection procedures: the potential role of calcium in the preservation of anthrax spores. *The Canadian Veterinary Journal (La revue veterinaire canadienne)*, 49(12): 1208–1210.

Hohenhaus, M.A. & Outteridge, P.M., 1995. The immunogenetics of resistance to *Trichostrongylus colubriformis* and *Haemonchus contortus* parasites in sheep. *British Veterinary Journal*, 151: 119–140.

Hunter, A.G., 1986. Urine odour in a camel suffering from surra (*T. evansi* infection). *Tropical Animal Health and Production*, 18: 146. https://doi.org/10.1007/BF02359524.

Knobloch, H., Köhler, H., Commander, N., Reinhold, P., Turner, C. & Chambers, M. 2009. Volatile organic compound (VOC) analysis for disease detection: proof of principles for field studies detecting paratuberculosis and brucellosis. In: Pardo, M. & Sberveglieri, G. (eds), *Proceedings of the 13th International Symposium of the American Institute of Physics*, 1137: 195–197.

Lans, C., Turner, N., Khan, T., Brauer, G. & Boepple, W., 2007. Ethnoveterinary medicines used for ruminants in British Columbia, Canada. *Journal of Ethnobiology and Ethnomedicine*, 3: 11.

Leeflang, P., 1993. Some observations on ethnoveterinary medicine in northern Nigeria. *Veterinary Quarterly*, 15(2): 72–74.

Leese, A.S., 1927. *A Treatise on the One-humped Camel in Health and Disease*. Stamford (Lincs): Haynes and Son.

Luseba, D. & van der Merwe, D., 2006. Ethnoveterinary medicine practices among Tsonga speaking people of South Africa. *Onderstepoort Journal of Veterinary Research*, 73: 115–122.

Makhubu, L., 1998. Bioprospecting in African context. *Science*, 282: 41–42.

Martin, M., Mathias, E. & McCorkle, C.M., 2001. *Ethnoveterinary Medicine: An Annotated Bibliography of Community Animal Healthcare*. London, UK: ITDG Publishing.

Mathias-Mundy, E. & McCorkle, C.M., 1989. *Ethnoveterinary Medicine: An Annotated Bibliography*. Washington, D.C.: Technology and Social Change Program, Iowa State University; Columbia, MO: Academy for Educational Development; Ames, IA: Small Ruminant Collaborative Research Support Program, Sociology Project, Dept. of Rural Sociology, University of Missouri.

Mattioli, R.C., Bah, M., Faye, J., Kora, S. & Cassama, M., 1993. A comparison of field tick infestation on N'Dama, Zebu and N'Dama × Zebu crossbred cattle. *Veterinary Parasitology*, 47: 139–148.

Mattioli, R.C., Bah, M., Kora, S., Cassama, M. & Clifford, D.J., 1995. Susceptibility to different tick genera in Gambian N'Dama and Gobra zebu cattle exposed to naturally occurring tick infection. *Tropical Animal Health and Production*, 27: 995–1005.

Mayer, M., Vogl, C.R., Amorena, M., Hamburger, M. & Walkenhorst, M., 2014. Treatment of organic livestock with medicinal plants: A systematic review of European ethnoveterinary research. *Forschende Komplementarmedizin*, 21: 375–386.

McCorkle, C.M. & Mathias-Mundy, E., 1992. Ethnoveterinary medicine in Africa. *Journal of the International African Institute*, 62: 59–93.

McGaw, L., 2013. Use of plant-derived extracts and essential oils against multidrug-resistant bacteria affecting animal health and production. In: Rai M.K. & Kon K.V. (eds) *Fighting Multidrug Resistance with Herbal Extracts, Essential Oils and Their Components*, 1st edn. London, UK: Academic Press, pp. 191–203.

Misra, K.K. & Kumar, K., 2004. Ethno-veterinary practices among the Konda Reddi of East Godavari district of Andhra Pradesh. *Studies of Tribes*, 2: 37–44.

Moran, M.C., Nigarura, G. & Pegram, R.G., 1996. An assessment of host resistance to ticks on cross-bred cattle in Burundi. *Medical and Veterinary Entomology*, 10: 12–18.

Niamir-Fuller, M. 1994. Women livestock managers in the Third World: Focus on technical issues related to gender roles in livestock production. Staff Working Paper 18, IFAD, Rome, Italy.

Ole-Miaron, J., Farah, K.O. & Ekaya, W.N., 2004. Indigenous knowledge: The basis of the Maasai Ethnoveterinary diagnostic skills. *Journal of Human Ecology*, 16: 43–48.

Ole-Miaron, J.O., 1997. Ethnoveterinary practice of the Loitokitok Maasai: Impact on the environment. *Tanzania Veterinary Journal*, 17: 159–167.

Ole-Miaron, J.O., 2003. The Maasai ethnodiagnostic skill of livestock diseases: A lead to traditional bioprospecting. *Journal of Ethnopharmacology*, 84: 79–83.

Peled, N., Ionescu, R., Nol, P., Barash, O., McCollum, M., VerCauteren, K., Koslow, M., Stahl, R., Rhyan, J. & Haick, H., 2012. Detection of volatile organic compounds in cattle naturally infected with *Mycobacterium bovis*. *Sensors Actuators B: Chemical*, 171–172: 588–594.

Purkhart, R., Köhler, H., Liebler-Tenorio, E., Meyer, M., Becher, G., Kikowatz, A. & Reinhold, P., 2011. Chronic intestinal mycobacteria infection: discrimination via VOC analysis in exhaled breath and headspace of feces using differential ion mobility spectrometry. *Journal of Breath Research*, 5: 027103.

Rajan, S. & Sethuraman, M., 1997. Traditional veterinary practices in rural areas of Dindigul district, Tamilnadu, India. *Indigenous Knowledge and Development Monitor*, 5: 15.

Roberts, J.A., Estuningsih, E., Widjayanti, S., Wiedosari, E., Partoutomo, S. & Spithill, T.W., 1997. Resistance of Indonesian thin tail sheep against *Fasciola gigantica* and *F. hepatica*. *Veterinary Parasitology*, 68(1–2): 69–78.

Ross, A.C. & Harrison, E.H., 2007. Vitamin A: Nutritional aspects of retinoids and carotenoids. In: Zempleni, J., Rucker, R.B., McCormick, D.B. & Suttie, J.W., (eds). *Handbook of Vitamins*, 4th edn. Boca Raton, FL: CRC Press, pp. 2–39.

Schillhorn van Veen, T.W., 1997. Sense or nonsense? Traditional methods of animal parasitic disease control. *Veterinary Parasitology*, 71: 177–194.

Sethi, S., Nanda, R. & Chakraborty, T., 2013. Clinical application of volatile organic compound analysis for detecting infectious diseases. *Clinical Microbiology Reviews*, 26: 462–475.

Spinhirne, J.P., Koziel, J.A. & Chirase, N.K., 2004 Sampling and analysis of volatile organic compounds in bovine breath by solid-phase microextraction and gas chromatography–mass spectrometry. *Journal of Chromatography*, 1025: 63–69.

Stear, M.J. & Murray M., 1994. Genetic resistance to parasitic disease: Particularly of resistance in ruminants to gastrointestinal nematodes. *Veterinary Parasitology*, 54: 161–176.

Sutherst, R.W., 1983. Management of arthropod parasitism in livestock. In: Dunsmore J.P. (ed) *Tropical Parasites and Parasitic Zoonosis*. Perth: Murdon University, pp. 41–56.

Toyang, N.J., Nuwanyakpa, M., Ndi, C., Django, S. & Kinyuy, W.C., 1995. Ethnoveterinary medicine practices in the Northwest Province of Cameroon. *Indigenous Knowledge and Development Monitor*, 3: 24.

Vogl, C.R., Vogl-Lukasser, B. & Walkenhorst, M., 2016. Local knowledge held by farmers in Eastern Tyrol (Austria) about the use of plants to maintain and improve animal health and welfare. *Journal of Ethnobiology and Ethnomedicine*, 12: 40.

Wanzala, W., 2017. Potential of traditional knowledge of plants in the management of arthropods in livestock industry with focus on (acari) ticks. *Evidence-Based Complementary and Alternative Medicine: eCAM*, 2017: 8647919. http://doi.org/10.1155/2017/8647919.

Wanzala, W., Zessin, K.H., Kyule, N.M., Baumann, M.P.O., Mathias, E. & Hassanali, A., 2005. Ethnoveterinary medicine: A critical review of its evolution, perception, understanding and the way forward. *Livestock Research and Rural Development*, 17: 119.
Wynn, S.G. & Fougère, B.J., 2007. Veterinary herbal medicine: A systems-based approach. In: Wynn, S.G. & Fougère, B.J. (eds), *Veterinary Herbal Medicine*. St. Louis, MO: Mosby, Elsevier, pp. 291–409.

CHAPTER 12

Local Ecological Knowledge on Climate Prediction and Adaptation: Agriculture-Wildlife Interface Perspectives from Africa*

Olga L. Kupika, Godwell Nhamo, Edson Gandiwa, and Shakkie Kativu

CONTENTS

ABSTRACT

Background: This study explores the role of local ecological knowledge (LEK) in climate change prediction and adaptation in Africa, including a case study of the Middle Zambezi Biosphere Reserve (MZBR), Zimbabwe. The study used both

* A version of this chapter was originally published as "Local ecological knowledge on climate change and ecosystem-based adaptation strategies promote resilience in the Middle Zambezi Biosphere Reserve, Zimbabwe" in the open-access journal *Scientifica, 2019*, 3069254.

quantitative and qualitative methods to collect data that included household surveys, desktop reviews of published journal articles, key informant interviews, and focus group discussions (FGDs). The household surveys, key informant interviews, and FGDs focused on climate change prediction as well as biodiversity-related coping and adaptation strategies. Both the literature review and the case study indicated that local communities at the agriculture-wildlife interface rely on both ethnobotanical and ethnozoological knowledge, particularly ethnophenology to predict and cope with the changing climate. By observing the phenology of animal and plant species and abiotic indicators, local communities are able to predict climatic events.

Relevance: Ecosystem-based adaptation strategies such as water conservation and harvesting of wildlife resources for consumption have enabled local communities to develop resilience to climate change. Local communities harvest wildlife resources to avert food shortages during drought periods. There is therefore a need to promote sustainable utilization of wildlife resources during drought periods to avoid disappearance of the species under a changing climate. LEK plays a vital role in promoting climate change resilience within terrestrial socio-ecological systems and consequently sustainable rural livelihoods and conservation of wildlife resources.

INTRODUCTION

The livelihoods of rural communities in developing countries are threatened by climate change (Brown et al., 2007). African rural communities have been documented as constructing climate change realities based on their experiences of the impacts and effects (Macherera and Chimbari, 2016). Although the observation of global climate change has been largely based on meteorological data, inadequate data had been generated from how local communities recognize and respond to such changes (Savo et al., 2016). This calls for research to explore the role of local culture in identifying and responding to threats imposed by the changing climate. Adger et al. (2013) suggested that local communities could interpret and construct climate change trends and local indicators within a cultural setting. The United Nations (UN) recognizes the significant role played by indigenous knowledge, cultures, and traditional practices in promoting sustainable development, equity, and management of the environment (United Nations, 2015). Boafa et al. (2016) noted that the Intergovernmental Panel on Climate Change (IPCC), since its establishment in 1988, has highlighted the role that traditional ecological knowledge can play in addressing the negative effects of global climate change and variability. Adger et al. (2013) argue that culture is embedded in societal modes of production, consumption, lifestyles, and social organization; this fact should be recognized in understanding both mitigation and adaptation to climate change. Thus, embracing indigenous knowledge in climate change adaptation is vital in order to enhance the resilience of communities to climate change (Nkomwa et al., 2014).

Fleischman and Briske (2016) identified four knowledge systems that contribute to ecological knowledge, namely, local/traditional/indigenous ecological knowledge, scientific, administrative/bureaucratic, and professional ecological knowledge, which are distinguished by the source, holder, content, and source of feedback of information. Several scholars have come up with definitions of indigenous knowledge systems (IKS). Nakashima et al. (2012) and Mapara (2009) defined indigenous or traditional knowledge as the intellectual behavior of indigenous societies or local information that exists and is developed through the experiences of the local community and know-how accumulated by successive generations. Indigenous ecological knowledge can also be defined as a cumulative body of knowledge and beliefs handed down through generations by cultural transmission about the relationship of living beings (including humans) with one another and with their environment (Gadgil et al., 1993). Indigenous knowledge is derived from IKS, the designation that is used to refer to the modus operandi and processes that indigenous peoples use to harness indigenous knowledge (Mapara, 2017). Also known as indigenous technical science, indigenous knowledge is dynamic and is informed by local communities' interactions with their local biophysical and social environment (Nakashima et al., 2012).

Indigenous knowledge relevant to climate change encompasses interrelated subsets of local ecological knowledge, seasonal knowledge, and phenological knowledge (Armatas et al., 2016). Local ecological knowledge (LEK) refers to knowledge, practices, and beliefs shared among local resource users regarding ecological interaction within ecosystems (Cook et al., 2014). Thus, LEK is composed of people's lived experiences and their dialectical interaction with the natural environment. Traditional phenological knowledge (TPK) refers to all forms of knowledge related to custom, observations, norms, beliefs, and baselines on the biological seasonality of life cycle stages of plant and animal species growth, abundance, and productivity (Fitchett et al., 2015). It falls within the study of ethnoecology, that is, the study of the human understanding of the relationships among plants, animals, ecological processes, and the environment (Berkes and Seixas, 2005). In this study, we investigated LEK, inclusive of seasonal ecological knowledge and TPK, on observations and responses to climate change impacts in the context of the agriculture–wildlife interface.

Local communities inhabiting the protected area margins or edge utilize LEK in their environment to develop ecosystem-based adaptation strategies as a means of coping with changes in climatic conditions. Ecosystem-based adaptation can be defined as the use of biodiversity and ecosystem services as part of an overall adaptation strategy to help people to adapt to the adverse effects of climate change (Doswald et al., 2014; Vignola et al., 2015). Adaptation has been understood differently by different scholars and practitioners. The IPCC (2001) defines adaptation as an adjustment in natural or human systems in response to actual or expected climatic stimuli or their effects, which moderates harm or exploits beneficial opportunities. Connolly-Boutin & Smit (2016:392) define "adaptation strategies as the actions that people individually or collectively undertake to adjust to changing conditions in order to maintain or improve their well-being for example through agricultural intensification, livelihood diversification and migration". Smit et al. (2001) define adaptation as the process through which actors adjust to changing conditions, hazards, risks, and opportunities posed by climate change. Actors refers to stakeholders such as individuals and civic, private, and public entities in the community (Agrawal, 2010). These definitions share fundamental similarities in that the concept of adaptation in the context of climate change refers to a climate-related stressor that exposes people to risk and causes communities to adjust their productive activities in order to survive or even improve their livelihoods. The process of adaptation to climate change involves two steps: (1) local communities notice that climate has changed, and (2) they identify and decide whether or not to adopt a useful strategy (Maddison, 2007; Gbetibouo, 2009). Knowledge about how climate is changing enables people to respond appropriately to current variability and changes to come (Pettengell, 2010), as supported by available natural, human, physical, social, and financial capital (Tompkins and Adger, 2004).

Indigenous knowledge has value not only to rural communities but also to scientists and planners who are determined to develop sustainable conditions in rural areas (Codjoe et al., 2014). Analysis of LEK is critical to understanding the impacts of and human responses to climate change. The growing body of information on local indigenous knowledge about climate change impacts on biophysical systems provides novel contributions towards our understanding of local climate change and people's responses (Nyong et al., 2007; Lebel, 2013; Reyes-García et al., 2016). Thus, LEK can be used to understand community adaptive practices in order to be resilient to the environmental changes (Nash et al., 2016) that have negative effects on local livelihoods and sustainable development (Nakashima et al., 2012). In this chapter, we focus on how local communities at the agriculture-wildlife interface utilize local knowledge to develop ecosystem-based adaptation strategies as a means of coping with changes in climatic conditions. The objectives of this chapter are to: (1) explore literature on the role of LEK on climate change prediction, adaptation, and mitigation from selected African countries, (2) determine local community knowledge on climate prediction in the Middle Zambezi Biosphere Reserve (MZBR), and (3) explore LEK on coping, adaptation, and mitigation strategies towards climate change in the MZBR.

MATERIALS AND METHODS

Description of Study Area

This chapter focuses on selected literature on LEK and climate change prediction and adaptation derived from across English-speaking African countries. In addition, a case study was conducted in Chundu Communal Area (CCA) and Nyamakate Resettlement Area (NRA), adjacent to the MZBR, in northern Zimbabwe (Figure 12.1). The study communities were purposively sampled due to their proximity to Mana Pools National Park and Hurungwe Safari area. Furthermore, the area was selected because it is made up of indigenous people who were forced to migrate from the Zambezi Valley prior to the establishment of the protected areas. The CCA was established in 1958 and is located adjacent to Mana Pools National Park (ZPWMA, 2009). The Nyamakate chiefdom was established in the Zambezi Valley

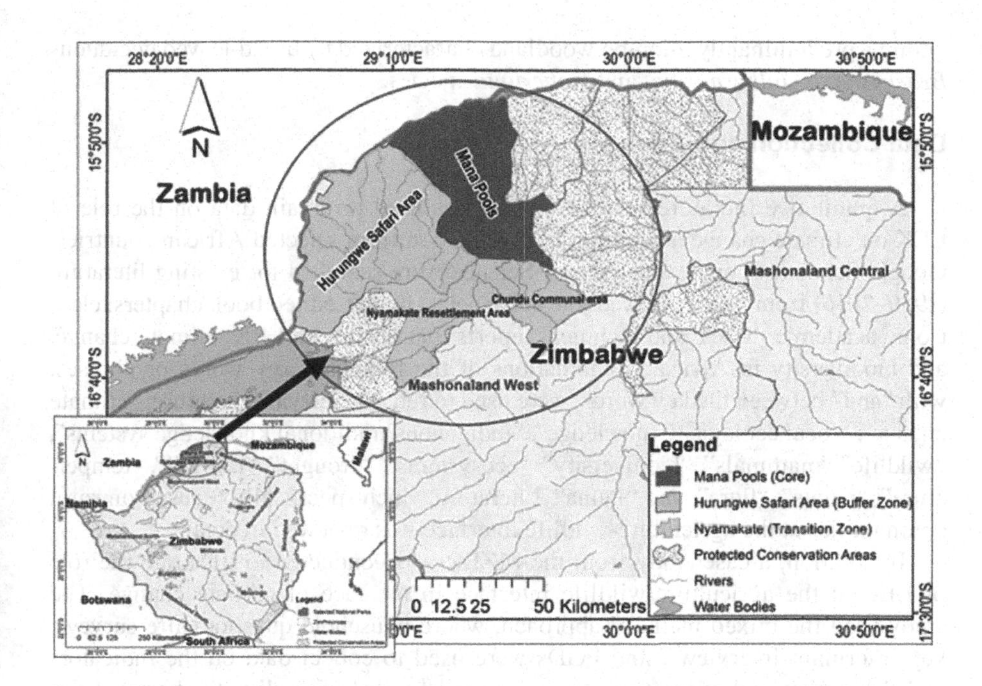

Figure 12.1 Location of Nyamakate Resettlement and Chundu Communal Area in the Middle Zambezi Biosphere Reserve (Source, Authors).

before colonization by the white men in 1890 (Mbereko et al., 2015). However, the chiefdom was dismantled by the colonial government, and the occupation of the contemporary Nyamakate resettlement was done in the 1980s (Chimhowu and Hulme, 2006; Mbereko et al, 2015).

Nyamakate resettlement and the CCA lie within agroecological region 3. Rainfall in the area varies seasonally, with an average annual rainfall of 700–800 mm (Chimhowu and Hulme, 2006; PWMA, 2010). Temperatures in the area are high, with the average maximum reaching 40°C and average minimum temperatures at around 10°C (Meteorological Services Department, 2012). Chimhowu and Hulme (2006) observed that most agricultural seasons experience drought in the area as indicated by the 1981–1982, 1982–1983, 1983–1984, 1986–1987, 1991–1992, 1994–1995, and 1996–1997 droughts. The 1991–1992 drought was the most severe, and it had profound effects on livelihoods (Chimhowu and Hulme, 2006).

The MZBR is the habitat of diverse and unique flora and fauna which contributes significantly towards the region's biodiversity. Terrestrial and aquatic flora and fauna species are found in the adjacent protected areas, i.e. Mana Pools National Park and the Hurungwe, Chewore, and Charara safari areas. The MZBR valley floor area is endowed with diverse wildlife species, including elephant (*Loxodonta africana*), buffalo (*Syncerus caffer*), black rhino (*Diceros bicornis*), painted wild dog (*Lycaon pictus*) and nyala (*Tragelaphus angasii*), impala, kudu, waterbuck, zebra, hyena, and on the escarpment, sable (ZPWMA, 2009). Nyamakate and the CCA are located within a predominantly miombo woodland characterized by broad-leaved deciduous *Brachystegia, Julbernardia*, and *Isoberlinia* species.

Data Collection and Analysis

A qualitative literature review method was used to obtain data on the role of LEK on climate change adaptation and mitigation from selected African countries. Google Scholar, Scopus, and Web of Science were searched for existing literature (1980–2016) from peer-reviewed journal articles, books, edited book chapters, electronic academic theses, and technical reports that dealt with LEK, climate change, and biodiversity in Africa. Combinations of the following key words or phrases, with "and" between the key words, were used to retrieve relevant literature: "climate change", "local ecological knowledge", "indigenous/traditional knowledge systems", "wildlife" "mammals", "biodiversity", "ecosystems", "drought", "rainfall", "temperature", "plants", "flora", and "fauna". Literature which specifically focused on rural communities at the agriculture-wildlife interface was reviewed in detail.

In addition, a case study from the MZBR was conducted to illustrate the role of LEK at the agriculture-wildlife interface in the face of climate change. The study used the mixed methods approach, where household questionnaire surveys, key informant interviews, and FGDs were used to collect data on the indicators and the coping and adaptation strategies used to deal with climate change. Prior to the survey, written annual permission to carry out the research was sought and granted from Hurungwe Rural District Council in 2015 and 2016. The research was approved by the Senate Research Council at the Chinhoyi University of Technology (CUT), Zimbabwe. In addition, all participants gave verbal informed consent to participate in the research. The study was also ethically cleared by the CUT ethics committee.

Household surveys were used to collect quantitative data on climate trends. The household questionnaire included questions related to (1) agriculture and land-use practices in relation to climate events, and (2) perceptions towards climate change impacts and coping and adaptation strategies. The questions comprised both close-ended and open-ended questions. The household survey was carried out in the two communities between August 2015 and October 2016. The questionnaire was pre-tested with 16 households from Lima village, located adjacent to the Charara Safari Area to improve validity and reliability of the instruments. The questionnaires were revised after a pilot test to remove ambiguities and misunderstandings. Villages which are located close to the Hurungwe Safari Area were purposively sampled for the survey. Village registers which were obtained from the village heads were used to come up with the representative sample for the survey. Every third household was systematically sampled on the ground for the survey. Interviews were conducted with the head of the household or their spouse if they were not available. The geographical location of each sampled household was captured using a Geographical Position System Garmin Model GPS Map 64 (2013) and recorded.

Household questionnaires were administered to 320 people, of which 30% (n = 96) were from five villages (India, Golf, Village 20, Murimbika, and Hotel) located in the NRA whilst the remaining 70% (n = 224) were from Kabidza and Mayamba villages in the CCA. The proportion of respondents from each area ward is proportional to the number of household in the sampled villages. We selected only respondents that were more than 20 years old, since we assumed that older persons would be more familiar with the local environment than younger people would be. The overall response rate was 100%, since research assistants administered the questionnaires. Many of the interviewees (85%) were local farmers in the two areas. A household survey, key informant interviews, FGDs, and field observations were used to gather information on the role of LEK in the study area. In addition, key informant interviews were held with traditional leaders and elders, government representatives, parks officials, and nongovernmental organizations. Information on traditional climate and weather indicators and prediction tools, as well as rituals and agronomic and agroforestry strategies, was solicited (Rivero-Romero et al., 2016). All participants from the household surveys, key informant interviews, and subsistence farmer interviews/focus groups gave verbal informed consent to participate in the research. For household surveys, key informant interviews, and focus groups, questions were asked either in Shona or English. Key informant interviews with traditional leadership and other local experts were done in Shona. However, the rest of the interviews with the Rural District Council officials were done in English. Responses were documented in field notes and audio-recorded with consent.

FGDs were held in August 2015 and November 2016 with participants who had previously participated in a broader household survey on impacts of climate change. Eight FGDs were organized in the NRA and the CCA. A random call for participation in FGDs was made among members in each village during the household surveys. Between eight and 15 farmers, including males and females aged 30 and above, participated in the FGDs in each village. During the FGDs, facilitators requested responses from both men and women; various opinions were documented as they arose, and a consensus was eventually reached (Meldrum et al., 2016). During the FGDs, participants were asked to describe the causes of climate change and how the community utilized LEK to respond to impacts of climate change.

Participant observation, on the other hand, allowed us to interact with people conducting their daily activities. During the 2015 and 2016 surveys, selected traditional leaders accompanied a researcher to their gardens, agricultural plots, and forest areas to identify wildlife species and to describe their use and their purpose in climate prediction, adaptation, and mitigation. This information, together with data from key informant interviews and FGDs, allowed the compilation of an inventory of flora and fauna species which are used in adaptation to climate change.

Data collected through the household questionnaire surveys were coded and entered into the Statistical Package for Social Sciences (SPSS software IBM Version 20, Chicago, IL) for analysis. Descriptive statistics were used to summarize demographic and socio-economic data from the questionnaire response data set. Data analysis involved the use of the sustainable livelihoods approach in order to get insights on livelihood strategies and to understand the adaptive capacities in the area. Household livelihood assets such as physical and natural capital were extracted from the household survey data. Where multiple responses were possible on an open-response question, data are presented as the percentage of respondents giving each response and may sum to over 100%.

Qualitative data collected from literature, key informant interviews, FGDs, and field observations were analyzed using deductive and/or theoretical thematic analysis (Corbin and Strauss, 2008; Ayal and Filho, 2017). Deductive and/or theoretical thematic analysis is a qualitative data analysis approach based on themes which are predetermined by the researcher's theoretical or analytic interest in the research area (Braun and Clarke, 2006). Transcriptions of key informant interviews and FGDs were translated to English. The content was inductively analyzed by drawing out recurring themes and key words (Mase et al., 2017). During this qualitative content analysis, information on local values and useful concepts related to LEK was extracted and synthesized. FGDs were recorded (when possible), transcribed, coded, and analyzed using Statistical Package for Social Scientists Version 21 (SPSS, 2013). Data obtained from literature, FGDs, key informant interviews, and field observations were categorized under the following themes: (1) local indicators and predictors of climate change, and (2) LEK about climate change adaptation and mitigation.

RESULTS

Use of LEK in Climate Prediction: Perspectives from Selected African Countries

Local communities from different parts of Africa have different interpretations of indicators of climate change, except for common ones such as shifting of the peak of rain season and sporadic rain, drier conditions, recurring weather variability, flooding, drought, and high temperatures (Chase, 2006). Gyampoh et al. (2009) stated that indigenous people in African communities "may not understand the concept of global warming or climate change, but they observe and feel the effects of decreasing rainfall, increasing air temperature, increasing sunshine intensity, and seasonal changes in rainfall patterns." Trees and animal behaviors form a basis for observing changing environmental conditions and indicate climate change in most communities in African landscapes.

Ecophysiological constraints, including high temperatures and low rainfall, have limited plants' resilience, affecting plant productivity and dispersal abilities. This leads to local (if not global) extinction of some species, with some failing to produce fruits (Kirilenko and Solomon, 1998). For example, in the Chiredzi District of Zimbabwe, trees, including *Bechermia discolor* and *Sclerocarya birrea* (amarula), are reported to produce fewer fruits, and some trees are no longer producing fruits, most probably due to climate change. Plants species, including grasses, now occur in low abundance and die before senescence (maturity age) due to high temperatures. In addition, the existence of habitat patches shows variability or change in environmental resource distribution, which indicates that climate is changing. Occurrence of new and invasive species that are more competitive and resistant to drought shows a change in environmental conditions in the African landscapes (Kirilenko and Solomon, 1998).

Local communities attribute increases in diseases such as foot and mouth, pests, ticks, wounds, and lump-skin disease to low rainfall and high temperatures (Simms and Murphy, 2005). Reduction in the abundance of some plant and wildlife species also indicates a change in climatic conditions. For example, snakes are reported to have reduced abundance since the 1992 drought due to prevailing harsh weather conditions. The extinction of two populations of a butterfly species (*Euphydryas editha bayensis*) has been attributed to variability in precipitation and habitat loss in Namibia (McLachlan et al., 2007). The low abundance of qualia birds, doves, and rats shows that there is limited food and unfavorable weather conditions (Ziervogel and Zermoglio, 2009). Moreover, mopane worms (*Gonimbrasia belina*) in South Africa and Zimbabwe used to occur in December and March, indicating abundant rain, but now might not occur at all or only occur in March, indicating drought conditions (Ziervogel and Zermoglio, 2009).

There is a wealth of local knowledge based on predicting weather and climate across various parts of Africa. African farmers, in particular, have developed intricate systems of natural resource gathering, weather prediction, interpretation, and decision making in relation to changing climatic conditions. Largely, these systems of climate forecasts have been helpful to the farmers in managing their vulnerability and promoting resilience. Farmers are known to make decisions on cropping patterns based on local predictions of climate and on planting dates based on complex cultural models of weather (Tazeze et al., 2012). "The appearance of certain birds, mating of certain animals, and flowering of certain plants are all important signals of changes in time and seasons that are well understood in traditional knowledge systems," Gyampoh et al. (2009) observe. The singing, nesting, and migration of some bird species, including southern hornbill (*Bucorvus abyssinicus*), appear to be useful indicators for the onset of the rains in African communities (Case, 2006). To adopt a specific strategy, local people need to have knowledge of the past, current, and expected rainfall in their area. Table 12.1 shows the use of LEK in weather prediction in selected African countries.

Role of LEK in Adaptation and Mitigation Strategies: Perspectives from Africa

LEK is playing an integral role in building climate resilience in developing countries. LEK contributes towards resilience by promoting grassroots adaptation and mitigation among smallholder farmers, pastoralists, fishing communities, and forest dwellers in African countries. LEK includes gender-defined knowledge of indigenous plant and animal species, especially drought-tolerant and pest-resistant varieties (Van Campenhout et al., 2016).

Savo et al. (2016) noted that local observations of climate-related events among subsistence farmers around the globe can contribute to the understanding of the frequency and magnitude of climate change impacts on ecosystems and societies. Such climatic changes can be grouped into three major categories, namely:

Table 12.1 Local Ecological Indicators for Weather Prediction

Category	Indicator	Meaning	Country	Reference
Vegetation characteristics	Reed grass (*Typha capensis*)	Presence of reed indicates consistent increase of rainfall over a specific season, whilst absence signifies consistent decrease of rainfall over specific season.	Ethiopia, Tanzania	Buizer et al. (2016); Tazeze et al. (2012)
	Cactus (*Opuntia ficus*)	Presence of cactus indicates consistent decrease of rainfall over long period of time, whilst absence signifies consistent increase of rainfall over long period.	Uganda, Kenya, Ethiopia	Orlove et al. (2010); Ifejika Speranza (2010); Tazeze et al. (2012)
	Senegalia nigrescens (formerly *Acacia nigrescens*)	High blooming means low rainfall to be received in that year	Botswana, Mozambique, South Africa, Zambia, Zimbabwe	Risiro et al. (2012); Tanyanyiwa and Chikwanha (2011)
	Peach tree (*Prunus persica*), apricot (*Prunus armeniaca*)	Flowering and budding of these plants indicate the beginning of rainy season	Botswana, Malawi, Zambia, Zimbabwe, South Africa	Risiro et al. (2012); Tanyanyiwa and Chikwanha (2011)
	Aloe ferox	Sprouting aloes indicate good rains	Botswana, Malawi, South Africa, Swaziland, Zimbabwe, Zambia	Mogotsi et al. (2011); Zuma-Netshiukhwi et al. (2013)
	Baobab (*Adansonia digitata*)	Germination of new leaves of baobab indicate good rains	Botswana, Malawi, South Africa, Swaziland, Zimbabwe, Zambia	Mogotsi et al. ((2011); Zuma-Netshiukhwi et al. (2013)
	Vangueria infausta, *Englerophytum natalense*, and *Sclerocarya caffra*	Abundance of these wild fruits during the months of December and February signify an imminent challenging farming season	Nigeria, Botswana	Buizer et al. (2016); Kalame et al. (2011)
	Mango tree (*Mangifera indica*)	Heavy flowering of the mango trees indicate a potential drought season	Malawi, Tanzania, Zimbabwe	Risiro et al. (2012); Mapfumo et al. (2016)
	Nandi flame tree (*Delonix regia*), *Adansona digitata*, *Strychnos innocua*, *Flacourtia indica*, *Diospyros mespiliformis*	Abundance of fruits indicates drought year	Botswana, Burkina Faso, DRC, Ghana, Nigeria, South Africa, Zimbabwe	Alvera (2013); Buizer et al. (2016)

(*Continued*)

Table 12.1 (Continued) Local Ecological Indicators for Weather Prediction

Category	Indicator	Meaning	Country	Reference
Animal behavior	Crocodiles (*Crocodylus* spp.), Hippos (*Hippopotamus amphibius*)	Movements indicate decrease or increase of rainfall amount	Ghana	Alvera (2013); Buizer et al. (2016)
	Elephants (*Loxodonta africana*)	Migrating to higher ground means above-normal rains expected, with the possibility of flooding	Mozambique, Tanzania, Zimbabwe	Buizer et al. (2016); Alvera (2013)
	Toads (*Xenopus laevis*)	Croaking a lot indicate coming of rains within a day or two	Zimbabwe	Alvera (2013)
Insect behavior	Red ants (*Solenopsis* spp.)	Frequent movements of red ants indicate rainfall onset	South Africa, Malawi, Tanzania, Zambia, Zimbabwe	Zuma-Netshiukhwi et al. (2013); Risiro et al. (2012)
	Termites (*Ancistrotermes* spp. and *Macrotermes* spp.)	Presence of termites in the evening during rainy season indicates end of rainfall	Botswana, Malawi, Zambia Zimbabwe	Mogotsi et al. (2011)
	Locust (*Schistocerca* spp.)	High abundance before rain season means it's a drought year	Zimbabwe	Mogotsi et al. (2011)
	Butterflies (*Danaus plexippus*), army worms (*Spodoptera exempta*)	Indicate imminent mid-season drought and possible famine	Swaziland, Botswana	Mogotsi et al. (2011)
Bird behaviours	Swallows (*Cecropis* spp.)	Movement from west to east indicates rainfall onset, whilst from east to west indicates end of rainfall season	Botswana, Malawi, South Africa, Swaziland, Tanzania, Zambia, Zimbabwe	Kalanda-Joshua et al. (2011); Mogotsi et al. (2011); Zuma-Netshiukhwi et al. (2013); Risiro et al. (2012)
	Owls (*Bubo* spp.), Pigeons (*Columba* spp.)	When singing frequently early in the morning and late in the evening rainfall is expected to begin	Botswana	Mogotsi et al. (2011)
	Abdim's stork (*Ciconia abdimii*)	Flying in circles at low altitude means that rain will fall within a day or two	South Africa, Malawi, Namibia Zimbabwe	Alvera (2013)
	Black storks (*Ciconia* spp.)	Foraging in fields means rain about to begin	Zimbabwe	Risiro et al. (2012)
	Doves (*Zenaida* spp.)	High breeding before rainy season means more rainfall	Zimbabwe South Africa	Risiro et al. (2012)

(1) changes in climate and weather, (2) changes in the biological components of the environment (plant and animal phenology and distribution), and (3) changes in the physical components of the environment (Savo et al., 2016). Codjoe et al. (2013) similarly observed that local knowledge acquired through direct experiences and observations of local communities may provide practical insights about seasonal trends and weather patterns. For example, farmers in sub-Saharan Africa use a suite of environmental indicators to predict local weather and climate (Macherera et al., 2016).

Common indicators used to predict rains and season quality include the behavior of animals, tree phenology, astronomical pointers, wind patterns, and cloud cover (Codjoe et al., 2013; Jiri et al., 2016). Understanding and building upon perceptions, experiences, and IK of climate change can contribute towards strengthening the resilience of poor societies with weak infrastructure and economic well-being (Codjoe et al., 2013). Table 12.2 summarizes local observations on the impacts of climate change across Africa and ecosystem-based adaptation strategies.

Findings from the literature review show that LEK can provide primary and comprehensive descriptions of the biophysical and socio-economic components of the environment in semi-arid areas which are experiencing climate change stresses (Alexander et al, 2011). In addition, local community traditional daily practices such as weather forecasting can provide a myriad of benefits, including making informed decisions to enhance agricultural food security (Armatas et al., 2016). The next section presents findings from a case study of local communities inhabiting the transitional zone of the MZBR in Zimbabwe.

Case Study of the Middle Zambezi Biosphere Reserve, Zimbabwe

Biodiversity Indicators of Weather Events in the MZBR

Local communities in the NRA and the CCA use LEK to predict and cope with climatic change. During the 2015–2016 surveys, key informants and FGD participants gave an account of plant and animal species which are used as climate predictors. Astronomical and other abiotic predictors were also mentioned as factors that informed the local community about changes in weather patterns. Thirty-two uses of plants, animals, or abiotic indicators to predict climate or, less commonly, to affect it by use in rainmaking ceremonies were reported (Figure 12.2; Tables 12.3, 12.4, and 12.5).

The NRA and the CCA are located within a predominantly miombo woodland characterized by broad-leaved deciduous *Brachystegia, Julbernardia,* and *Isoberlinia* species. The terrestrial flora is one of the most valuable natural resource components, providing goods and services to the rural communities. Findings from FGDs and key informant interviews indicate that communities use both wild and exotic tree species for climate prediction. For instance, the mango tree (*Mangifera indica*), among the exotic species commonly found around homesteads as an agroforestry species, is a well-known indicator species. A total of 19 uses of plants to predict (or modify) weather were reported. Respondents generally stated that heavy

Table 12.2 Summary of Climate Change Impacts and Biodiversity-Related Mitigation and Adaptation in Africa

Climate change impact	Adaptation/mitigation strategy	Significance	Country	Reference
Low agricultural output	Organic agriculture, mixed cropping or intercropping, irrigation, growing drought-resistant crops, dry planting, soil conservation through zero tillage	Prevents nutrient and water loss, increasing soil resilience to floods, drought, and land degradation processes and preserving soil fertility; minimal tillage keeps top soil strong enough not to be washed away by flood water	Botswana, DRC, Angola, Malawi, Mozambique, Tanzania, South Africa, Zimbabwe, Zambia	Van Campenhout et al. (2016); Mwaura et al. (2008); Domfeh (2007); Mapfumo et al. (2016);
Livestock loss due to forage or water shortage	Use of zhombwe (*Neorautanenia brachypus* (Harms) C.A.Sm.)	Zhombwe acts as an antioxidant, antibiotic, and dewormer, and has a high nutrient value and high water content; *N.brachypus* contains adequate nutrients to mantain ruminant during drought	Zimbabwe	Murungweni et al. (2012); Tanyanyiwa and Chikwanha (2011); Mapfumo et al. (2016)
	Planting fodder crops	Use of alternative plants as supplementary cattle feed during drought improves survival	Botswana, South Africa	Zuma-Netshiukhwi et al. (2013)
	Using plant stalks (maize) as feed	Use of alternative plants as supplementary cattle feed during drought improves survival	Uganda, Kenya, Zimbabwe	Orindi et al. (2007)
Water shortages	Digging of wells around homesteads; drilling boreholes and wells; zero tilling practices in cultivation mulching crops; digging infiltration pits along contours	Water conservation for domestic use and to sustain crops during drought	Botswana, Burkina Faso, Ethiopia, Zimbabwe, Nigeria, Namibia, South Africa, Bangladesh	Mapfumo et al. (2016); Ziervogel et al. (2006); Dockerty et al. (2005); Ifejika Speranza (2010); Van Campenhout et al. (2016); Kalame et al. (2011)
Excessive temperatures	Agroforestry; organic farming	Promotes carbon sequestration; conserves carbon in soils, reducing greenhouse gas emissions	Nigeria, South Africa, Zimbabwe, Sahel region	Tanyanyiwa and Chikwanha (2011); Osunade (1994)

(Continued)

Table 12.2 (Continued) Summary of Climate Change Impacts and Biodiversity-Related Mitigation and Adaptation in Africa

Climate change impact	Adaptation/mitigation strategy	Significance	Country	Reference
Food shortages	Home gardens; Domestication of edible plants; Hunting, fishing and sustainable harvesting of wild food plants and animals such as amarula, vegetables, ivory fruit (*Hyphaene petersiana*), muzhanje, mutohwe, insects, medicinal plants, including *Ziziphus mucronata*, mopane worms, sherpard tree; Use of *Hyphaene petersiana)* leaves for basket weaving and production of wine from the sap; Producing crafts for selling for income generation; Bartering; Food-for-work; Storing food in granaries; Livestock culling	Diversification of food or income sources; Zimbabwe, njemani production has become a source of living for many people in Sengwe area; Employment creation for alternative source of income ensured survival of thousands of starving pastoralists	Botswana, Burkina Faso, Ethiopia, Kenya, Tanzania, Ghana, South Africa, Malawi, Mozambique, Zimbabwe	Mapfumo et al. (2016); Mogotsi et al. (2011); Kalanda-Joshua et al. (2011); Mapfumo et al. (2016); Tanyanyiwa and Chikwanha (2011); Dockerty et al. (2005); Gyampoh et al. (2009)
Increase in pests and diseases	Burning dry grass; Use of aloe, chiwololo	Kills pests and ticks to minimize disease risk in livestock	South Africa	Zuma-Netshiukhwi et al. (2013)

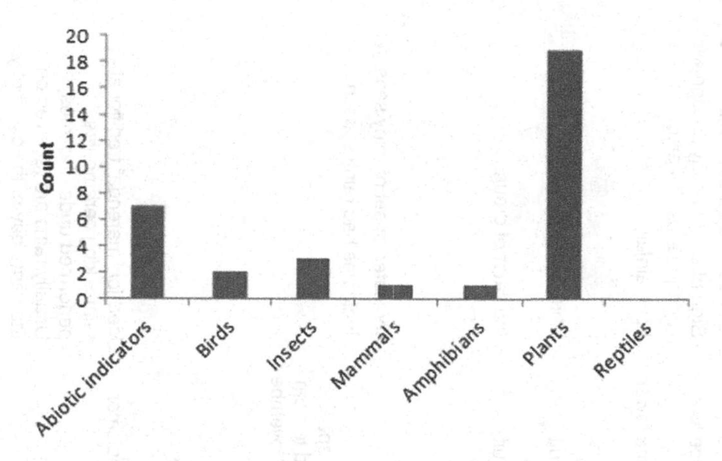

Figure 12.2 Taxonomic distributions of climate indicators in MZBR.

fruit production among indigenous fruit trees signals drought whilst heavy fruit production among exotic trees would signal a good rainy season.

Nineteen uses of plants as climate predictors or in traditional rainmaking ceremonies were mentioned. These included 15 species belonging to 11 families. Respondents mentioned that observation of these trees was most commonly done during and towards the start of the rainy season, which is normally from August through March. The Fabaceae included the highest number of species (4) cited as key indicators for the status of the rainy season. Other species which were commonly mentioned were *Mangifera indica* and *Lannea discolor.* Rainmaking trees, including *Ficus sycomorus* (Figure 12.3), *Parinari curatellifolia, Pseudolachnostylis maprouneifolia,* and *Khaya anthotheca,* were commonly cited by the participants as being specially protected by traditional laws. Cutting down of these trees is a punishable offense, although unlike other species that are multipurpose, these species cannot be used for household purposes nor do they provide fruits or medicine.

Experience and personal observation of faunal species provide knowledge pertaining to climate and weather events. Key informants and FGD participants stated that historically, animal abundance and behavior such as communication sounds were critical in predicting weather events. Animal predictors included vertebrates (birds, mammals) and invertebrates (Table 12.4). For instance, one key informant (a traditional elder) confirmed that in the past, people would hear sounds of a roaring lion from across the Hurungwe Safari Area, signalling the coming of abundant rains. However, FGD participants stated that these sounds are no longer heard, which they attribute to the high incidence of violation of traditional taboos in the area.

In addition to plant and animal indicators, FGD participants and key informants mentioned abiotic indicators associated with changing weather and climate patterns in the area. Astronomical indicators are related to the intensity and direction of wind, appearance of cloud cover, ambient temperature conditions, and winter precipitation (Table 12.5).

Table 12.3 Plant Indicators and Uses in Relation to Rainfall in MZBR. Common Names Are Marked as English (E) or Shona (S).

Scientific name	Common name and vernacular name	Family	Indicator	Significance
Mangifera indica	Mango (E)	Anacardiaceae	Trees produce lots of fruits	Sign of a good rainy season with adequate rainfall amounts
Mangifera indica	Mango (E)	Anacardiaceae	Trees produce fewer fruits	Low rainfall
Lannea edulis	Tsambati	Anacardiaceae		
Lannea discolor	Live-long (E), mugan'acha (S)	Anacardiaceae	Abundant fruit production	Abundant rainfall
Parinari curatellifolia	Hissing tree/mobola plum (E), muhacha/ muchakata (S)	Chrysobalanaceae	Abundant fruit production	Indicator of drought
Uapaca kirkiana	Mazhanje(S), wild loquat (E)	Euphorbiaceae		
Diospyros mespiliformis	Mushuma(S), jackalberry (E)	Ebenaceae		
Faidherbia albida	Apple ring acacia (E)	Fabaceae	Flowering	Indicates onset of rainy season
Brachystegia boehmii	Mupfuti (S)	Fabaceae	Leaf fall for deciduous species now shifted to from August to November	Indicates bad rainy season
Brachystegia spiciformis	Musasa (S)	Fabaceae		
Julbernardia globiflora	Munondo (S)	Fabaceae		
Pseudolachnostylis maprouneifolia	Mutsonzowa (S), duiker/kudu berry (E)	Phyllanthaceae		
Terminalia sericea	Mukonono (S)	Combretaceae		
Diplorhynchus condylocarpon	Mutowa (S)	Apocynaceae		
Ficus sycomorus	Mukuyu (S)	Moraceae	No specific indicator	Used for "marenda" traditional rainmaking ceremonies performed under the trees; Usually rains are received on the same day of the ceremony
Pseudolachnostylis maprouneifolia	Mutsonzowa (S)	Phyllanthaceae		
Parinari curatellifolia	Muhacha (S)	Chrysobalanaceae		
Khaya anthotheca	Mururu (S)	Meliaceae		

Table 12.4 Local Ethnozoological Knowledge on Local Animal Species Used As Weather Predictors

Scientific name or general category	Vernacular name	Family	Indicator	Significance
White storks	Mashuramurove	Ciconiidae	Appearance or disappearance	Signal imminent abundant rains or end of rainy season
Cicada	Nyenze	Cicadoidea	Persistent sounds	Signal high temperatures and hot season
Panthera leo	Lion, shumba	Felidae	Roaring lions	Indicator of abundant rains in the next season
Apis mellifera	Honeybee, nyuchi	Apidae	Production of abundant honey	Indicate low rainfall
Buceros bicornis	Dendera(S), hornbill (E)	Bucerotidae	Singing early in the morning	Imminent change of weather
Locusta migratoria	mhashu	Acrididae	High rate of reproduction	Signal low rainfall
Frogs			Persistent croaking	Coming of the rain season

Table 12.5 Local Abiotic Climatic Indicators

Indicator	Significance
Excessive winds and clear sky	Indicates dry spell for that season
Calm conditions followed by a heavy cloud appearance	Indicates heavy rains
Excessive heat	Indicates good rains
"Bumharutsva" and "gukurahundi"	Rains would start falling in November
When it gets too hot for 3 days before rainfall comes we will also know that our crops will be damaged	Signals oncoming excessive rains which can damage rains
High temperatures	Signal high rainfall
Wind movement from west to east	Signals rainfall coming

LEK on Climate Change Impacts, Adaptation and Mitigation in the MZBR

Household respondents, key informants, and FGDs were aware of the impacts of climate change on socio-economic and biophysical components of the environment (Table 12.6). The changing climate has resulted in a general decline in agricultural productivity, including changes in the availability of ecosystem goods and services. About 43% ($n = 139$) of the respondents indicate that they had experienced livestock diseases such as red water due to climate change. Approximately 59% ($n = 189$) attributed the decrease of pastures to drought, which ranked third after increase in

Figure 12.3 (a,b) Examples of two tree species (a) *Uapaca kirkiana*, and (b) *Ficus sycomorus* used as climate indicators or rainmaking plants in the MZBR. (Photo Credit: O.L. Kupika.)

livestock and population increase. About 8% (26) households had lost at least two cattle (currently valued at approximately US$ 800) due to drought.

A large proportion of household respondents (78%; n = 250); cited declining rainfall (53%; n = 168) as the major factor contributing to water shortages whilst (43%; n=136) also indicated that disappearance of wetlands (68%; n = 218) were also mentioned to be suffering from decreasing rainfall. Field observations revealed that some farmers cultivated in wetlands and valley bottoms, taking advantage of residual moisture in the soils. Personal field observations revealed boreholes, rivers, streams, and wells as the key water sources in the area. Key informants indicated that there have been substantial changes to the flow regime of streams such as Chitake, Chewore, Kabidza, Chitake, Mvurameshi, Mvuramachena, Samhofu, and Hodobe. For instance, key informants reported that the Rukomechi river flow regime had changed from perennial to seasonal since the 2015–2016 season. The study villages used to have several water sources which would supply water throughout the year via natural springs and wetlands, but most of them have dried up.

Field observations and FGD findings also revealed that the local communities have resorted to digging of wells along riverbeds and wetlands (Figure 12.4) to access underground water which they now use for the greater part of the year.

Evidence from field observations shows that the terrain has also influenced the manner in which farmers adjust their farming ways. Generally, the NRA is characterized by fairly undulating terrain, whilst the CCA is typically mountainous. In the CCA, farmers tend to cultivate in low-lying valley areas and foot slopes (Figure 12.5a) where moisture gathers for a longer time, hence sustaining plant growth despite the shorter rainy seasons. Farmers in Nyamakate have adapted through utilization of wetlands such as the Nyamakate River (Figure 12.5b-c) for cultivation of seasonal vegetables all year round. In some selected villages, nutritional gardens associated with non-governmental organizations (NGOs) such as Carbon Green initiatives are also present.

Table 12.6 Local Community Perceptions of Climate Change Impacts on Livelihood Systems

Perceived climate trend	Impacts on livelihood system	Coping and adaptation strategy
	Agricultural activities	
Declining and erratic rainfall Shifting rainy season	Inadequate moisture for plants production (BI); Decrease in crop productivity e.g. maize (SE); Changes in crops/varieties; Short and unpredictable planting season (SE); Increased prevalence of new pests and diseases (BI/SE)	Cultivate in wetlands and low-lying areas; Change crop variety from long season to short season (A). Use stored grain as seed reduced the overall area under cultivation
Extreme temperatures (heat waves and very cold winters)	Wilting of maize and tobacco, mostly affected by excessive heat (BI/SE)	Water conservation techniques such as conservation agriculture, mulching
Persistent droughts	Household food shortages due to poor harvest/low agricultural output (SE)	Harvest wild fruits, e.g. *muchekecha*, and wild legumes such as *Dioscorea praehensilis*, Benth (mupama), and the air potato, *Dioscorea bulbifera (manyanya)*, and *Rhynchosia venulosa (mukoyo) during* drought periods; Harvest wild animals, e.g. rabbits, warthogs, and mice, community/ nutritional gardens; Off-farm jobs in nearby commercial farms; Cooking bananas and mangoes; Gold panning; Selling livestock; Bartering, for example, exchanging two gallons of maize for a goat
	Reduced household income (SE)	Rearing of domestic guinea fowls (*Numida meleagris* f. *domestica*) and rock hyrax (*Procavia capensis*) for sale; Beekeeping informal trading; food-for-work Weaving and hand crafting; Off-farm activities, e.g. seek employment in nearby farms and migrant labour from Zambia
Reduced rainfall and excessive heat	Deterioration of quantity and quality of livestock grazing areas (BI/SE); Reduced livestock, e.g. cattle, goats, reproductive rate and capacity has been affected (SE);	Livestock grazing within wetlands and adjacent protected area
Drought	Increase in roadrunner mortalities and reduced reproduction (SE);	Store maize crop residue for cattle feed

(*Continued*)

Table 12.6 (Continued) Local Community Perceptions of Climate Change Impacts on Livelihood Systems

Perceived climate trend	Impacts on livelihood system	Coping and adaptation strategy
	Increase in cases of climate-induced disease outbreaks (SE); Shortage of water for livestock (BI); Human, wildlife, livestock, affected; Livestock mortalities (cattle, goats, sheep); Livestock diseases, e.g. red water in sheep, goats, cattle; Deteriorating health (all animals)	Reduce livestock numbers
	Water resources	
Reduced rainfall and high temperatures	Reduction in water sources due to drying up of water sources; Boreholes, domestic wells drying up before the end of the next rainy season (BI/SE); Change in river flow from perennial to seasonal (BI); Disease outbreaks: headache, malaria, diarrhoea (SE); Lack of water for setting up tobacco seed beds (BI/SE); Wetlands drying up	Dig deep wells along river beds and on wetlands; Women travel long distances to fetch water; Several households (e.g. up to 44) share same borehole
	Forest resources	
Reduced rainfall	Changes in tree phenology (both domestic and exotic tree species), e.g mazhanje, mango, (BI); Prolonged leaf senescence time in leaf fall for deciduous trees from August to October, tree leaves would be green but now the leaves have actually fallen off when they used to fall off only in August, e.g. Mupfuti, Munondo, Mutsonzowa, Mukonono, Mutowa (BI/SE);	Planting of indigenous and exotic tree species

(*Continued*)

Table 12.6 (Continued) Local Community Perceptions of Climate Change Impacts on Livelihood Systems

Perceived climate trend	Impacts on livelihood system	Coping and adaptation strategy
	Most tree indigenous and exotic fruits, including fruit trees, are no longer producing fruits; Disappearance of or reduced fruit production e.g. *Diospyros mespiliformis* and muhacha (BE/SE)	
Extreme temperatures (heat waves and excessive cold)	Premature drying up of fruits like nhunguru (BI)	Planting indigenous and exotic trees
	Soil resources	
Reduced rainfall and excessive temperatures	Disturbance of soil carbon stocks (BI); Change in soil quality over time (BI)	Conserve our soils through the use of "madhunduru" in applying fertilizers
	Wildlife resources	
Drought (2002–2008)	Habitat encroachment, e.g. human expansion of cultivation into buffer zone (BI); Livestock depredation during the prolonged dry season; lions mostly follow after donkeys and cows; hyenas target goats; wild animals, lions and hyenas, attack livestock (BI); Crop destruction (buffaloes and elands destroy tobacco and eat maize in the fields during March, April, and May; bush pigs and baboons are a menace during the planting and harvesting season from December to May) (SE)	Illegal hunting and harvesting

Key: Socio-economic impact (SE); Biophysical impact (BI).

Key informants and FGD participants were asked to identify wild fruits or forest products which were used to sustain livelihoods during periods of drought. Participants indicated that generally, periods of food shortage coincide with a reduction in the availability of indigenous fruits available in adjacent forests. Despite this, *Piliostigma thonningii* fruits have been harvested and pounded into powder to cook porridge, while *Diospyros mespiliformis* (mushuma) and *Parinari curatellifolia* (muchakata) fruits are used as an alternative food source during drought. In addition, exotic tree species such as raw mangoes were harvested prematurely and cooked for consumption during the 1991–1992 and 2007–2008 drought periods.

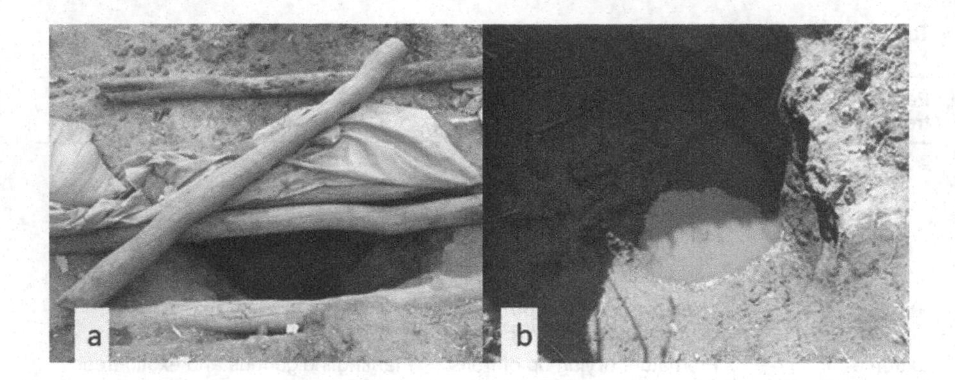

Figure 12.4 Adaptation options include digging deep wells in response to water shortages: (a) in Kabidza, Chundu Communal Area; (b) along Nyamakate River. (Photo credits: O.L. Kupika.)

Figure 12.5 (a) Cultivation on the foot of the mountain to take advantage of moisture which accumulates at the base of the foot slope; (b) and (c) Community gardens along wetlands used as coping strategies in response to soil moisture stress and shortage. (Photo Credits: O.L. Kupika.)

During drought, local communities also rely on indigenous shrubs and underground tubers such as *Dioscorea bulbifera* (air potato, manyanya) and *Dioscorea praehensilis* (mupama), although even these only do well when there is good rainfall. Both of these are woody perennial plants, belonging to the family Dioscoreaceae, which produce edible underground organs (tubers). The air potato (manyanya) is one of the most widely consumed yam species. The mupama tree is also used during drought periods. For example, respondents mentioned that this species provided food relief to the most vulnerable and poor households during the most severe 1991–1992 and 2007–2008 drought periods. Unlike ordinary potato and sweet potato tubers, the air potato tubers have to be thoroughly boiled to remove the bitter taste before consumption. Other key informants reported that if the tubers are underprepared, their consumption can lead to severe stomach ailments and eventually death.

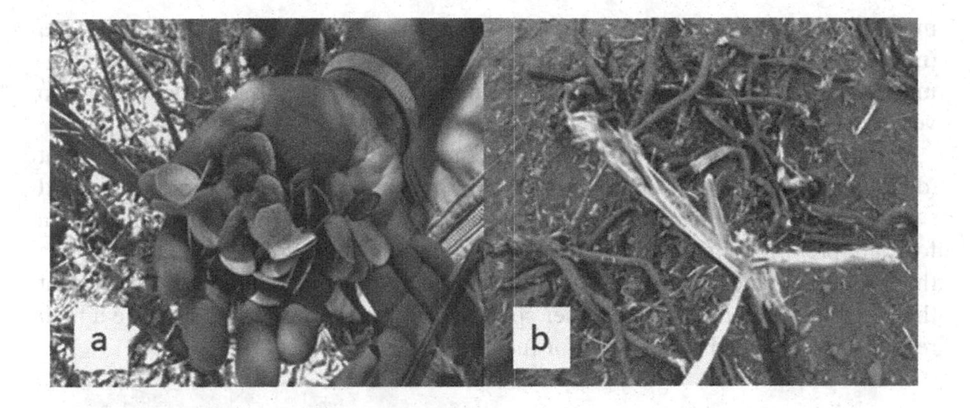

Figure 12.6 (a) Fruits, and (b) Roots of *Rhynchosia venulosa* (mukoyo).

Respondents indicated that *Rhynchosia venulosa* (mukoyo) is one of the popular drought relief plant species among local community members. Key informants and FGD participants stated that traditional beer brewing experts use the roots (Figure 12.6) of the legume to brew beer, which can be sold locally or taken to the border town of Chirundu for sale. Key informants suggested that commercialization of the by-product from the legume could be critical in contributing towards income for selected households during drought periods. The shrub has been cited by key informants as one of the under-utilized legumes which have tremendous potential for commercial exploitation in the area. Participants reported that they even illegally harvest the plant in the adjacent protected area since the species is already disappearing in nearby community forests.

A few respondents cited bee farming (Figure 12.7a) and rearing of wild guinea fowl (Figure 12.7b) as key coping strategies in response to shortages. Beehives are

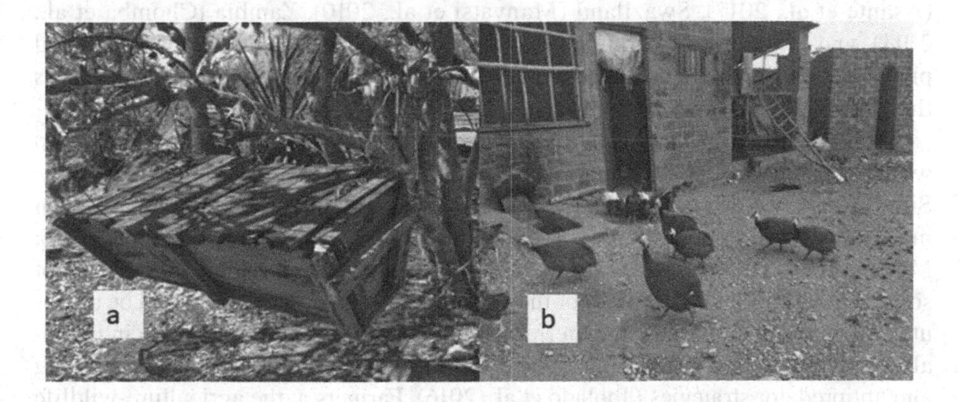

Figure 12.7 (a) Beekeeping and (b) Breeding of guinea fowls as alternative livelihood sources (livelihood diversification).

made from timber from mupfuti trees. They are placed in the *Bauhinia petersiana* (mupondo) tree because it has flowers that attract bees towards the hives. Local community members highlighted that bee farming requires patience and dedication, so very few people use it as a coping strategy.

Domestication of wild animal species such as guinea fowl and the rock rabbit (dassie) was observed to be one of the livelihood strategies used to promote food security at the household level. Respondents stated that wild species such as guinea fowl are preferred since they are resistant to drought and diseases, as compared to the traditional chicken. Hence, they can provide a protein source in the event that the latter succumb to drought. Other species bred include mbira, which have a high reproductive rate and can be sold for income.

DISCUSSION

Local Knowledge of Climate Change Prediction in Africa and the MZBR

Most African countries rely on LEK in biodiversity conservation, disaster prediction, forecasting agricultural seasons, and determining weather patterns. This chapter shows that farmers use LEK of flora, fauna, and abiotic events to understand trends in climate patterns. Our findings are in line with other studies, which have documented the use of plants, animals, and abiotic factors as climate predictors from Southern Africa (Jiri et al., 2016; Macherera and Chimbari, 2016) and South America (Rivero-Romero et al, 2016). Findings related to tree phenology such as variations in fruit production in mango trees, *Parinari curatellifolia* (muhacha), *Lannea discolor* (mugan'acha), *Uapaca kirkiana* (mushuku) to indicate low rainfall have been observed in Botswana, (Batisani and Yarnal, 2010), Malawi (Nkomwa et al., 2014), Mozambique (Arnall et al., 2013), Tanzania (Asante et al., 2017), Swaziland (Manyatsi et al., 2010), Zambia (Chomba et al., 2011), and other parts of Zimbabwe (Jiri et al., 2015). Knowledge of traditional plant species which are used for rainmaking ceremonies tend to be uniform across the Lowveld regions of Zimbabwe.

Several animal species belonging to different taxa (mammals, birds, and insects) which can be used as indicators of climatic events were recorded in this study. Similar observations on the taxa which are used for climate prediction have been made in other parts of Zimbabwe (Unganai and Murwira, 2010; Soropa et al., 2015; Mugabe et al., 2010; Jiri et al., 2015). Farmers also recognize that the home range of some animal species has shifted, or they could be disappearing. This could be attributed to variation in movement patterns and home range formation between individuals, populations, and species in response to intrinsic characteristics such as feeding and antipredator strategies (Tablado et al., 2016). Farmers at the agriculture-wildlife interface also indicated that disturbances such as habitat encroachment due to the expansion of land for cultivation (Baudron et al., 2011) might have resulted in the migration of some animal species.

Climate change has led to highly variable yields in arable agriculture (both rain-fed and irrigated) in African countries (Cooper et al., 2008). Livestock production faces the problem of poor and variable rangeland productivity and desertification processes (Yayeh and Leal, 2017; Murungweni et al., 2014; Thornton et al., 2009). The recruitment of plants has been constrained by low rainfall and high temperature, leading to low productivity (Jagtap and Nagle, 2007). Low productivity of primary producers (plants) had severe cascading effects on both domestic and wildlife species along the food chain. Climate change has led to the creation of strong tensions between humans and wildlife species due to a shortage of resources (especially food and water). The occurrence of habitat patches, invasive species (which are inedible to animals), led to animals exceeding their home ranges and encroaching on human habitats, thereby causing damage to crops and property and paving the way to human-wildlife conflict. Climate change has the potential to alter migratory routes (and timings) of species that use seasonal wetlands (migratory birds) and track seasonal changes in vegetation (herbivores) (Orlove et al., 2010). This increases the conflicts between people and large mammals such as elephants, particularly in areas where rainfall is low.

In the MZBR case study, we recorded several animal species, including birds, terrestrial vertebrates, and insects, that can be used as indicators of climatic events. Similar observations of taxa which are used for climate prediction have been made in other parts of Southern Africa, including Botswana, Malawi, South Africa, Zambia, Swaziland, and Tanzania, as well as other regions in the world (Leonard et al., 2013; Rivero-Romero et al., 2016). However, unlike in other studies, farmers in this study did not mention reptiles as indicators of climate; instead, they mentioned amphibians. Recognition of animal species could be broadly influenced by different factors that affect the geographical distribution of species and their habitat in a particular locality. Farmers also recognize that the home range of some of the animal species that are used for weather prediction has shifted, or they may be disappearing. This could be the reason why farmers in the study area can no longer hear the sounds of roaring lions as an indicator of the rainy season. Since farmers in this study are at the agriculture-wildlife interface, disturbance such as habitat encroachment due to the expansion of land for cultivation (Baudron et al., 2012) might have resulted in the migration of some animal species.

LEK on Coping and Adaptation Strategies in the MZBR

Findings from this study revealed that local communities use LEK related to agricultural practices, wildlife resources, water conservation, and seeking alternative food sources and income in response to food shortages and to adapt and cope with changing rainfall patterns, extreme temperatures and droughts. Results from the study revealed that farmers have adjusted their farming ways, such as adjusting planting dates, planting drought-resistant crops, diversifying crops, and using water conservation techniques in response to changing rainfall patterns. Growing of drought-resistant crops is now practised widely in African countries (Davis, 2011). Crop diversification and planting crops that are drought-tolerant or resistant to temperature stresses serves as an important form of insurance against rainfall variability by reducing the risk of complete crop failure, since different crops are affected differently by climate changes (Case, 2006). Similar adaptation strategies have been observed in Nigeria, where farmers vary the planting dates and increase the use of water and soil conservation techniques (Ifejika Speranza, 2010). Adoption of farm-level adaptations aim at increasing productivity and dealing with existing climatic conditions and draw on farmers' knowledge and farming experience (Van Campenhout et al., 2016).

Local farmers indicated that they use soil conservation techniques such as zero tillage and agroforestry as part of agricultural technologies to conserve water and boost agricultural production. Similar observations have been made in the Sahel and have been known to conserve carbon in soils through the use of zero tilling practices in cultivation and mulching (Osunade, 1994). Natural mulches moderate soil temperatures and extremes, suppress diseases and harmful pests, and conserve soil moisture. Before the advent of chemical fertilizers, local farmers largely depended on organic farming, which is a balanced land-use planning system that tries to balance the raising of food crops and forests whilst at the same time promoting carbon sequestration and thereby reducing greenhouse gas emissions (Kalame et al., 2011). Moreover, domestication of edible plant species is being practised by most African rural people. Domestication of plants increases carbon sequestration and soil fertility, and reduces food insecurity during drought periods. Most farmers grow trees on their farms to shade their crops from intense sunshine (Kalame et al., 2011). Local communities also reported that they kept farm residues for cattle feed during the dry season. Storing farm residues for livestock feeding during the dry season is also an adaptation strategy practised by many African countries. In South Africa, farmers burn grass or moribund material during the dry season to kill pests and ticks to minimize risk of climate-induced diseases in livestock (Zuma-Netshiukhwi et al., 2013).

Local communities use water harvesting such as digging wells, which collect water within wetlands during the rainy season, as one of the key strategies of coping with water shortages. The water is then used for supplementary irrigation of vegetables and crops during the dry season. Similar observations have been made by Van Campenhout et al. (2016); in the dry season, farmers use the water conserved in the wells and basins for irrigation. Van Campenhout also reports (2016) that home gardens are widely used in Southern Africa, especially in Zimbabwe, Mozambique, Botswana, and South Africa, where there is growing of food crops like maize, rice, vegetables, and fruits for bartering and income generation. For example, in South Africa, indigenous people grow cash crops like vegetables, tomatoes, and maize in home gardens to increase household income (Ziervogel et al., 2006).

This study revealed that local communities engage in off-farm activities such as food-for-work programmes as alternative coping strategies during periods of drought. For example, small-scale farmers often provide labour to large-scale commercial farmers or even engage in food-for-work schemes in return for food. Food-for-work schemes are also used in many African countries in order to minimize starvation risks (Bryan et al., 2009). Moreover, in Southeast Lowveld, local Zimbabwe communities sell cattle or barter with those that have better harvest in order to get food

(Risiro et al., 2012). Rural women are involved in both farm and non-farm occupations in order to cope with the challenges of climate change (Van Campenhout et al., 2016). Similarly, in this study, women engage on informal trading along the Harare-Chirundu highway or even within villages, whilst others seek migrant labour in neighbouring countries, i.e. Zambia and South Africa.

This study established that harvesting of wildlife and consumption of wild fruits and legumes alleviate food shortages during drought. Similar findings have been observed by researchers (Udoto, 2012; Hazzah et al., 2013) who noted that off-farm income derived from exploitation of wildlife resources is critical to livelihoods and overall adaptive capacity. In coping with risk due to excessive or low rainfall, drought, and crop failure, some traditional people in Ghana also supplement their food by hunting, fishing, and gathering wild food plants (Gyampoh et al., 2009). In this study, few individuals indicated that they resort to production of handcrafts for sale as strategy to increase household income. Mogotsi et al. (2011) concur that local communities have engaged in different living strategies, like producing crafts for selling for income generation in order to reduce poverty and starvation during drought periods. For example, in Zimbabwe and South Africa, residents use murara (*Hyphaene petersiana*) leaves for basket weaving and murara sap for the production of wine (Murungweni et al., 2014; Chari et al., 2016). In Zimbabwe, njemani production has become a source of living for many people in the Sengwe area, and some of the people no longer are involved in farming (Mapfumo et al., 2016). Similarly, in this study, some farmers indicated that they used local plant resources to weave mats, carve wood objects, and brew beer to boost household income but that they still engage in farming. However, unlike other communities who use palms like *Hyphaene petersiana* to brew local beer (Mapfumo et al, 2016), in this study they used the legume *Rhynchosia venulosa* to brew beer.

Findings from this study on the use of edible tubers from the family Dioscoreaceae are similar to findings elsewhere in Africa. For example, Bruschi et al. (2014) found that some plant species like *Dioscorea cochleariapiculata* and *Dioscorea dumetorum* have been collected from a miombo woodland and eaten as a means of averting food shortages during drought periods in a rural community of Muda-Serraçã, central Mozambique. Similarly in this study, the same species provide alternative food during drought periods. In addition, both studies acknowledge that utilization of the tubers requires one to be thoroughly acquainted with the skills and techniques for making some of the poisonous wild plants edible. Apart from Southern Africa, the edible legumes are widely known as famine foods in East Africa and have also been reported as cultivated in some parts of West Africa (Maundu et al., 2013).

Key informants also emphasized that *Dioscorea cochleariapiculata* and *D. dumetorum* may be eaten only after they have undergone appropriate preparation. Basically, the preparation procedure for the tubers involves the following: peeling the tuber, cutting it into thin slices, drying and washing several hours in a river, always changing the place, then boiling thoroughly for a prolonged period of time until they are cooked. Failure to do this may cause vomiting and even death, as revealed by observations from East Africa (Maundu et al., 1999). Such traditional expertise can be complemented by scientific knowledge in order to enhance food sources. Thus, farmers suggested the need for research institutions to assist in supplying innovative scientific knowledge on how to utilize natural resources commercially. In this study, respondents cited economic and policy constraints coupled with a lack of scientific knowledge as factors which hamper such initiatives. Similarly, Charnley et al. (2007) are of the view that social, economic, and policy constraints have prevented the flourishing and application of LEK.

CONCLUSION

Findings from this review and case study indicate LEK is vital for understanding biodiversity options available for climate prediction and adaptation, particularly at the agriculture-wildlife interface. Understanding local people's knowledge on climate change prediction, impacts, coping, and adaptation strategies is fundamental to addressing the dual challenge of threats to biodiversity conservation and human livelihoods at the agriculture-wildlife interface. However due to a lack of empirical data on the status of wildlife resources under changing climate, our results highlight the significance of harnessing local knowledge to assist agroecological communities in developing effective adaptive management. LEK can therefore shed light on the status of biophysical resources under a changing climate, especially in under-researched areas such as the MZBR.

Integrating LEK into climate change adaptation and biodiversity conservation is possible if the knowledge holders are directly engaged as active participants in these efforts. LEK can complement scientific data to inform policy on best practices to build the adaptive capacity of rural communities within biosphere reserves in a semi-arid tropical savannah. To develop appropriate policies and responses, it will be important to anticipate not only the nature of expected changes but also how they are perceived, interpreted, and adapted to by local communities, thereby promoting ecosystem-based strategies and community resilience.

REFERENCES

Adger, W.N., Barnett, J., Brown, K., Marshall, N. and O'Brien, K., 2013. Cultural dimensions of climate change impacts and adaptation. *Nature Climate Change*, *3*, pp. 112–117.

Agrawal, A., 2010. Local institutions and adaptation to climate change. In: Mearns R, Norton A (eds) *Social Dimensions of Climate Change: Equity and Vulnerability in a Warming World*. World Bank, Washington, D.C., pp. 173–198.

Alexander, C., Bynum, N., Johnson, E., King, U., Mustonen, T., Neofotis, P., Oettlé, N., Rosenzweig, C., Sakakibara, C., Shadrin, V. and Vicarelli, M., 2011. Linking indigenous and scientific knowledge of climate change. *BioScience*, *61*(6), pp. 477–484.

Alvera, P., 2013. The role of Indigenous Knowledge Systems in coping with food security and climate challenges in Mbire District, Zimbabwe (Doctoral dissertation, University of Zimbabwe).

Armatas, C.A., Venn, T.J., McBride, B.B., Watson, A.E. and Carver, S.J., 2016. Opportunities to utilize traditional phenological knowledge to support adaptive management of social-ecological systems vulnerable to changes in climate and fire regimes. *Ecology and Society*, *21*(1), 16.

Arnall, A., Thomas, D.S., Twyman, C. and Liverman, D., 2013. Flooding, resettlement, and change in livelihoods: evidence from rural Mozambique. *Disasters*, *37*(3), pp. 468–488.

Asante, E.A., Ababio, S. and Boadu, K.B., 2017. The use of indigenous cultural practices by the Ashantis for the conservation of forests in Ghana. *SAGE Open*, *7*(1), 215824401668761. http://journals.sagepub.com/doi/10.1177/2158244016687611.

Ayal, D.Y. and Leal Filho, W., 2017. Farmers' perceptions of climate variability and its adverse impacts on crop and livestock production in Ethiopia. *Journal of Arid Environments*, *140*, pp. 20–28.

Batisani, N. and Yarnal, B., 2010. Rainfall variability and trends in semi-arid Botswana: Implications for climate change adaptation policy. *Applied Geography*, *30*(4), pp. 483–489.

Baudron, F., Corbeels, M., Andersson, J.A., Sibanda, M. and Giller, K.E., 2011. Delineating the drivers of waning wildlife habitat: the predominance of cotton farming on the fringe of protected areas in the Middle Zambezi Valley, Zimbabwe. *Biological Conservation*, *144*(5), pp. 1481–1493.

Baudron, F., Tittonell, P., Corbeels, M., Letourmy, P. and Giller, K.E., 2012. Comparative performance of conservation agriculture and current smallholder farming practices in semi-arid Zimbabwe. *Field Crops Research*, *132*, pp.117–128.

Berkes, F. and Seixas, C.S., 2005. Building resilience in lagoon social-ecological systems: a local-level perspective. *Ecosystems*, *8*(8), pp. 967–974.

Boafo, Y.A., Saito, O., Kato, S., Kamiyama, C., Takeuchi, K. and Nakahara, M., 2016. The role of traditional ecological knowledge in ecosystem services management: the case of four rural communities in Northern Ghana. *International Journal of Biodiversity Science, Ecosystem Services & Management*, *12*(1–2), pp. 24–38.

Braun, V. and Clarke, V., 2006. Using thematic analysis in psychology. *Qualitative Research in Psychology*, *3*, pp. 77–101.

Brown, O.L.I., Hammill, A. and Mcleman, R., 2007. Climate change as the 'new' security threat : implications for Africa. *International Affairs*, *83*(6), pp. 1141–1154.

Bruschi, P., Mancini, M., Mattioli, E., Morganti, M. and Signorini, M.A., 2014. Traditional uses of plants in a rural community of Mozambique and possible links with Miombo degradation and harvesting sustainability. *Journal of Ethnobiology and Ethnomedicine*, 10(1), 59.

Bryan, E., Deressa, T.T., Gbetibouo, G.A. and Ringler, C., 2009. Adaptation to climate change in Ethiopia and South Africa: options and constraints. *Environmental Science and Policy*, *12*(4), pp. 413–426.

Buizer, J., Jacobs, K. and Cash, D., 2016. Making short-term climate forecasts useful: linking science and action. *Proceedings of the National Academy of Sciences*, *113*(17), pp. 4597–4602.

Chari, T., Mulaudzi, M. and Masoga, M., 2016. Introduction: towards the integration of indigenous knowledge systems into climate change science. *Indilinga African Journal of Indigenous Knowledge Systems*, *15*(2), pp. v–xii.

Charnley, S., Fischer, A.P. and Jones, E.T., 2007. Integrating traditional and local ecological knowledge into forest biodiversity conservation in the Pacific Northwest. *Forest Ecology and Management*, *246*(1 SPEC. ISS.), pp. 14–28.

Chase, L.E., 2006. Climate change impacts on dairy cattle. *Fact sheet, Climate Change and Agriculture: Promoting Practical and Profitable Responses.* Online at http://climateandfarming.org/pdfs/FactSheets/III.3Cattle.pdf.

Chimhowu, A. and Hulme, D., 2006. Livelihood dynamics in planned and spontaneous resettlement in Zimbabwe: converging and vulnerable. *World Development*, *34*(4), pp. 728–750.

Chomba, C., Mwenya, A.N. and Nyirenda, N., 2011. Wildlife legislation and institutional reforms in Zambia for the period 1912–2011. *Journal of Sustainable Development in Africa*, *13*, pp. 218–236.

Codjoe, F.N.Y., Ocansey, C.K., Boateng, D.O. and Ofori, J., 2013. Climate change awareness and coping strategies of cocoa farmers in rural Ghana. *Journal of Biology, Agriculture and Healthcare*, *3*(11), pp. 19–30.

Codjoe, S.N.A., Owusu, G. and Burkett, V., 2014. Perception, experience, and indigenous knowledge of climate change and variability : the case of Accra, a sub-Saharan African city. *Regional Environmental Change*, *14*(1), pp. 369–383.

Connolly-Boutin, L. and Smit, B., 2016. Climate change, food security, and livelihoods in sub-Saharan Africa. *Regional Environmental Change*, *16*(2), pp. 385–399.

Cook, C., Wardell-Johnson, G., Carter, R. and Hockings, M., 2014. How accurate is the local ecological knowledge of protected area practitioners? *Ecology and Society*, *19*(2), 32.

Cooper, P.J.M., Dimes, J., Rao, K.P.C., Shapiro, B., Shiferaw, B. and Twomlow, S., 2008. Coping better with current climatic variability in the rain-fed farming systems of sub-Saharan Africa: an essential first step in adapting to future climate change? *Agriculture, Ecosystems and Environment*, *126*(1–2), pp. 24–35.

Corbin, J. and Strauss, A., 2008. *Basics of Qualitative Research: Techniques and Procedures for Developing Grounded Theory*. Sage Publications, London, UK.

Davis, C.L., 2011. *Climate Risk and Vulnerability: A Handbook for Southern Africa*. Council for Scientific and Industrial Research, Pretoria, South Africa.

Dockerty, T., Lovett, A., Sünnenberg, G., Appleton, K. and Parry, M., 2005. Visualising the potential impacts of climate change on rural landscapes. *Computers, Environment and Urban Systems*, *29*(3), pp. 297–320.

Domfeh, K.A., 2007. Indigenous knowledge systems and the need for policy and institutional reforms. *Tribes and Tribals*, *1*, pp. 41–52.

Doswald, N., Munroe, R., Roe, D., Giuliani, A., Castelli, I., Stephens, J., Möller, I., Spencer, T., Vira, B. and Reid, H., 2014. Effectiveness of ecosystem-based approaches for adaptation: review of the evidence-base. *Climate and Development*, *6*(2), pp. 185–201.

Fitchett, J.M., Grab, S.W. and Thompson, D.I., 2015. Plant phenology and climate change: Progress in methodological approaches and application. *Progress in Physical Geography*, *39*(4), pp. 460–482.

Fleischman, F. and Briske, D.D., 2016. Professional ecological knowledge: an unrecognized knowledge domain within natural resource management. *Ecology and Society*, *21*(1), p. 32.

Gadgil, M., Berkes, F. and Folke, C., 1993. Indigenous knowledge for biodiversity conservation. *Ambio*, *22*(2/3), pp. 151–156.

Gbetibouo, G.A., 2009. Understanding Farmers Perceptions and Adaptations to Climate Change and variability: The case of the Limpopo Basin farmers South Africa. *International Food Policy Research institute (IFPRI) Discussion Paper 849*. IFPRI, Washington, D.C.

Gyampoh, B.A., Amisah, S., Idinoba, M. and Nkem, J., 2009. Using traditional knowledge to cope with climate change in rural Ghana. *Unasylva*, *60*(281/232), pp. 70–74.

Hazzah, L., Dolrenry, S., Kaplan, D. and Frank, L., 2013. The influence of park access during drought on attitudes toward wildlife and lion killing behaviour in Maasailand, Kenya. *Environmental Conservation*, *40*(3), pp. 266–276.

Ifejika Speranza, C., 2010. *Resilient Adaptation to Climate Change in African Agriculture.* German Institute of Development, Bonn, Germany.

IPCC, 2001: Climate Change 2001: Impacts, Adaptation and Vulnerability, Contribution of Working Group II to the Third Assessment Assessment Report of the Intergovernmental Panel on Climate Change. Cambridge University Press, Cambridge, UK, and New York.

Jagtap, T.G. and Nagle, V.L., 2007. Response and adaptability of mangrove habitats from the Indian subcontinent to changing climate. *AMBIO: A Journal of the Human Environment, 36*(4), pp. 328–334.

Jiri, O., Mafongoya, P.L. and Chivenge, P., 2015. Indigenous knowledge systems, seasonal 'quality' and climate change adaptation in Zimbabwe. *Climate Research, 66*(2), pp. 103–111.

Jiri, O., Mafongoya, P.L., Mubaya, C. and Mafongoya, O., 2016. Seasonal climate prediction and adaptation using indigenous knowledge systems in agriculture systems in Southern Africa: a review. *Journal of Agricultural Science, 8*(5), p. 156.

Kalame, F.B., Aidoo, R., Nkem, J., Ajayie, O.C., Kanninen, M., Luukkanen, O. and Idinoba, M., 2011. Modified taungya system in Ghana: a win–win practice for forestry and adaptation to climate change?. *Environmental Science and Policy, 14*(5), pp. 519–530.

Kalanda-Joshua, M., Ngongondo, C., Chipeta, L. and Mpembeka, F., 2011. Integrating indigenous knowledge with conventional science: enhancing localised climate and weather forecasts in Nessa, Mulanje, Malawi. *Physics and Chemistry of the Earth Parts A/B/C, 36*(14), pp. 996–1003.

Kirilenko, A.P. and Solomon, A.M., 1998. Modeling dynamic vegetation response to rapid climate change using bioclimatic classification. *Climatic Change, 38*(1), pp. 15–49.

Lebel, L., 2013. Local knowledge and adaptation to climate change in natural resource-based societies of the Asia-Pacific. *Mitigation and Adaptation Strategies for Global Change, 18*(7), pp. 1057–1076.

Leonard, S., Parsons, M., Olawsky, K. and Kofod, F., 2013. The role of culture and traditional knowledge in climate change adaptation: insights from East Kimberley, Australia. *Global Environmental Change, 23*(3), pp. 623–632.

Macherera, M. and Chimbari, M.J., 2016. A review of studies on community based early warning systems. *Jàmbá: Journal of Disaster Risk Studies, 8*(1), pp. e1–e11.

Macherera, M., Chimbari, M.J. and Mukaratirwa, S., 2016. Indigenous environmental indicators for malaria: a district study in Zimbabwe. *Acta Tropica, 175*, pp. 50–59.

Maddison, D., 2007. The Perception of and Adaptation to Climate Change in Africa, pp. 1–51. http://papers.ssrn.com/sol3/papers.cfm?abstract_id=1005547.

Manyatsi, A.M., Mhazo, N. and Masarirambi, M.T., 2010. Climate variability and change as perceived by rural communities in Swaziland. *Research Journal of Environmental and Earth Sciences, 2*(3), pp. 164–169.

Mapara, J., 2009. Indigenous knowledge systems in Zimbabwe: juxtaposing postcolonial theory. *Journal of Pan African Studies, 3*(1), pp. 139–155.

Mapara, J., 2017. Binarism as a recipe for lukewarm research into indigenous knowledge systems in Zimbabwe. In: Ngulube, P. (ed), *Handbook of Research on Theoretical Perspectives on Indigenous Knowledge Systems in Developing Countries*. IGI Global, pp. 1–21.

Mapfumo, P., Mtambanengwe, F. and Chikowo, R., 2016. Building on indigenous knowledge to strengthen the capacity of smallholder farming communities to adapt to climate change and variability in southern Africa. *Climate and Development, 8*(1), pp. 72–82.

Mase, A.S., Gramig, B.M., and Prokopy, L. S. 2017. Climate change beliefs, risk perceptions, and adaptation behavior among Midwestern US crop farmers. *Climate Risk Management, 15*, pp. 8–17.

Maundu, P., Bosibori, E., Kibet, S., Morimoto, Y., Odubo, A., Kapeta, Brian, Muiruri, Patel, Adeka, Ruth and Ombonya, Julia, 2013. *Safeguarding Intangible Cultural Heritage: A Practical Guide to Documenting Traditional Foodways*. UNESCO, Nairobi, Kenya.

Maundu, P.M., Ngugi, M.G. and Kabuye, C.H.S., 1999. *Traditional Food Plants of Kenya*. National Museum of Kenya, Nairobi, Kenya.

Meteorological Services Department 2012. Climate Data Services, Zimbabwe.

Mbereko, A., Scott, D., and Kupika, O.L., 2015. First generation land reform in Zimbabwe: historical and institutional dynamics informing household's vulnerability in the Nyamakate Resettlement community. *Journal of Sustainable Development in Africa, 17*(3), pp. 21–40.

McLachlan, J.S., Hellmann, J.J. and Schwartz, M.W., 2007. A framework for debate of assisted migration in an era of climate change. *Conservation Biology*, *21*(2), pp. 297–302.

Meldrum, G., Mijatović, D., Rojas, W., Flores, J., Pinto, M., Mamani, G., Condori, E., Hilaquita, D., Gruberg, H. and Padulosi, S. 2017. Climate change and crop diversity: farmers' perceptions and adaptation on the Bolivian Altiplano. *Environment, Development and Sustainability*, *20*(2), pp. 703–730.

Mogotsi, K., Moroka, A.B., Sitang, O. and Chibua, R., 2011. Seasonal precipitation forecasts: agro-ecological knowledge among rural Kalahari communities. *African Journal of Agricultural Research*, *6*(4), pp. 916–922.

Mugabe, F.T., Mubaya, C.P., Nanja, D., Gondwe, P., Munodawafa, A., Mutswangwa, E., Chagonda, I., Masere, P., Dimes, J. and Murewi, C., 2010. Use of indigenous knowledge systems and scientific methods for climate forecasting in Southern Zambia and North Western Zimbabwe. *Zimbabwe Journal of Technological Sciences*, *1*(1), pp. 19–30.

Murungweni, C., Andersson, J.A., Van Wijk, M.T., Gwitira, I. and Giller, K.E., 2012. Zhombwe (*Neorautanenia brachypus* (Harms) CA Sm.) – a recent discovery for mitigating effects of drought on livestock in semi-arid areas of Southern Africa. *Ethnobotany Research and Applications*, *10*, pp. 199–212.

Murungweni, C., van Wijk, M.T., Giller, K.E., Andersson, J.A. and Smaling, E.M., 2014. Adaptive livelihood strategies employed by farmers to close the food gap in semi-arid South Eastern Zimbabwe. *Food Security*, *6*(3), pp. 313–326.

Mwaura, P.A., Sylva, K. and Malmberg, L.E., 2008. Evaluating the Madrasa preschool programme in East Africa: a quasi-experimental study. *International Journal of Early Years Education*, *16*(3), pp. 237–255.

Nakashima, D.J., Galloway McLean, K., Thulstrup, H.D., Ramos Castillo, A. and Rubis, J.T., 2012. *Weathering Uncertainty: Traditional Knowledge for Climate Change Assessment and Adaptation*. UNESCO, Paris and UNU, Darwin.

Nash, H.C., Wong, M.H.G. and Turvey, S.T., 2016. Using local ecological knowledge to determine status and threats of the critically endangered Chinese pangolin (*Manis pentadactyla*) in Hainan, China. *Biological Conservation*, *196*, pp. 189–195.

Nkomwa, E.C., Joshua, M.K., Ngongondo, C., Monjerezi, M. and Chipungu, F., 2014. Assessing indigenous knowledge systems and climate change adaptation strategies in agriculture: a case study of Chagaka Village, Chikhwawa, Southern Malawi. *Physics and Chemistry of the Earth*, *67–69*, pp. 164–172.

Nyong, A., Adesina, F. and Osman Elasha, B., 2007. The value of indigenous knowledge in climate change mitigation and adaptation strategies in the African Sahel. *Mitigation and Adaptation Strategies for Global Change*, *12*(5), pp. 787–797.

Orindi, V., Nyong, A. and Herrero, M., 2007. Pastoral livelihood adaptation to drought and institutional interventions in Kenya. UNDP Human Development Report Office Occasional paper.

Orlove, B., Roncoli, C., Kabugo, M. and Majugu, A., 2010. Indigenous climate knowledge in southern Uganda: the multiple components of a dynamic regional system. *Climatic Change*, *100*(2), pp. 243–265.

Osunade, M.A., 1994. Indigenous climate knowledge and agricultural practice in Southwestern Nigeria. *Malaysian Journal of Tropical Geography*, *25*(1), pp. 21–28.

Pettengell, C., 2010. Climate change adaptation: enabling people living in poverty to adapt. *Oxfam Policy and Practice: Climate Change and Resilience*, *6*(2), pp. 1–48.

Reyes-García, V., Fernández-Llamazares, Á., Guèze, M., Garcés, A., Mallo, M., Vila-Gómez, M. and Vilaseca, M., 2016. Local indicators of climate change: the potential contribution of local knowledge to climate research. *Wiley Interdisciplinary Reviews: Climate Change*, *7*(1), pp. 109–124.

Risiro, J., Mashoko, D., Tshuma, T. and Rurinda, E., 2012. Weather forecasting and indigenous knowledge systems in Chimanimani District of Manicaland, Zimbabwe. *Journal of Emerging Trends in Educational Research and Policy Studies*, *3*(4), pp.561–566.

Rivero-Romero, A.D., Moreno-Calles, A.I., Casas, A., Castillo, A. and Camou-Guerrero, A., 2016. Traditional climate knowledge: a case study in a peasant community of Tlaxcala, Mexico. *Journal of Ethnobiology and Ethnomedicine*, *12*(1), 33.

Savo, V., Lepofsky, D., Benner, J.P., Kohfeld, K.E., Bailey, J. and Lertzman, K., 2016. Observations of climate change among subsistence-oriented communities around the world. *Nature Climate Change*, *6*(5), pp. 462–473.

Simms, A. and Murphy, M., 2005. Africa up in smoke: the second report from the working group on climate change and development. *Oxfam Policy and Practice: Climate Change and Resilience*, *1*(1), pp. 58–101.

Smit, B., Pilifosova, O., Burton, I., Challenger, B., Huq, S., Klein, R., and Yohe, G., 2001. Adaptation to climate change in the context of sustainable development and equity. "IPCC Third Assessment Report, Working Group II".

Soropa, G., Gwatibaya, S., Musiyiwa, K., Rusere, F., Mavima, G.A. and Kasasa, P., 2015. Indigenous knowledge system weather forecasts as a climate change adaptation strategy in smallholder farming systems of Zimbabwe: case study of Murehwa, Tsholotsho and Chiredzi districts. *African Journal of Agricultural Research*, *10*(10), pp. 1067–1075.

Tablado, Z., Revilla, E., Dubray, D., Saïd, S., Maillard, D. and Loison, A., 2016. From steps to home range formation: species-specific movement upscaling among sympatric ungulates. *Functional Ecology*, *30*(8), pp. 1384–1396.

Tanyanyiwa, V.I. and Chikwanha, M., 2011. The role of indigenous knowledge systems in the management of forest resources in Mugabe area, Masvingo, Zimbabwe. *Journal of Sustainable Development in Africa*, *3*(3), pp. 13–149.

Tazeze, A., Haji, J. and Ketema, M., 2012. Climate change adaptation strategies of smallholder farmers: the case of Babilie District, East Harerghe Zone of Oromia Regional State of Ethiopia. *Journal of Economics and Sustainable Development*, *3*, pp. 1–12.

Thornton, P.K., Van de Steeg, J., Notenbaert, A. and Herrero, M., 2009. The impacts of climate change on livestock and livestock systems in developing countries: a review of what we know and what we need to know. *Agricultural Systems*, *101*(3), pp. 113–127.

Tompkins, E.L. and Adger, W.N., 2004. Does adaptive management of natural resources enhance resilience to climate change? *Ecology and Society* *9*(2), 10. http://www.ecologyandsociety.org/vol9/iss2/art10/.

Udoto, P., 2012. Wildlife as a lifeline to Kenya's economy: making memorable visitor experiences. *The George Wright Forum*, *29*(1), pp. 51–58.

Unganai, L.S. and Murwira, A., 2010. Challenges and opportunities for climate change adaptation among small-holder farmers in southeast Zimbabwe. Second International Conference: Climate, Sustainability and Development in Semi-Arid Regions, 2009.

United Nations, 2015. The road to dignity by 2030: ending poverty, transforming all lives and protecting the planet. Synthesis report of the Secretary-General on the post-2015 sustainable development agenda. United Nations, A/69/700 (December 2014).

Van Campenhout, B., Bizimungu, E. and Birungi, D. 2016. Risk and Sustainable Crop Intensification: The Case of Smallholder Rice and Potato Farmers in Uganda. (IFPRI Discussion Paper 1521). IFPRI, Washington, DC.

Van Campenhout, B., Vandevelde, S., Walukano, W. and Van Asten, P., 2016. Agricultural extension messages using video on portable devices: Increase knowledge about seed selection and seed storage and handling among smallholder potato farmers in southwestern Uganda. *PLoS ONE 12*(1): e0169557.

Vignola, R., Harvey, C.A., Bautista-Solis, P., Avelino, J., Rapidel, B., Donatti, C. and Martinez, R., 2015. Ecosystem-based adaptation for smallholder farmers: definitions, opportunities and constraints. *Agriculture, Ecosystems & Environment*, *211*, pp. 126–132.

Yayeh, D. and Leal, W., 2017. Farmers perceptions of climate variability and its adverse impacts on crop and livestock production in Ethiopia. *Journal of Arid Environments*, *140*, pp. 20–28.

Ziervogel, G., Bharwani, S. and Downing, T.E., 2006. November. Adapting to climate variability: pumpkins, people and policy. *Natural resources forum, 30*(4), pp. 294–305.

Ziervogel, G. and Zermoglio, F., 2009. Climate change scenarios and the development of adaptation strategies in Africa: challenges and opportunities. *Climate Research*, *40*(2–3), pp. 133–146.

Zimbabwe Parks and Wildlife Management Authority (ZPWMA), 2009. Mana Pools National Park General Management Plan part 2: Background.

Zimbabwe Parks and Wildlife Management Authority (ZPWMA), 2010. Gonarezhou National Park General Management Plant 2011–2021, Harare.

Zuma-Netshiukhwi, G., Stigter, K. and Walker, S., 2013. Use of traditional weather/climate knowledge by farmers in the South-western Free State of South Africa: Agrometeorological learning by scientists. *Atmosphere*, *4*(4), pp. 383–410.

CHAPTER 13

Food and Nutrition Innovation in the Context of Indigenous Knowledge Systems

Roseanna Avento

CONTENTS

ABSTRACT

Background The combination of indigenous knowledge on underutilized and local food plants with information on consumer markets, regulations, and entrepreneurship can give rise to new food products that support community development and benefit health and nutrition. Workshops held in Peru, Finland, Kenya, and South Africa brought together stakeholders, including consumers, researchers, business and legal experts, and potential entrepreneurs to brainstorm and test new products in a process of co-creation. Participants reported gaining knowledge and feeling more comfortable with the process of innovation, and some proceeded to development of products for market as a result of the workshops.

Relevance Institutions of higher learning can be suitable organizers and facilitators of such processes due to their role in implementing interdisciplinary work and communicating with a variety of communities. For best results, it is necessary that the agenda, goals, and suitable partnerships be determined in advance.

INNOVATION IN THE CONTEXT OF INDIGENOUS KNOWLEDGE SYSTEMS

To be instrumental in creating enterprise and knowledge societies, creative engagement with new partners and the building of unique learning, idea-generation environments are increasingly necessary. A storm propelled by increased internationalization and a renewed interest in indigenous knowledge and its potential to developing a variety of sectors, especially food, nutrition, and health, is leading a paradigm change through innovation in consumer society, which encompasses cultures, values, people, and behaviors.

Contemporary views on innovation (Fagerberg et al., 2010) present a more holistic approach than the more traditional views, which treat innovation as a high-level R&D activity performed only by leading research institutes. Innovation, understood as a set of solutions developed and implemented in response to problems, challenges, or opportunities, provides a foundation for utilizing indigenous knowledge to create new products and services. Schumpeter, the Founding Father of Innovation Theory, enumerated five types of innovation (Schumpeter, 1934) (Figure 13.1), which are characterized by both novelty and implementation (Śledzik, 2013). Hence, the holistic perspective emphasizes the entire cycle from ideation to implementation and finally dissemination.

This holistic approach is useful when considering the use of indigenous knowledge in food and nutrition innovation. Indigenous knowledge is rarely scientific, but

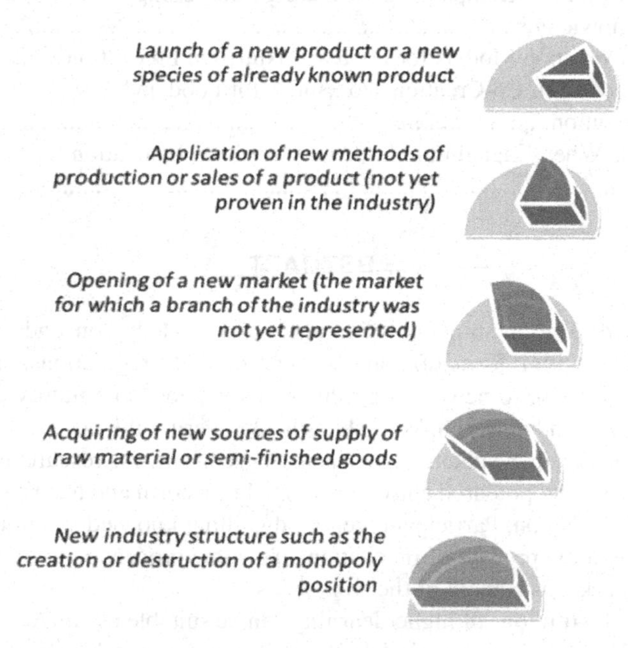

Launch of a new product or a new species of already known product
Application of new methods of production or sales of a product (not yet proven in the industry)
Opening of a new market (the market for which a branch of the industry was not yet represented)
Acquiring of new sources of supply of raw material or semi-finished goods
New industry structure such as the creation or destruction of a monopoly position

Figure 13.1 Schumpeter's five types of innovation. (Source: Śledzik, 2013.)

rather, more practical and contextual, accumulated over time, locally embedded, and unique to a given culture, society, or local community. It is also transmitted orally or through demonstration and imitation, and learned by repetition (Subba Rao, 2006; Sillitoe and Marzano, 2009).

Using indigenous knowledge to innovate requires, as in any innovation process, a road map for creating value and an atmosphere that encourages creativity and collaboration. Collaboration essentially refers to networking, through which more and more innovation takes place. Knowledge exchange cannot happen in a vacuum. It requires communication between people. As the globe becomes smaller, and as people have greater access to information networks, knowledge exchange has become less expensive and more accessible and flexible. The potential for accessing and utilizing indigenous knowledge in food and nutrition innovation has therefore increased.

Utilizing indigenous knowledge in food and nutrition innovation (Figure 13.2) can be looked upon, within product development, as a means of modernizing traditions, strengthening culinary institutions in different communities and societies, and creating an inclusive innovation environment where rural communities are given the opportunity to participate in the development of their own society through income generation and entrepreneurship. This can be done in two ways: (1) food and nutrition innovation processes may take external knowledge and localize it through indigenous knowledge, or (2) indigenous knowledge may become a part of an external knowledge-based innovation. Both processes combine indigenous and external knowledge and apply them to experiences.

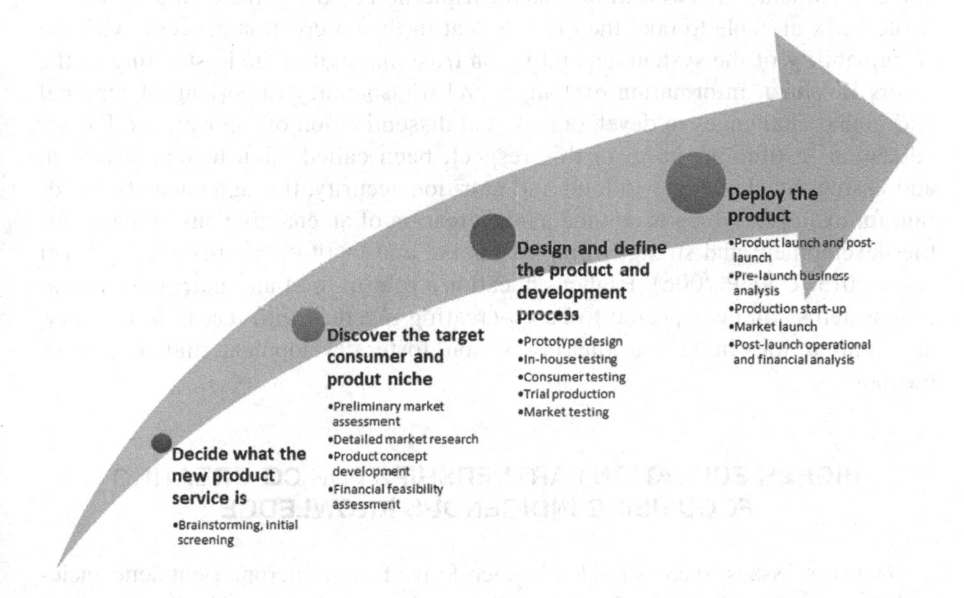

Figure 13.2 Food product development process.

CO-CREATION FOR FOOD AND NUTRITION INNOVATION

When using indigenous knowledge in food and nutrition innovation, it is essential not only to understand how different consumer groups may respond to the new innovations, but also to work together with consumers to create innovations that are acceptable to them. Food and nutrition innovation can thus, on one hand, be perceived as a process where value is created for consumers from different ingredients to ensure health and well-being. On the other hand, the value created must also come from consumers. This process of partnering with consumers to ideate, problem-solve, or create a new product, service, or business is called co-creation (Guiné et al. 2016). The food and beverage industry has already traditionally looked into consumer trends, which influence food and beverage purchases, behavior, beliefs, and innovation in the industry (Figure 13.3).

Co-creation, however, takes the dialogue much deeper by consulting with consumers to generate new ideas. It is a form of open innovation engagement that recognizes the importance of consumer knowledge. Co-creation requires real scrutiny into human experiences, letting go of one's inhibitions, and including actors who may have alternative views into the process.

Co-creation is thus founded in interactive engagement, which, as Prahalad and Ramaswamy (2004) say, is the crux of our emerging reality of creative dialogue. Utilizing the co-creation model in food and nutrition innovation implies integrating existing knowledge from different organizations, for example, through a triad of networks of operation involving universities, communities (including industry), and government, otherwise known as the triple-helix. Any of the actors within the triple-helix are able to take the driver's seat in the co-creation process, with the sustainability of the system depending on trust and mutual understanding of the actors involved, information exchange, and transparency in solving of regional and global challenges to development, and dissemination of innovations. Higher education institutions have, in this respect, been called upon to play a role in addressing the challenges to food and nutrition security, through capacity building, for example, which is defined as the creation of an enabling environment for the development and strengthening of human and institutional resources (Aerni et al. 2015; UNDP 2006). Higher education's role in food and nutrition innovation systems can be expected to be co-creating together with research, industry, and communities in order to fuel innovation, foster development, and create new business.

HIGHER EDUCATION PARTNERSHIPS FOR CO-CREATING FOOD USING INDIGENOUS KNOWLEDGE

Nutrition issues, such as hidden hunger derived from micronutrient deficiencies and diseases of "affluence" (obesity, cardiovascular, and degenerative diseases), represent an enormous challenge for the future (FAO, IFAD, and WFP, 2014). At the same time, Shand (1993) reports that about 75% of plant genetic diversity has been

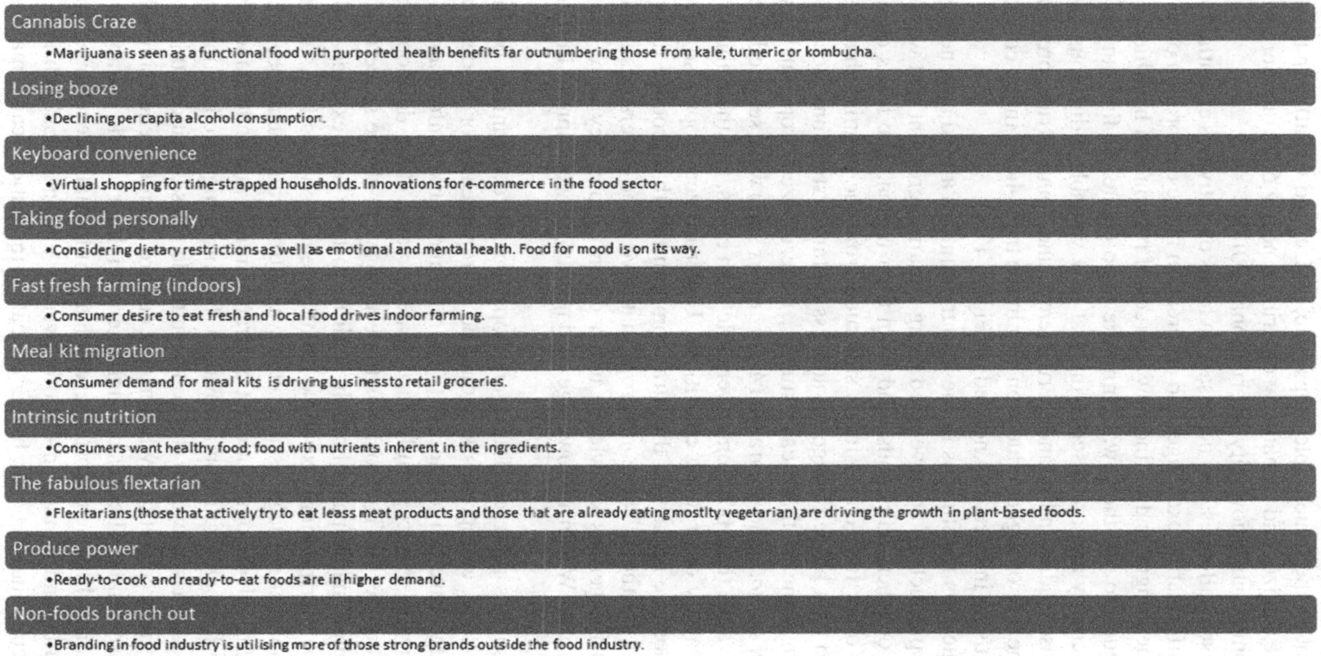

Cannabis Craze

- Marijuana is seen as a functional food with purported health benefits far outnumbering those from kale, turmeric or kombucha.

Losing booze

- Declining per capita alcohol consumption.

Keyboard convenience

- Virtual shopping for time-strapped households. Innovations for e-commerce in the food sector

Taking food personally

- Considering dietary restrictions as well as emotional and mental health. Food for mood is on its way.

Fast fresh farming (indoors)

- Consumer desire to eat fresh and local food drives indoor farming.

Meal kit migration

- Consumer demand for meal kits is driving business to retail groceries.

Intrinsic nutrition

- Consumers want healthy food; food with nutrients inherent in the ingredients.

The fabulous flextarian

- Flexitarians (those that actively try to eat leass meat products and those that are already eating mostly vegetarian) are driving the growth in plant-based foods.

Produce power

- Ready-to-cook and ready-to-eat foods are in higher demand.

Non-foods branch out

- Branding in food industry is utilising more of those strong brands outside the food industry.

Figure 13.3 Macro trends driving food and beverage innovation. (Adapted from Mattson, 2018.)

abandoned, as farmers worldwide have left their countless local varieties and landraces for genetically uniform, high-yielding varieties. Further, countries are today 36% more dependent on the same staple crops (rice, maize, and wheat) than they were 50 years ago. Rice, maize, and wheat now contribute nearly 50% of calories obtained by humans from plants (Bioversity International, 2017).

To face these challenges, it is necessary to focus on activities essential to accelerate dialogue, information exchange, and co-creation in the exploration of novel solutions to feed the hungry and to find new sources of protein and healthier, nutritious foods. In our quest to find the new, to innovate or co-create, we find ourselves in the winds of rediscovery and are discovering just how valuable indigenous knowledge can be. Four case studies in Finland, Peru, Kenya, and South Africa, presented here, demonstrate the role of higher education within the triple-helix in the co-creation of food products using indigenous knowledge (Figure 13.4).

Higher education institutions in these four countries aimed at generating solid ideas for food products to be developed using a main ingredient that was an underutilized crop or other indigenous food from their region. The focus on underutilized crops or other indigenous foods stemmed from the premise that these have the potential to fight hidden hunger or address dietary restrictions faced by certain populations. In some of these case studies, underutilized crops such as quinoa (*Chenopodium quinoa*) and amaranth (*Amaranthus* spp.) were selected due to their high nutritive value, their position as versatile cash crops, and their ability to withstand a wide variety of growing conditions. Underutilized crops also have a high potential for positive impact on the rural communities that produce them and the consumers that use them. These are crops that have faced adversity similar to that faced by most indigenous knowledge under colonization; they were banned and shunned in favor of Western knowledge, and new crops were spread in their native regions.

The case studies represent four workshops that were organized on the premise that small-scale entrepreneurs, would-be entrepreneurs, or holders of indigenous knowledge have limited access to training and market information that they can use holistically to successfully participate in value chain development in the food and nutrition sector. The goal was to develop a knowledge space in a spirit of co-creation, where participants brought with them their own existing and indigenous knowledge. Furthermore, expertise from the region on specific issues (business development, marketing, legal issues, etc.) was introduced to enhance the co-creation process through dialogue, interaction, transparency, and understanding of risks and benefits in food and nutrition innovation. Having a higher education institution in the driver's seat has relevancy due to the multiple roles that it can take on. These include economic acceleration by fostering a culture of entrepreneurship relevant to solving global societal challenges through interdisciplinary work, collaboration with the world of work which helps foster research and information exchange within and outside academia, and promotion of both diversity and inclusion. While disruption of economies continues to occur, higher education institutions can remain as strong foundations and guardians of knowledge economies, propelling the development of society.

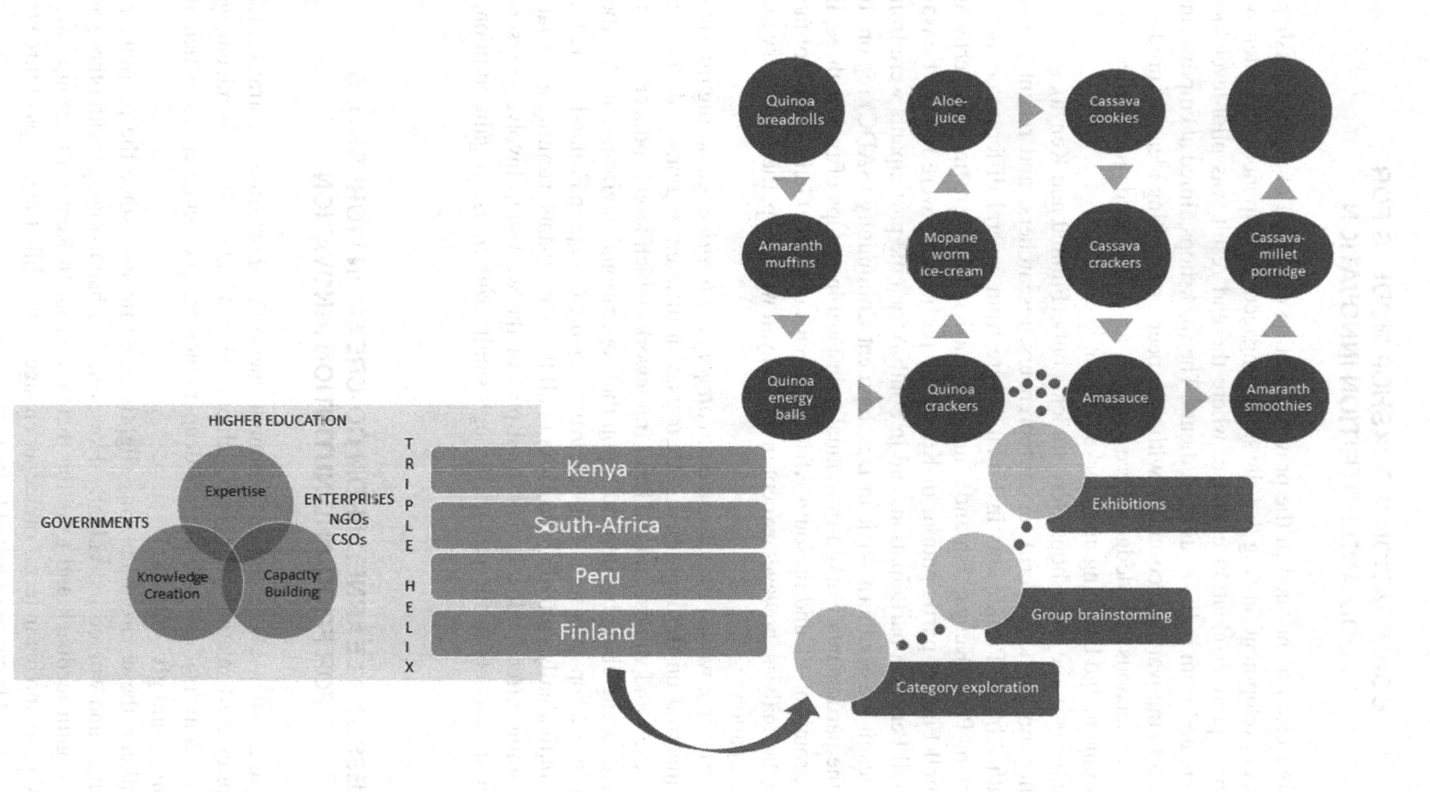

Figure 13.4 Co-creating food innovations from indigenous foods.

CO-CREATION WORKSHOP MODELS FOR FOOD AND NUTRITION INNOVATION

The four cases mentioned in the previous section consisted of short workshops on product development, lasting 3-4 days, that engaged participants in an intensive and vigorous product ideation exercise, where the end result was pilot-level food products created from indigenous ingredients. The workshops aimed at empowering participants to innovate and co-create with their peers, generating ideas for products using their indigenous knowledge where possible, assessing needs for further product development, and branding and market launch.

Participants in the workshops were students only (Finland and Kenya), students and teachers/researchers (Peru), or entrepreneurs, researchers, and recent graduates (South Africa). Participants in Finland were from several different countries (Kazakhstan, Peru, Kenya, Finland), in Peru participants were mostly Peruvian with a small Finnish participation, in Kenya the participants were mostly Kenyan with a small Peruvian participation, and in South Africa the participants were from 10 countries in the Southern African Development Community (SADC) region. In Finland, the participants worked individually, in Peru in groups of two or three, in Kenya in groups of 10, and in South Africa in groups of three. Gender was not factored into the analysis; however, except in the Kenyan workshop, there was very low male participation.

One part of the workshop consisted of *category exploration*, gathering information and material on the ingredients, their uses in different regions and different communities, brand values, and the legal frameworks which food producers have to operate within and abide by, depending on their operational environment. Another part of the workshop consisted of *group brainstorming* (except in Finland where this was done individually), pitching the ideas to all the participants, running test-trials, and fine-tuning after feedback. The final part of the workshop involved presenting the pilot products to the public, through small *exhibitions*, to gain additional feedback.

LESSONS LEARNED FROM CO-CREATION WORKSHOPS FOR FOOD AND NUTRITION INNOVATION

Despite an intense working environment due to the short time constraints, the workshops created an atmosphere in which participants had to quickly bond, engage in authentic conversations, and explore deeper emotions and perspectives, which all resulted in meaningful outputs.

In Finland, the environment was slightly more intense, since the participants brainstormed and worked individually. However, they had to engage with other participants to gain feedback and to refine their products. In Kenya, the groups were deemed slightly too big in terms of deeper engagement, while the groups of three (in Peru and South Africa) were found to be optimal.

The South African group worked faster than any of the other groups and in general seemed to be more mature in terms of brainstorming and delivery, as they needed less advice and guidance. This can be attributed to the age of the participants (older than the other groups) and experience (entrepreneurs, academics, and recent graduates, with no students). However, despite the different backgrounds of the participants and modes of working (e.g. group size), the outputs and aims were very well achieved in all the workshops.

In all four cases, participants commended group work, brainstorming exercises, and the information exchange that resulted. The participants were highly appreciative of the new networks they formed and general information exchange. As one participant from Kenya commented, "It was interesting to exchange information and ideas on the scale Finland–Kenya–Peru." The participants appreciated the opportunity to gain information on how other cultures and communities used ingredients they normally used in an alternative manner.

Indigenous knowledge on the use of different ingredients was found valuable during the discussions and information exchange. For example, a participant from Zambia said, "I have learnt more about indigenous food, labelling, and networking." The participants further commended the information they received regarding product development, marketing, and business development. A participant from South Africa commented, "I have got knowledge and skills on producing a product that can be marketed. Also to network for further business opportunities."

The participants were generally satisfied with the information on legal issues around innovation in the food industry; however, more in-depth regional expertise would have greatly improved the workshops. The challenge was most apparent in the African region, where discussions on intellectual property rights (IPR) and legalities surrounding indigenous knowledge, not to mention the food industry, are still more or less in the fledgling stage. The differences between the more regulated markets in the EU and less regulated markets in Africa and Latin America were also discussed, with highlights being the difficulty of market entry into the EU due to restrictions on size, form, and labeling of products, whereas EU products generally enter African and Latin American markets more easily. Labeling of products was deemed an area for development. Legal issues remain a challenge as well. For many small-scale entrepreneurs, it is not practical or possible to obtain reliable legal advice, and many have to survive with false or outdated information. Especially when dealing with indigenous knowledge, clarity on what can be done with which ingredient, for whom, and by whom is needed. In addition, gaps exist between indigenous knowledge and Western-validated information on the health benefits of certain ingredients. Health claims regarding ingredients and products have to be carefully considered, preferably with an expert in the field.

Actually engaging in kitchen trials and showcasing the products to the public for feedback was well-commended by the participants. A participant from Botswana said that she appreciated the workshop as it gave her new opportunities: "Having my product being tried by those I could never have imagined was an excellent part of the course", while a participant from Kenya said, "The practical aspects of the course made learning concrete."

There were complaints of the shortage of time in all workshops. However, they were intended to create intensity through the time constraint. The South African case had actually had a day less (3 days) than the other cases (4 days), leading the organizers to the conclusion that the outputs would be the same regardless of the time factor. Given more time, the participants may have had more time to brainstorm and refine their products. However, the idea was to engage with consumers before further refining of products and thus more time was not necessary for the exercise. In terms of creating value, the workshops were successful. The participants made initial marketing decisions, based on their understanding of their consumer value needs, and then created a value proposition through their pilot products.

During the workshops, the participants had the opportunity to engage with different experts from academia (researchers), from business incubators, from businesses, and from legal experts. The inclusion of different actors in the program allowed for category exploration, in which they were given advice and insight for their product ideation and development.

Furthermore, during the exhibition of the pilot products, engagement with stakeholders and consumers provided valuable feedback to the participants on how to further refine their products. There was also interest in commercialization of some of the products directly from industry (Kenya and Finland), while other participants were eager to test their products on the market and proceed to market launch individually. This was inherent especially in the South African workshop, where a good number of the participants were already active entrepreneurs. In fact, while some of the products were still in pilot stage, a good many were ready for scaling up and branding considerations.

Further development of the innovations should focus not only on scaling up and branding, but also on business development. As many of the participants were holders of indigenous knowledge, academics, small-scale entrepreneurs, and mostly women entrepreneurs, it was inherent that they faced many challenges in terms of how to develop their business, price their product, and take it to market. Many questions were raised during the co-creation exercises on with whom to partner, especially when considering market entry points. Supporting participants on these aspects is therefore of the utmost importance and a second natural phase in the co-creation process for food and nutrition innovation.

CONSIDERATIONS WHEN ENGAGING IN FOOD AND NUTRITION INNOVATION

To conclude, partnerships propelled by higher education and formed through the co-creation process provide a platform for dialogue, technology development, and transfer of indigenous knowledge. Partnerships also work as an interface to translate outputs and innovations into policies, practices, and products. Experiential learning, such as seen through these co-creation workshops, plays an integral part in dissemination of indigenous knowledge. It is, however, important to ask the following questions when engaging in food and nutrition innovation utilizing indigenous knowledge within the co-creation process:

- What do we want to achieve? What do our partners want to achieve?
- Whose agenda is the driver? Whose voice drives the agenda?
- Are our goals legitimate?
- Do we know and understand the local environment?
- Is the necessary scaling up or down feasible?
- Are we collaborating with the right people and institutions? Do we really trust our partners?

Harnessing creativity and indigenous knowledge for food and nutrition innovation through collaborative networks has great potential for creating value for customers. This collaborative networking is not just about exchanging contacts but about co-creating and cultivating unique relationships that result in diverse opportunities.

REFERENCES

Aerni P, Nichterlein K, Rudgard S, Sonnini A (2015) Making agricultural innovation systems (AIS) work for development in tropical countries. *Sustainability* 7:831–850. https://doi.org/10.3390/su7010831.

Bioversity International (2017) Mainstreaming agrobiodiversity in sustainable food systems: scientific foundations for an agrobiodiversity index. Bioversity International, Rome, Italy.

Fagerberg J, Srholec M, Verspagen B (2010) The role of innovation in development. *Revista de Economía Institucional* 1(2):2. https://doi.org/10.5202/rei.v1i2.2

FAO, IFAD, WFP (2014) The state of food insecurity in the world 2014. Strengthening the enabling environment for food security and nutrition. FAO, Rome, Italy. http://www.fao.org/3/a-i4030e.pdf.

Guiné RPF, Ramalhosa ECD, Valente LP (2016) New foods, new consumers: innovation in food product development. *Current Nutrition & Food Science* 12(3):175–189.

Mattson (2018) 10 Macrotrends driving food & beverage innovation. FoodNavigator. 18.01.2018. https://www.foodnavigator-usa.com/Article/2018/01/18/Mattson-unveils-10-macro-trends-driving-food-and-beverage-innovation

Prahalad CK, Ramaswamy V (2004) Co-creating unique value with customers. *Strategy Leadership* 32(3):4–9.

Schumpeter JA (1934) The theory of economic development: an inquiry into profits, capital, credit, interest and the business cycle. Harvard Economic Studies 46. Harvard College, Cambridge, MA.

Shand H (1993) *Harvesting Nature's Diversity*. FAO, Rome, Italy.

Sillitoe P, Marzano M (2009) Future of indigenous knowledge research in development. *Futures* 41(1):13–23.

Śledzik K (2013) Schumpeter's view on innovation and entrepreneurship. In: Hittmar S (ed.) *Management Trends in Theory and Practice*. University Publishing House, University of Zilina, Zilina, Slovakia, pp. 89–95.

Subba Rao S (2006) Indigenous knowledge organization: an Indian scenario. *International Journal of Information Management* 26(3):224–233.

United Nations Development Programme (2006) Definition of basic concepts and terminologies in governance and public administration. E/C.16.2006.4. Committee of Experts on Public Administration, 5th session, New York.

CHAPTER 14

Hurdles in the Commercialization of Tribal and Indigenous Knowledge-Derived Technologies

David R. Katerere, Wendy Applequist, and Trish Flaster

CONTENTS

ABSTRACT

Background Technologies derived from indigenous knowledge are typically user-innovations produced by the communities that use them via affordable, low-capital means of production; variability is considered acceptable. Economic, logistical, cultural, and legal barriers to broader-scale sale of these products can be substantial, especially in the field of traditional medicine, where the potential value of indigenous knowledge is particularly well known. Among the challenges are ensuring sustainable supply of material, ensuring consistent product quality by the methods expected in a given market, and complying with often very strict and expensive regulations in high income-potential markets. These burdens are often too expensive for indigenous knowledge holders to meet alone; to develop marketable products from their knowledge, partnerships with larger companies or fair-trade organizations are necessary.

Relevance This chapter reviews the challenges to developing and commercializing innovative products from indigenous knowledge.

INTRODUCTION: THE STATUS OF INDIGENOUS TECHNOLOGIES

Traditional knowledge is a major contributor to global knowledge and innovation, particularly in natural resource management, agriculture, and healthcare (Joshi and Chelliah 2013). Indigenous people in biodiversity-rich regions have learnt to innovate products from their local natural resources. For instance, Himalayan tribes continue to use ancient fermentation technologies to enhance aroma, flavour, and texture of foods and aid in preservation (Nehal 2013). These technologies use microbial cultures to ferment leafy vegetables, cereals and pulses, fish, bamboo, and milk beverages, among other foods, which can potentially be industrialized into symbiotic formulations and products. Most such indigenous products can be termed as user-innovations (Baldwin et al. 2006), products that are manufactured by the users for their own use using low-capital-cost production technologies that are often highly variable.

User-innovation products do not follow a clear development process. They often come into the mainstream market by accident or as niche products. Should producers decide to commercialize, they will have a high variable cost and an initially limited market (Baldwin et al. 2006). As the product grows in popularity and market size, established manufacturers move to invest in high-volume standardized products leading to unit cost reduction and the elbowing of the innovators out of the marketplace.

For example, as shea butter from West Africa has become a popular commodity, income from its production increasingly goes to urban residents who have the capital to purchase nuts and costly milling equipment and who have better access to fair-trade markets, rather than rural women (Pouliot and Elias 2013). Price volatility of raw commodities due to large fluctuations in demand and supply puts small farmers at considerable risk, which also may affect product quality (e.g. for turmeric: Booker et al. 2016).

Commercialization of technologies derived from indigenous knowledge requires that multiple challenges be overcome. The contextual nature of such knowledge is the first challenge and creates tensions at the global market level (Joshi and Chelliah 2013). Other important challenges include economic, regulatory, cultural, and ecological barriers to participation in the market economy. To start any business takes some level of capital, and if the goal is to participate not only in local markets but in the global marketplace, potential buyers often expect a scale of production that requires very large upfront investments. Small-scale local businesses may be subject to the same expensive regulations applicable to globalized industry, depending upon the choice of national governments. To market the products of tribal and indigenous knowledge to consumers outside the source group, cultural resistance to the valuation of that knowledge must be overcome, which itself often requires significant financial investment. Finally, if entry into larger markets is successfully negotiated, producers of goods derived from indigenous plant, animal, or fungal species must face the challenge of ensuring a sustainable supply of high-quality raw materials.

In this chapter, we emphasize the situation of indigenous traditional medicine (ITM) or complementary medicine (CM), the latter defined as health and healing practices that are not inherent to a country and that are considered to lie outside the country's conventional healthcare system (Falkenberg 2012). Dietary and/or health supplements are also generally classified as complementary medicines. In the minds of many consumers, traditional knowledge is virtually equated with traditional and complementary medicine, while ITM more than any other traditional technology faces all of the barriers to commercialization enumerated above, and more. There is a global marketplace dominated by relatively large companies; governments in many nations stringently regulate the conditions of manufacture and marketing, and sometimes prohibit or severely restrict the introduction of species not previously used within the country; consumers are often suspicious of unfamiliar medicines from other cultures, and often favor those that have been the subject of substantial research, including costly clinical trials; and wild harvest of medicinal species for global commerce has already been shown to cause scarcity of many species and indeed threaten their extinction. Hence, we will discuss ITM as a case in which both the potential and the challenges are at a maximum, though most of these challenges apply, to varying degrees, to the commercialization of other indigenous technologies.

CHALLENGES TO THE MARKETING OF INDIGENOUS TRADITIONAL MEDICINE

In most countries of the world, conventional (allopathic) medicine coexists with ITM. There has been rapid growth of ITM in the West since the 1980s, which has resulted in the commercialization of products of non-Western origin, particularly from China and the Indian subcontinent. This growth is largely fueled by the need to self-medicate and disillusionment with the medical/pharmaceutical model of disease and health management. For the past several years, the dietary supplement industry has been on a steady upward trajectory; for example, retail sales of supplements in the United States totaled an estimated $6.92 billion in 2015, the 12th consecutive year of growth (Smith et al. 2016). More people are recognizing the value of treating themselves for selected conditions. With this maturing of the industry, its scientific integrity and the possibility for innovation increase. Contrarily, so do the demands on would-be participants in the market.

Previously, ITM products were generally made extemporaneously by the traditional knowledge holder/healer in small quantities with limited distribution to individual patients, or raw materials were purchased by patients themselves for preparation of remedies. Strict quality management is neither feasible nor necessary in those circumstances. In light of commercialization and globalization, raw materials are frequently produced in one part of the world and processed in another part, or processed in one region and sold in another. Because this increases both opportunities for quality problems and the potential repercussions, such products must meet strict quality standards for correct identification (to exclude deliberate or inadvertent adulterants), purity (to exclude contaminants, including mold), often quality (as measured by the content of marker compounds), and stability (to ensure product integrity after transportation and use until the end of shelf life). The gap between these two models presents ongoing challenges to the commercialization of products.

The safety and activity or efficacy of herbal medicines largely depend on their quality (WHO 2017). Therefore, large-scale commercialization should be accompanied by quality management systems which track products from sourcing of the raw materials (whether farmed or wild harvested), through processing, formulation, and manufacturing, all the way to the finished product. The World Health Organization (WHO) has over the years developed guidelines for stakeholders in this area, covering quality control methods (WHO 1998), good agricultural and collection practices (GACP) for medicinal plants (WHO 2003), assessing quality of herbal medicines with reference to contaminants and residues (WHO 2007), and most recently, in draft form, good herbal processing practices (GHPP) for medicinal products (WHO 2017). Various national or trade organizations also have guidelines whose purpose is to attempt to maintain quality throughout the entire value chain of herbal medicinal plants. This is something which has become indispensable as indigenous knowledge has been turned into commercial products. Quality systems can, however, be onerous and expensive and may not always be feasible or appropriate for implementation by small-holder farmers or small manufacturers (e.g. Booker et al. 2016).

The development of a marketing plan for a botanical or other natural product should always include consideration of sustainable sourcing. Wild collection of medicinal species for use by local populations can be sustainable. Unfortunately, too often a product that becomes fashionable in global commerce continues to be collected by traditional means until (and after) it becomes apparent that wild harvest cannot sustainably provide the quantities of material needed to supply a marketplace of over seven billion potential consumers. Many plants and some fungi are already threatened by overharvesting for commerce (e.g. Nantel et al. 1996; Mulligan and Gorchov 2004; Bodeker et al. 2014; Castle et al. 2014; Baral et al. 2015). Further, the expansion of commercial harvesting may itself reduce economically marginalized collectors', especially women's, access to suitable land, as powerful individuals see more profit to be made from restricting access (e.g. Elias and Carney 2005).

For products that are still largely collected in the wild, commercial traders can help to arrange access to land and can train collectors in sustainable harvest practices, including field identification, while ensuring that scarcity does not lead to the substitution of other species. However, to supply material sustainably, it may be necessary to develop cultivation methods, when feasible, and provide financial support to enable farmers to begin cultivation, especially of perennials and species that require infrastructure such as shade houses. Such contract farming is well developed in the food and spice industry but less so in the medicinal and cosmetic value chain. For example, the supply of organic paprika in Zimbabwe is being stimulated via contract farming agreements. However, contract farming has challenges, including low prices paid to local producers, the frequent disappearance of contract schemes, and the need for farmers to be able to afford and maintain locally appropriate cultivation technologies if they are to produce material in quantity (e.g. Hanlon and Smart 2013).

For commercial use, government regulations in many countries require that each ingredient batch must be tested for identity and purity, which includes not only the absence of excessive material of the wrong species or incorrect plant parts, but of excessive levels of bacteria, mold, and often heavy metals or pesticides. Identity will often be adequately determined by morphology if material is minimally processed at the time of acquisition, but otherwise, some combination of chemical and perhaps validated molecular methods will be necessary. Methods range from the relatively low-tech and familiar, such as High-Performance Thin Layer Chromatography (HPTLC), to the elaborate and very costly, such as Nuclear Magnetic Resonance (NMR) (Reynertson and Mahmood 2015). Market demands for a specified content of putative active compounds or popular marker compounds also create increased need for chemical testing.

In the United States, Good Manufacturing Practices regulations for dietary supplements require manufacturers to conduct double identity testing, i.e. of ingredients when they are received and then again of the finished product whenever that is scientifically feasible. Even when raw materials have been adequately identified by classical botany or macromorphology, the second round of identity tests on the finished product will typically be required to include chemical fingerprinting, as morphological features and often DNA are no longer present. If a small business is not capable of developing in-house lab methods for the necessary tests, substantial fees will have to be paid to contract labs even if minimal methods, rather the most high-tech and costly methods, are used. This creates a substantial burden on start-up businesses. Further, many contract labs are not really familiar with botanical identification or taxonomy and lack adequate reference standards or well-validated methodologies, raising doubts about the value of those tests.

Opportunities for success in the marketplace are greatest when, rather than simply producing "me-too" products, a company can engage in successful innovation. Innovation can mean many things, but in this context, it refers to the development of new ingredients not previously commodified (at least in the region where it will be marketed), or of formulas that combine familiar ingredients in new combinations or for new uses. However, with greater requirements or public expectations for quality standards, innovation is reduced and the use of lesser-known plants is reduced as money is spent preferentially on larger, better-known commodities. Cooper (1999) has identified several critical factors for successful product innovation in many industries, which we think can be applied in the present context. These include conducting in-depth upfront homework; detailed market studies, including customer tests and field trials; and developing differentiated products with unique customer benefits, among others. In addition, strategic organizational factors that support corporate innovation, such as management, cross-functional and multidisciplinary teams, and long-term vision and strategy, have been identified (Sun and Wing 2005). Those factors are generally lacking in the development of bioresource products because they are launched onto the market initially by user-innovators (as stated earlier) or by small enterprises or lone entrepreneurs. Enterprises which have followed well-thought-out pathways to market are few and frequently require large corporate partners with capital to invest. For example, L'Oréal has invested in a research and development facility in South Africa to develop culturally appropriate cosmetic products (Pitman 2016).

Introduction of a new ingredient to a marketplace is particularly challenging. First, there may be high regulatory barriers to overcome. For example, in the United States, the Dietary Supplement Health and Education Act grandfathered plants that were "marketed in the United States before October 15, 1994" (DSHEA 1994), so that they could continue to be used unless safety risks were demonstrated, whereas manufacturers who wished to use ingredients not proven to be so marketed had to provide the Food and Drug Administration (FDA) with evidence of their safety, which FDA has interpreted to mean safety as a conventional food. The Native American traditional practitioners whose knowledge is recorded in ethnobotanical studies usually did not sell their medicines to members of their own tribes for cash money, nor did they own and run stores of the kind that produced potentially surviving proofs of sale such as printed advertisements. Hence, species that are well documented as being widely used by indigenous people are legally treated as if they had never been used in the United States. If the history of human use as conventional food is not viewed as adequate, expensive long-term animal toxicological studies and chemical analyses demonstrating inter-batch consistency may be demanded.

European regulations are even stricter, making the legal introduction of species not previously used in Europe into a major category of finished products almost impossible. Products that qualify to be registered as traditional herbal medicinal products (THMP) are required to show usage for a generation (30 years) anywhere in the world, including at least 15 years in the European Union (European Parliament 2004). For products that meet all requirements for registration except the last, the state receiving the application can forward it to a European Union committee that may (or may not) choose to create a monograph for the product; after its completion, the member state would be free to choose to register the product. This severely limits the introduction of non-Western medicinal products into the large European market, making innovation less lucrative.

For ingredients that have not previously been commercialized, much more effort is required to develop methods of both sourcing and quality control. It is likely that such species have not been widely cultivated and that neither potential cultivation methods nor the levels of harvest that can be sustained from wild populations are well understood. There will be no validated chemical or molecular methods of authenticating raw material, and chemical fingerprinting as a proxy method of identification or an estimate of quality may initially be unfeasible because, if any characteristic compounds have been identified, reference standards are not likely to be commercially available. Hence, quality control will initially have to depend upon examination of unprocessed material by manufacturers. This may be beneficial for local small businesses, as shortening the supply chain reduces opportunities for contamination or adulteration of materials, and if the companies or cooperatives that supply the raw materials also process it, thus adding value, while documenting its identity, they will receive a much fairer share of the profits. It will be particularly important to understand traditional knowledge associated with the species' use, e.g. must it be collected at a particular season for maximum potency? Until chemistry and mechanisms of action are understood, such guides should be respected.

A case in point is *Hoodia gordonii*, a small succulent perennial that was traditionally used by the Khoisan people of Southern Africa to suppress hunger. Its efficacy as an appetite suppressant was established through research beginning in the 1960s by South African scientists who patented an active compound initially without acknowledging their debt to traditional knowledge or offering benefit-sharing, leading to charges of biopiracy (e.g. Amusan 2017). Its modern application has been in weight-loss products marketed as dietary supplements or the equivalent. However, several issues soon arose. Because of supply limitations, overharvesting became a problem, and cultivation in monoculture was not successful. Many products contained little if any of the marker compounds typically used for *Hoodia* (Avula et al. 2008; Vermaak et al. 2010); there was suspicion that products could be adulterated with related or unrelated succulents (e.g. Vermaak et al. 2010), while some unquestionably diluted the costly raw material down to a small dose that could not be expected to have much effect (Pereira et al. 2011). Additionally, consumers may have had unrealistic expectations of the products, and Western media emphasized potential safety risks and reported side effects (Blom et al. 2011). Soon, *Hoodia* was discredited as a weight-loss product and sales figures declined precipitously. If investments in sustainable supply had preceded widespread marketing, this might not have happened.

If the history of human use is not sufficient to support a presumption that the ingredient is likely to be safe under the circumstances of commercial use (considering preparation methods and dose, but also expected potential users and duration of use), then animal toxicology tests should be done before a product is marketed, as regulations will often require (Naidoo and Seier 2010). Further, the marketing for a novel ingredient or formulation will be greatly enhanced if there are animal studies demonstrating benefit for the intended use, if animal models are available, or ideally, in nations where it is feasible, one or more human clinical trials. Even in nations that do not permit manufacturers to make disease claims or acknowledge the existence of such studies, consumers do hear about and value them. The existence of such data also increases the expectation that methods will be developed to screen batches of raw material for adequate potency, either by identification and quantification of marker compounds or by use of a bioassay considered relevant to the in vivo activity.

Clinical trials are particularly expensive, and pose legal as well as practical and economic challenges. Among these are the fact that a manufacturer who funds a clinical trial of a traditional product will not be rewarded with a period of exclusive marketing, because its use cannot be patented, although testing of a proprietary formula may benefit a manufacturer. In the United States, an Investigational New Drug application (IND) must be filed with the FDA for any substance to be used in a clinical trial with disease end points. The IND must characterize the substance and sourcing in sufficient detail to demonstrate the ability to obtain virtually identical batches in future. Worse, US law permanently prohibits the use in foods or dietary supplements of a substance not previously marketed as such in the US that has ever been the subject of clinical investigation; for dietary supplements, the law also includes the words "authorized for investigation". The FDA chooses to interpret this to mean that on the date when an IND becomes active, if the product is not already legally sold as a dietary supplement it may never be marketed as such by any business. This will apply even if no human study is ever conducted or a a drug developed (FDA 2013). Botanical drugs are almost never approved in the United States, so very few herbal medicines will ever be saleable except as conventional foods, such as teas, or as dietary supplements. Thus it is imperative that any new medicinal species or product be legally introduced into US commerce as a food or dietary supplement before any steps towards conduct of human studies in the United States are taken. An attempt to place high-quality research before marketing, which some would consider most logical, would usually result in the permanent prohibition of the product. (This legal regime also creates the opportunity for malevolent actors to file INDs that they do not even intend to utilize, merely to bar other businesses from marketing imported botanicals that foreign research has shown to be useful alternatives to drugs marketed by the IND holders; it is not clear whether such stratagems are yet being adopted.)

Thus, in an industry with minimal intellectual property protections and low return on investment, it is a rare executive team that sees beyond familiar products. The discovery of a new ingredient or new formula takes a large investment of time, money, and human resources. However, ITM is an invaluable source of new ingredients and formulas that would never be developed *de novo* by industry itself. Multispecies formulas enjoy an increasingly large share of the market (Smith et al. 2016), buttressed by the increasing number of Western clinical trials examining such products (e.g. Echinaforce® Hotdrink [Rauš et al. 2015]) as well as the ongoing interest in and scientific study of Traditional Chinese Medicine and Ayurveda. A brand-named novel combination thus has the potential to attract substantial interest.

THE SUPPORT REQUIRED FOR THE COMMERCIALIZATION OF INDIGENOUS PRODUCTS

The successful introduction of a new product into international commerce requires upfront expenditures that are simply not feasible under most circumstances for local holders of traditional knowledge. Therefore, the single greatest need is for links to be formed between indigenous knowledge systems (IKS) holders and commercial partners or universities. It is essential that these links be equal partnerships rather than relationships in which knowledge holders are viewed primarily as raw material suppliers and have little influence over business activities. If they are not equal partners, local suspicion of outside participants will persist. Recommendations for priority action include the following:

- Develop regulatory regimes that protect both knowledge holders and commercializing partners. As a backlash against past exploitative practices, recent regulatory regimes are often overprotective, which has resulted in the sector being highly politicized, overly restrictive, and unattractive to commercial interests. An example of this is the promulgation of Access and Benefit Sharing legislation in South Africa that made research and development difficult and complicated (Crouch et al. 2008).

- Provide education and capacity building to both partners in a potential relationship so that each has an understanding of what the other does and of each party's roles and responsibilities. This will lessen suspicion and manage expectations.
- Invest in modern science to be used in quality management of products developed from indigenous resources. This can be done through centres of excellence where collaboration between IKS practitioners and scientists or industry can be nurtured. For example, indigenous partners should receive assistance to develop methods for quality assurance, authentication, and maintaining a chain of custody. However, demanding standards that would be too burdensome for many knowledge holders to meet should not be imposed without clear need.
- Invest in the resources and knowledge necessary to ensure sustainable supply. Collection or cultivation should be beneficial to communities of traditional knowledge holders. Care must be taken to avoid encouraging farmers to spend time and money cultivating more material than the market needs, leading to serious financial losses when the price of raw material drops. Local value-added processing should be developed when possible to increase the percentage of income going to local communities and reduce the percentage of their income that is reliant upon production of raw materials.
- Create market access for IKS products. Linking products and IKS knowledge to key markets in the West through ethical/fair-trade partners is important to ensure a just distribution of profits.

CONCLUSION

The commercialization of indigenous knowledge and/or resources remains a highly controversial area. Many activists and IK practitioners contend that this knowledge is sacred and cannot/should not be commercialized. However, it can also be argued that if IK is not commercialized it is likely to be lost and no benefit will accrue to the community. There is therefore a need to ethically navigate this sensitive terrain for the good of the community.

REFERENCES

Amusan L (2017) Politics of biopiracy: An adventure into *Hoodia*/Xhoba patenting in southern Africa. *Afr J Tradit Complement Altern Med* 14:103–109.

Avula B, Wang YH, Pawar RS, Shukla YJ, Smillie TJ, Khan IA (2008) A rapid method for chemical fingerprint analysis of *Hoodia* species, related genera, and dietary supplements using UPLC-UV-MS. *J Pharm Biomed Anal* 48:722–731.

Baldwin C, Hienerth C, von Hippel E (2006) How user innovations become commercial products: A theoretical investigation and case study. *Res Policy* 25:1291–1313.

Baral B, Shrestha B, Teixeira da Silva JA (2015) A review of Chinese *Cordyceps* with special reference to Nepal, focusing on conservation. *Environ Exper Biol* 13:61–73.

Blom WA, Abrahamse SL, Bradford R, et al. (2011) Effects of 15-d repeated consumption of *Hoodia gordonii* purified extract on safety, ad libitum energy intake, and body weight in healthy, overweight women: A randomized controlled trial. *Am J Clin Nutr* 94:1171–1181.

Bodeker G, van't Klooster C, Weisbord E (2014) Prunus africana (Hook. f.) Kalkman: The overexploitation of a medicinal plant species and its legal context. *J Altern Complement Med* 20:810–822.

Booker A, Johnston D, Heinrich M (2016) The welfare effects of trade in phytomedicines: A multi-disciplinary analysis of turmeric production. *World Dev* 77:221–230.

Castle LM, Leopold S, Craft R, Kindscher K (2014) Ranking tool created for medicinal plants at risk of being overharvested in the wild. *Ethnobiol Lett* 5:77–88.

Cooper RG (1999) The invisible success factors in product innovation. *J Prod Innov Manag* 16:115–133.

Crouch NR, Douwes E, Wolfson MM, Smith GE, Edwards TJ (2008) South Africa's bioprospecting, access and benefit-sharing legislation: Current realities, future complications, and a proposed alternative. *S Afr J Sci* 104:355–366.

Dietary Supplement Health and Education Act of 1994 (DSHEA) (1994) Dietary Supplement Health and Education Act of 1984. Public Law 103-417. 103rd Congress. Available via Office of Dietary Supplements. https://ods.od.nih.gov/About/DSHEA_Wording.aspx. Accessed 14 December 2017.

Elias M, Carney J (2005) Shea butter, globalization, and women of Burkina Faso. In: Nelson L, Seager J (eds), *A Companion to Feminist Geography*. Blackwell Publishing Ltd., Oxford, UK, pp. 93–108.

European Parliament (2004) Directive 2004/24/EC of the European Parliament and of the Council of 31 March 2004 amending, as regards traditional herbal medicinal products, Directive 2001/83/EC on the Community code relating to medicinal products for human use. Available via the European Commission, Directorate-General of Health and Food Safety. https://ec.europa.eu/health/human-use/herbal-medicines_en. Accessed 14 December 2017.

Falkenberg T (2012) Traditional and complementary medicine policy. In: Management Sciences for Health (ed.), *MDS-3: Managing Access to Medicines and Health Technologies*. Management Sciences for Health, Arlington, VA, pp. 5.1–5.17.

Food and Drug Administration (FDA) (2013) Investigational new drug applications (INDs) – Determining whether human research studies can be conducted without an IND. Guidance for clinical investigators, sponsors, and IRBs. Available via U.S. Food and Drug Administration. https://www.fda.gov/Drugs/GuidanceComplianceRegulatoryInformation/Guidances/ucm064981.htm. Accessed 14 December 2017.

Hanlon J, Smart T (2013) Making money farming in Manica. Research Report 2. Available via The Open University. http://www.open.ac.uk/technology/mozambique/mozambique-reports-and-documents. Accessed 13 Decemeber 2017.

Joshi RG, Chelliah J (2013) Sharing the benefits of commercialization of traditional knowledge: What are the key success factors? *Intell Prop Forum* 93:60–66.

Mulligan MR, Gorchov DL (2004) Population loss of goldenseal, *Hydrastis canadensis* L. (Ranunculaceae), in Ohio. *J Torrey Bot Soc* 131:305–310.

Naidoo V, Seier J (2010) Preclinical safety testing of herbal remedies. In: Katerere DR, Luseba D (eds), *Ethnoveterinary Botanical Medicine: Herbal Medicines for Animal Health*. CRC Press, Boca Raton, FL, pp. 69–94.

Nantel P, Gagnon D, Nault A (1996) Population viability analysis of American ginseng and wild leek harvested in stochastic environments. *Conserv Biol* 10:608–621.

Nehal N (2013) Knowledge of traditional fermented food products harbored by the tribal folks of the Indian Himalayan belt. *Int J Agric Food Sci Technol* 4(5):401–414.

Pereira CA, Pereira LLS, Corrêa AD (2011) High performance liquid chromatography (HPLC) of *Hoodia gordonii* commercial powder. *J Med Plants Res* 5:5766–5722.

Pitman S (2016) L'Oréal raises the bar in Africa with new R&D facility. Available via https://www.cosmeticsdesign-europe.com/Article/2016/05/24/L-Oreal-raises-the-bar-in-Africa-with-new-R-D-facility. Accessed 15 December 2017.

Pouliot M, Elias M (2013) To process or not to process? Factors enabling and constraining shea butter production and income in Burkina Faso. *Geoforum* 50:211–220.

Rauš K, Pleschka S, Klein P, Schoop R, Fischer P (2015) Effect of an echinacea-based hot drink versus oseltamivir in influenza treatment: A randomized, double-blind, double-dummy, multicenter, noninferiority clinical trial. *Curr Therapeut Res* 77:66–72.

Reynertson K, Mahmood K (eds) (2015) *Botanicals: Materials and Techniques for Quality and Authenticity*. CRC Press, Boca Raton, FL.

Smith T, Kawa K, Eckl V, Johnson J (2016) Sales of herbal dietary supplements in the U.S. increased 7.5% in 2015. *HerbalGram* 111:67–73.

Sun H, Wing WC (2005) Critical success factors for new product development in the Hong Kong toy industry. *Technovation* 25:293–303.

Vermaak I, Hamman JH, Viljoen AM (2010) High performance thin layer chromatography as a method to authenticate *Hoodia gordonii* raw material and products. *S Afr J Bot* 76:119–124.

World Health Organization (WHO) (1998) *Quality Control Methods for Medicinal Plant Materials*. World Health Organization, Geneva, Switzerland.

World Health Organization (WHO) (2003) *WHO Guidelines on Good Agricultural and Collection Practices (GACP) for Medicinal Plants*. World Health Organization, Geneva, Switzerland.

World Health Organization (WHO) (2007) *WHO Guidelines for Assessing Quality of Herbal Medicines with Reference to Contaminants and Residues*. World Health Organization, Geneva, Switzerland.

World Health Organization (WHO) (2017) R-draft: WHO guidelines on good herbal processing practices for herbal medicines – revised draft for comments, March 2017. World Health Organization, Geneva, Switzerland. Available via http://www.who.int/medicines/areas/quality_safety/quality_assurance/projects/en/. Accessed 13 December 2017.

CHAPTER 15

The State of Traditional Leadership in South Africa from Colonialism and Apartheid to Democracy

Mothusiotsile E. Maditsi, David R. Katerere, and Phillip F. Iya

CONTENTS

ABSTRACT

Background Do traditional leaders have a role to play in local governance of today's constitutional democratic South African society? Due to South Africa's history, and the power that the apartheid regime bestowed on traditional leaders, the country is in a unique situation. This article is based on the premise that the customary system of African traditional leadership is facing extinction and is not fully acknowledged by government. The overall aim of this article is, therefore, to analyse the role of traditional leaders and to understand and analyse the legislative framework surrounding the institution of traditional leadership in South Africa.

The authors contend that traditional leaders are part of Africa's long-standing heritage and that their system of governance is important to many rural communities on the continent and in South Africa in particular. This is because there is a huge link between indigenous knowledge systems (IKS) and governance. IKS is crucial in terms of decision making in rural communities, thus indigenous knowledge (IK) plays a significant role in justifying and legitimising the role of traditional leaders in local governance, especially in rural communities.

A qualitative approach to investigating the problem was applied. This was mainly because studying traditional governance systems requires in-depth research. A case study utilising a qualitative approach was employed, as case studies bind the sequence of events together to form a cohesive story.

Relevance This chapter reveals that many people, especially those still residing in rural communities, have confidence in traditional leaders and still view their institution as relevant even in modern times. To many rural communities, these leaders are a source of ethnic and cultural identity and development. Therefore, the chapter recommends developing capacity building for traditional leaders in order to equip them with necessary governance, administrative, and legal knowledge and skills.

INTRODUCTION

For many centuries – and to this day – the institution of traditional leadership has been a big part of the lives of people, not just in Africa. It is a form of governance found in traditional societies across the world. One of the most successful democracies, the United Kingdom, is anchored by traditional leadership in the form of the British Monarchy. To date, Europe has 12 monarchies (traditional leaders) of which 10 are hereditary. Thus, universally, traditional leaders are established by virtue of ancestry, and the person who occupies the "throne" (the "stool" in the case of Africa) leads the ethnic community. Traditional leaders were and are still appointed in accordance with the customs and traditions of a particular area and are responsible for performing the cultural, customary, and traditional functions (Weber, 2007:5).

The origin of traditional leadership in Africa can be traced to pre-colonial times. While the organisation structure did not remain intact through colonialisation and post-colonisation, traditional leadership worked and is still relevant to most indigenous African communities, especially those resident in rural areas where customary law systems still exist (Cele, 2011:5). Furthermore, as in other parts of Africa, colonialism and apartheid resulted in South Africa's traditional leaders being co-opted by the colonial powers to govern rural areas. From the 1950s, under the apartheid government, development of legislative and administrative structures in the so-called Bantustans (or black homelands) saw traditional leaders being used in cynical ways, which implicated chiefs deeply in the apartheid government (Cele, 2011:5).

As an institution, traditional leadership represents the early form of societal organisation and good governance, and embodies the preservation of culture, customs, traditions, and values. However, the introduction of colonialism, and especially the apartheid era, legalised and institutionalised racial discrimination. As a result, the apartheid government created Bantustans based on the language and culture of a particular ethnic group. For example, Bophuthatswana was established as a cultural state under the leadership of Chief Lucas Mangope for Setswana speakers (Khunou, 2009:1).

In light of the co-opting of traditional leaders to prop up apartheid, this chapter aims to explore the status and role of the institution of traditional leadership in modern, democratic, and constitutional South Africa. It also seeks to analyze contemporary problems such as the negative impact of legislation and policies towards traditional leadership and the undefined role that traditional leaders have to play in governance (Logan, 2008:3). Therefore, the following questions arise: What are the challenges and prospects of traditional leaders in local governance? How do policies and legislation at national, provincial, and local government impact on the role of traditional leaders in the governance of local communities? How is the relationship between traditional leaders and local government defined and practised in South Africa? And what should be the role of traditional leaders in the new democratic South Africa? The following discussion will attempt to provide some answers to the above questions and will focus on:

- Understanding the concept of traditional leadership in South Africa
- Application of the concept of traditional leadership towards the transition to constitutional democracy in South Africa
- The emerging challenges of the application of traditional leadership in South Africa: an overview of socio-political, legal and practical application.

UNDERSTANDING THE CONCEPT OF TRADITIONAL LEADERSHIP IN AFRICA

Nature and Characteristics

Traditional leaders include village headsmen, chiefs, and kings. They have been an integral part of the lives of Africans for centuries. In the main, they were male and would serve as political, social, judicial, and religious heads of the group (Ayittey, 1999). The literature reveals information on the emergence of traditional leaders and the institution of traditional leadership and how these institutions have transformed over the years. Throughout the history of Africa, traditional leaders have been the basis of local governance, especially in peripheral rural areas of Africa. Traditional leaders have served through wars, slavery, famine, freedom struggles, economic and political restructurings, and during pre-colonial and colonial eras (Tlhoale, 2012:7).

Role of the Institution of Traditional Leadership in Pre-Colonial Africa

Before colonialism was introduced to Africans, in general and South Africa in particular, social organisations were characterised by patriarchal systems of tribal regimes. However, women played advisory roles at both family and political levels (Machingura, 2012). There is much evidence to demonstrate that during that period, a significant proportion of the Southern African population was organised into political groupings with centralised authority vested in hereditary leaders known in Setswana as "Kgosi" (Khunou, 2009:94). During this period, the roles of traditional leaders were to serve as political, military, spiritual, and cultural leaders. They were responsible for looking out for the interests of the people and the community at large. The welfare of the entire community depended on traditional leaders, especially on issues such as land acquisition and agriculture, in order to sustain the livelihood of communities (Rugege, 2003:172).

Traditional leaders and traditional authorities were important institutions which gave effect to life and played an essential role in the day-to-day administration of their areas and the lives of their people. There was an important relationship between the leader and community, and the normal functioning of the traditional community was the responsibility of the traditional authority. The traditional authorities (traditional leaders and various councils) were the governments of their respective indigenous communities, and a traditional leader was accountable to his subjects and the ancestors (Khunou, 2009:94). Political organisation began at village level, which was made up of extended families who chose a village head. Spiegel and Boonzaier (1988:49) argue that during the pre-colonial period, the power of communities (several villages) was centralised and vested in hereditary rulers known then as "chiefs". This meant that the chief was the highest authority in the territory. The chief performed various functions with the cooperation of several advisory councils – inner (privy council), council of elders, and village assembly of all residents (commoners) (Ayittey, 1999). For example, among the Zulu, the king was assisted in decision making by the *ibandla* (general council) and the Swazi king utilised the *liqoqo* (inner council) and the *Libandla* (general council) (Olivier, 1969).

Decisions would be made by consensus, therefore inculcating participatory democracy as opposed to parliamentary democracy, which the Europeans came with. The system also included checks and balances to prevent abuse of power (Ayittey, 1999; Düsing, 2002).

Davison (1992) and Tlhoale (2012), in their studies, identify the three important principles which stressed and underpinned the functions and existence of traditional leaders during the pre-colonial era, as follows:

- The unity of purpose acknowledging the supremacy of the "golden stool" (throne);
- The unifying force depended on participation which must be publicly seen as working; and
- The principle of systematic distrust of power with systems that had built-in mechanisms to prevent the abuse of power.

TRADITIONAL LEADERSHIP IN SOUTH AFRICA'S POLITICAL TRANSITION

Impact of Colonialism on Traditional Leadership in South Africa

The arrival of Jan van Riebeeck in 1652 marked the initial stage of colonialism in South Africa. Although history records Riebeeck as an explorer, he initially opened the floodgates to colonialism and subsequently apartheid in South Africa. For outsiders at that time, the local people had no apparent form of formal governance; they instead vested all power and authority in their traditional leaders (Tlhoale, 2012:20). The people lived in small tribal formations under tribal authority, but this system changed when the colonisers landed on the shores of what was later known as the Cape Colony (Roodt, Rusch and Tandy, 1993:19).

As previously alluded to, in the pre-colonial period traditional authorities were accountable to their communities. However, this system was either abolished or co-opted to extend colonial and apartheid rule, and the power of traditional leaders was significantly reduced by the colonial masters. This meant that they were no longer accountable to the communities but were now accountable to the colonial authorities (Khan and Lootvoet, 2001:3). Those that resisted, like Bambatha, were removed. The traditional leaders who were left had little power to allocate and distribute land that they still had under their control, since most of it had been taken by the colonizers. The responsibility of these traditional leaders was minimised to the point where they no longer had authority or the capacity to address developmental issues in their respective areas (Khan and Lootvoet, 2001:3).

The colonialists, therefore, co-opted traditional leaders to govern rural areas. The legislative and administrative structures in Bantustans saw traditional leaders being used in cynical ways from the 1950s, which deeply implicated traditional leaders even more in the apartheid government. This meant that traditional leaders turned from serving their people to serving their colonial masters or governments. The apartheid system turned traditional leaders into civil servants, which meant that they could be fired and hired and were paid by the colonial government (Cele, 2011:5–6).

African traditional government was systematically weakened, and the bond and relationship between traditional leaders and the people went sour. The colonial masters made sure that the Africans were subjugated and conquered in order to manipulate them (Maubane, 2007:4). The aspect of divide and rule was a major campaign, which was basically aimed at disrupting and destroying the African people and their way of life. The colonial masters made it a point that they turned the people against their traditional leaders and traditional leaders against their people, by making them agents of the colonial government, as long as they served the interest of their colonisers (Seiler, 2000:10).

A good example is the case of what happened in the homelands (Bantustans) whereby traditional leaders adopted and promoted the cause of colonisers. This resulted in a perversion of their roles (Spiegel and Boonzaier, 1988:49). They became enforcers of apartheid laws and policies, and helped to contain the nationalists. In the Eastern Cape, for example, Kaiser Matanzima became a key ally and collaborator of the apartheid government in suppressing political activity and silencing of numerous opponents to the regime (Koelble, 2004).

Roodt et al. (1993:19) gives an overview of the Black Administration Act 38 of 1927, and according to Section Two (2) of the Act, "the Governor-General was made supreme chief of all traditional leaders in the Union of South Africa". Through this Act not only were the colonial masters able to control the indigenous population but the colonial government was also able to implement more legislation and policies which changed the pre-colonial organs and structures of rural communities and the roles and powers of traditional leaders in general.

It is clear from the discussion above that the colonial era disorganised, conquered, and subverted traditional leaders. The implementation of many legislative measures that basically removed the roles, power, and authority of traditional leaders in their rural communities eroded the foundation by which the institution of traditional leadership at large was established and built upon (Ntsebeza, 2003:69).

After the enactment of the Self-Governing Territories Constitution Act 21 of 1971, traditional leaders were given direct recognition in the legislative assembly. Khunou (2009:94) notes that Bophuthatswana, as one of the Bantustans during the colonial era, became a self-governing homeland in 1972. After achieving this milestone, Bophuthatswana also gained its nominal independence on 6 December 1977. Bophuthatswana consisted of tribal land administered by so-called tribal authorities. Khunou further maintains the gaining of its independence was achieved through the enactment of the status of the Bophuthatswana Act 89 of 1977. However, the Act did not directly stipulate or outline and define the roles, functions, and powers of traditional leaders.

In order for the institution of traditional leadership to be regulated in Bophuthatswana, the Bophuthatswana Traditional Authorities Act was introduced. The Act prescribed the roles, powers, and functions of traditional authorities. Through this Act, traditional leaders were also made ex-officio members of the Bophuthatswana Parliament (section 27 of Bophuthatswana Traditional Authorities Act 23 of 1978). Even though traditional leaders were recognised through legislative measures on a large scale in the Bophuthatswana area, many argue that the Bophuthatswana Traditional Authorities Act was basically a replica of the Black Administration Act, as it stipulated that the Bophuthatswana President had the power to depose and appoint a traditional leader (Khunou, 2009:95). Chief Lucas Mangope was the president of Bophuthatswana until 1994, and during his term as president, he emphasised the ethnic origin of the Batswana nation (Khunou, 2009:94).

A good example as argued by Khunou is the case of the Bafokeng traditional leader, Chief Lebone, who defied Chief Mangope and refused to hoist the Bophuthatswana flag at the Bafokeng Tribal Offices (Khunou, 2009:95). Kgosi Lebone also instructed the Bafokeng to relinquish their Bophuthatswana citizenship.

From Apartheid to Constitutional Development

The convening of the Conference for a Democratic South Africa (CODESA) on 25 December 1991 pinpointed the entrance of South Africa to the last mile, which was aimed at extending political rights to all citizens of the country. On 17 March 1992, a Whites-only referendum voted in favour of continuation of the negotiations. The progress was increased by the convergence of opinions from two major parties, namely, the African National Congress (ANC) and the National Party government (Currie and de Waal, 2001).

CODESA was divided into five working groups for the purpose of substantive negotiations. The groups were established to negotiate and present arguments to the plenary sessions of CODESA. The terms of reference included the following: the reincorporation of the Transkei, Bophuthatswana, Venda, and Ciskei (TBVC) Bantustans into mainland South Africa; creation of a transitional government to lead the country to democracy; a set of constitutional principles for drafting and adopting a new constitution; and the creation of a climate for political activity (Currie and de Waal, 2001).

South Africa's transition ground to a halt in mid-1992 after nearly two-and-a-half years of slow progress. On 22 December 1993, the tricameral Parliament adopted the Interim Constitution after months of negotiations at the World Trade Centre in Kempton Park, Johannesburg (Currie and de Waal, 2001).

After the transition into democracy in 1994, a great deal of effort was paid to issues surrounding traditional leaders and their role in the new democratic South Africa. The roles performed before and those they were supposed to perform in the new South Africa became a key discussion. As an indication that the government was committed to accommodating the institution of traditional leadership and securing a role for traditional leaders in the affairs of their communities, the National Department of Traditional Affairs (DTA) was established. To legitimise the Department of Traditional Affairs, legislation was enacted to strengthen the system. The Council of Traditional Leaders Act of 1997, the House of Traditional Leaders, both provincial and national, and the Traditional Courts Bill, 2017, were enacted. Furthermore, the 1996 Constitution of South Africa (Chapter 12) acknowledges the institution of traditional leadership and its place in the democratic government.

The role of the Congress of Traditional Leaders of South Africa (CONTRALESA) was, at this point, critical for establishing the role of traditional authorities in both the 1993 Interim and the Final 1996 Constitution of the country (Oomen, 1996). The 1996 Constitution, in Sections 211 and 212, provides that the institution, status, and role of traditional leadership, according to indigenous law, shall be recognised and protected in the Constitution. It also provides that indigenous law, such as common law, shall be recognised and applied by the courts.

EMERGING CHALLENGES FOR THE APPLICATION OF TRADITIONAL LEADERSHIP IN SOUTH AFRICA

South Africa's transition to democracy brought about a number of measures that were meant to include and involve traditional leaders in the governance of their communities. Their participation in governance was valued, and in order for the government to show their commitment to the institution, a number of legislative and policy provisions have been developed and implemented. This is one strategy employed to legitimise traditional leaders' and their institutions' roles in governance, most especially in rural communities.

Socio-Political Perspectives

The Congress of Traditional Leaders of South Africa (CONTRALESA) is a non-governmental pressure group that was formed in 1987 by some traditional leaders of the homeland of KwaNdebele, with the support of the United Democratic Front (UDF) and the ANC. CONTRALESA assisted in the formation of the anti-apartheid front in the homelands, and continues to advocate greater rights for traditional leaders in the country post-apartheid (CONTRALESA, 2011). It also participates in both national and provincial gatherings of traditional leaders to garner support for legislation from non-CONTRALESA members.

CONTRALESA has also disagreed with many recent pieces of legislation which seek to advance the rights of women, but that in the process, tend to undermine traditional chiefs and leaders. Since 2008, when the Traditional Courts Bill (TCB) was shelved, much criticism had been levelled against the ANC by CONTRALESA for the stance it took in matters pertaining to traditional justice. CONTRALESA seeks a reform to the Bill, which it considers to be a threat to the authority and power of chiefs and to lack respect for traditional customs. They also see the Bill as deeply flawed and incorrect on many aspects of traditional justice, such as the representation of women within it (CONTRALESA, 2011).

Hweshe (2010) argues that while certain groups have enjoyed much benefit from the strong power of CONTRALESA, others have been ignored. For example, the Zulu Kingdom has had a very strong voice in political matters pertaining to the role of traditional leadership. The Khoi-San, on the other hand, have struggled to find a voice and have constantly been met with disregard for their demands to national government. In 2010, the Khoi-San decided to take legal action against the government for historic wrongs done to them and ongoing discrimination which they consider as "cultural genocide and discrimination against the Khoi-San Nation" (Hweshe, 2010).

The Impact of the Legal Framework

The transition from apartheid to democracy in South Africa saw a number of policies and legislation about traditional leadership being enacted. It is argued that some of these legal frameworks are not favourable to traditional leaders and their institution at large. Prominent traditional leaders across South Africa have argued that these legal frameworks give them too little power and minimise their role in governance.

The South African Constitution of 1996

Before the final Constitution of the country, there was an Interim Constitution of 1993 by which the Republic of South Africa was governed after the 1994 general elections. This Constitution required the Constitutional Assembly (CA) to draft and approve a permanent Constitution. After the new Constitution was drafted and certified by the Constitutional Court, the president of the country at the time, Nelson Mandela, signed it into law. The Constitution came into force on 3 February 1997 (Tlhoale, 2012).

The 1993 Interim Constitution paid more attention to the role and functions of traditional leaders compared to the 1996 Constitution, and provided them with more powers, as it retained some of the old pieces of legislation. Chapter 11 of the 1993 Interim Constitution (section 181), on the recognition of traditional leaders and indigenous law, stipulated that a traditional authority who observed a system of indigenous law and was recognised by law immediately before the commencement of the Constitution, should continue as such an authority and perform the powers and functions vested in it. Section 182 further stipulated that a traditional leader of a community observing a system of indigenous law and residing on land within the area of jurisdiction of an elected local government, should be entitled ex-officio to be a member of that local government and be eligible to be elected to any office of such local government. It was during that time that the controversial Ingonyama Trust was approved.

On the other hand, Chapter 12 of the 1996 Constitution devotes only two sections to traditional leaders (sections 211 and 212). These sections provide very little insight into the future of traditional leaders. The 1996 Constitution (section 211) stipulates as follows:

- That the institution, status and role of traditional leadership, according to customary law, are recognised, subject to the Constitution;
- That a traditional authority that observes a system of customary law may function subject to any applicable legislation and customs, which includes amendments to, or repeal of, that legislation or those customs; and
- That the courts must apply customary law when that law is applicable, subject to the Constitution and any legislation that specifically deals with customary law.

Section 212 of the Constitution, on the role of traditional leaders, also outlines the following:

- National legislation may provide for a role for traditional leadership as an institution at the local level on matters affecting local communities; and
- To deal with matters relating to traditional leadership, the role of traditional leaders, customary law and customs of communities observing a system of customary law:
  - National or provincial legislation may provide for the establishment of houses of traditional leaders; and
  - National legislation may establish a council of traditional leaders.

After the Constitution went into effect, many traditional leaders saw it as not being consistent with their traditional roles. As outlined above, the 1996 Constitution of the Republic of South Africa fails to give clarity on how traditional leaders fit into the governance system of local areas and how they are to participate in decision making at the local level. The Constitution appears to fall short in meeting the wishes and aspirations of traditional leaders nationwide. This shortfall is not viewed and noticed only by traditional leaders but has been recognised by scholars.

Nkosi Phatekile Holomisa, a prominent traditional leader and stalwart of CONTRALESA, wrote a memorandum on 6 May 1998 to President Nelson Mandela (republished in Holomisa, 2011) on matters of concern to traditional leaders. In that memorandum, he argued that the National House of Traditional Leaders should control its own finances and resources allocated to them:

> "CONTRALESA sees the government as not taking these structures with the seriousness they deserve. We see the Council and Houses as organs of state whose functions and responsibilities are of a full time nature. Yet, they are not given budgets of their own. They are made appendages of the Department of Constitutional Development and of one or other of the provincial governments" (Holomisa, 2011:152).

A close reading of Chapter 12 of the Constitution reveals that there is no sense of commitment which binds the government to any action whatsoever except for section 211 on matters of the courts. If the government was committed to ensuring a place for traditional leaders, it could have, at least, stipulated what the government was ready and willing to do for traditional leaders, subject to the Constitution, what the laws permit, and what resources can afford.

As indicated earlier, it is not only traditional leaders who are concerned about the role of the custodians of tradition; scholars also believe that Chapter 12 of the Constitution has inadequately addressed the roles and powers of traditional leaders. The Constitution, in sections addressing issues of local governance, also fails to indicate the roles of traditional leaders at that level, forgetting that local municipalities exist and local governance takes place in territories governed by traditional leaders themselves. In his memorandum to Dr Mandela, Holomisa argued for the incorporation of traditional leaders into democratic local governance:

> "There is currently a stalemate between traditional authorities and rural local councils. We have been, and still are, advocating for the recognition and transformation of

traditional authorities into democratic rural local councils. This can be achieved by having the majority of members democratically and directly elected by members of the community. The hereditary members, i.e. chief and headmen, would remain ex-officio members with full powers and privileges like everyone else" (Holomisa, 2011:51).

Craythorn (2003) observes that the aim of Chapter 12 of the Constitution is to prevent traditional leaders from inflicting traditional punishment. The aim was to greatly weaken the power of traditional leaders. Tlhoale (2012) also supports Craythorn by arguing that the 1996 Constitution of the Republic of South Africa fell far short of the expectations of traditional leaders. He maintains this is because only two sections are devoted to matters of traditional leadership, compared to Chapter 11 of the Interim 1993 Constitution, which contained four sections specifically for traditional leadership (Sections 181–184) He further observes that the two relevant sections of the 1996 Constitution give very little insight into the future of traditional leadership (Tlhoale, 2012).

The National House of Traditional Leaders Act 10 of 1997

The National House of Traditional Leaders Act 10 was established through the provision of the 1996 Constitution that stated that traditional leaders were to be recognised as a customary system of governance. The objectives of the Act were mainly to promote the role of traditional leaders within a democratic constitutional dispensation. Fostering unity and understanding among traditional leaders was one of the aims of the Act, as was enhancement of cooperation between the traditional council and various legislative houses with a view to addressing common matters of interest. The functions of the National House of Traditional Leaders include the following:

- It may advise the national government and make recommendations regarding:
  - Matters relating to traditional leadership
  - The role of traditional leaders
  - Customary law
  - The customs of communities observing a system of customary law
- It may investigate and disseminate information on the above-mentioned matters.
- At the request of the President, it has to advise him or her on any matter referred to it.
- It has to submit an annual report to Parliament (Du Plessis and Scheepers, 2000).

The Act had its shortfalls as well. There is no clear stipulation of the roles that traditional leaders are to play in the governance of local communities. This legislation also clashes with the Municipal Structures Act (discussed further below), which limits the participation of traditional leaders in municipal councils and their ability to vote on matters affecting their communities. The legislation above, however, stipulates that traditional leaders should submit an annual report to Parliament. What will go into that report if traditional leaders cannot fully participate in the governance of local areas?

National House of Traditional Leaders Act 22 of 1999

Following the implementation of the 1996 Constitution, the establishment of the National House of Traditional Leaders was subsequently followed by a second piece of legislation known today as the National House of Traditional Leaders Act. This Act was implemented after numerous letters and addresses were made by Nkosi Phatekile Holomisa at a number of events and formal gatherings. One of the papers presented by Holomisa was on accommodating the role of traditional leadership in the new dispensation, and one of the burning issues was addressed as follows:

> "Traditional authorities, in reality, constitute a form of local government. Both in terms of indigenous law and legislation, these authorities perform functions at local government with regard to development. The construction of roads, schools, clinics and other similar social amenities falls within the area of competence of traditional authorities" (Holomisa, 2011:120).

According to Section 11(1) of the Act, the powers and functions of the House are:

- Cooperation with provincial houses of traditional leaders, to promote the role of traditional leadership within a democratic constitutional dispensation;
- Nation building;
- Peace, stability and cohesiveness of communities;
- Preservation of the moral fibre and regeneration of society;
- Preservation of the culture and traditions of communities;
- Socio-economic development and service delivery;
- The social well-being and welfare of communities;
- The transformation and adaptation of customary law and custom so as to comply with the provisions of the Bill of Rights in the Constitution, in particular by:
  - Preventing unfair discrimination;
  - Promoting equality;
  - Seeking to progressively advance gender representation in the succession to traditional leadership positions; and
- Enhancing co-operation between the House and the various provincial houses with a view to addressing matters of common interest.

The Act, however, fails to enumerate specific roles of traditional leaders, especially when one looks at the issue of local governance. Traditional leaders are viewed by many people residing in rural areas as being legitimate leaders who are best suited to govern the communities; however, legislation and policies do not provide an active role for them and in fact appear to exclude them from constitutionally recognised governance structures.

The above Act stipulates that the function of traditional leaders is to promote socio-economic development and service delivery. This thus raises the issue of the type of resources which they are supposed to use to promote and carry out such functions. Local municipalities, together with the provincial and national governments, do not allocate resources that are controlled by traditional leaders. All resources, especially financial, are controlled by government departments, which in the long run, fail to work hand in hand with traditional leaders for the development of local areas. As noted above, Holomisa (2011) argued that the National House of Traditional Leaders and individual leaders should control the resources allocated to them so that they would have the power to improve delivery of services to rural communities.

Municipal Structures Act 117 of 1998

The implementation of the Municipal Structures Act was another means for the government to show their commitment to traditional leaders and their institution at large. The Act was enacted in order for traditional leaders to participate in the local governance of rural communities. Section 81(1) of the Act stipulates that traditional authorities that observe a system of customary law in the area of a municipality may participate through their leaders, identified in terms of subsection (2), in the proceedings of the council of that municipality, and those traditional leaders must be allowed to attend and participate in any meeting of the council.

Section 81(2) (a) and (b) further stipulate that the Member of the Executive Council (MEC) for local government in a province, in accordance with Schedule 6 and by notice in the Provincial Gazette, must identify traditional leaders, who in terms of subsection (1) may participate in the proceedings of a municipal council. The number of traditional leaders who may participate in the proceedings of a municipal council may not exceed 10% of the total number of councillors in that council, but if the council has fewer than 10 councillors, only one traditional leader may participate. Section 81(3) states that before a municipal council takes a decision on any matter directly affecting the area of a traditional authority, the council must give the leader of that authority the opportunity to express a view on that matter.

As found in the principles of this Act, traditional leaders must be given an opportunity to participate in municipal councils. This means that the role to be played by traditional leaders must be clarified so that they are able to make input on any service delivery enhancement processes of the municipality. However, the Act stipulates that the MEC for local government is the person responsible for identifying traditional leaders who can participate in the proceedings of a municipal council. This, therefore, marginalises other traditional leaders, who in many cases may not be very close to the MEC. Corruption in the formal government in South Africa is on the rise, and that is also one of the issues that makes other traditional leaders find themselves in the margins of government and local governance, in particular.

The quota of 10% representation in the municipal council reserved for traditional leaders by the Act is potentially problematic. This low representation limits the role and participation of traditional leaders in local governance of South Africa in general, as they may be overruled by elected municipal councillors should voting need to take place. It is evident that municipal councillors commonly side with the national government on issues affecting local areas, and only a few might support the views of traditional leaders. Setting up a traditional council at local level which interacts with the council on an equal footing might be advisable.

Traditional Leadership Governance and Framework (TLGF) Act 41 of 2003

The enactment of this Act was in line with section 212(1) of the Constitution, which stipulates that "national legislation may provide for a role for traditional leadership as an institution at local level on matters affecting the community". The aim of the Act was to harmonise the institution of traditional leadership with the new constitutional democracy created in 1996. The objectives of the Act are not merely to democratise traditional institutions but to constitutionalise them.

The preamble of the TLGF Act 41 of 2003 clearly and bluntly states those objectives (see below). It aims to shape traditional leadership in accordance with the Constitution. The Act also intends to reconcile customary law and practices with human rights and democratic imperatives (as acknowledged in the Preamble of the Act). In stating this, it creates the perception that customary law and traditional norms are, by their nature, undemocratic and inimical to human rights.

The objectives of the Traditional Leadership Governance and Framework Act 41 of 2003 are to:

- Set out a national framework and norms and standards that will define the place and role of traditional leadership within the new system of democratic governance;
- Transform the institution in line with constitutional imperatives; and
- Restore the integrity and legitimacy of the institution of traditional leadership in line with customary law and practices.

The Act complies with Section 211 of the South African Constitution in that it emphasises recognition of the status and role that the institution of traditional leadership plays. The Act aims to provide for the recognition of traditional communities around the country who still observe their customary judicial systems. Provision for the recognition of traditional leaders is also addressed by the Act as well as the establishment of traditional councils in communities living under customary systems. Provision for a statutory framework for leadership positions within the institution of traditional leadership is among the objectives set out by the Act. The legislation also provides for statutory frameworks that also address issues concerning the removal of traditional leaders from office (TLGFA, 2003).

The legislation also provides and advocates for houses of traditional leaders to be established and stipulates the functions and roles played by traditional leaders. Dispute resolution mechanisms and the establishment of commissions on leadership disputes and claims are also provided for. The Act also provides for matters connected to the institution of traditional leadership at large, including the provision for amendments to the Remuneration of Public Office Bearers Act of 1998.

There is great concern among supporters of traditional leadership that the legislation does not really allocate a proper role for traditional leaders. The burning concern is that the new law overlooks the powers of traditional leaders to rule and govern, and limits them to becoming ceremonial leaders who are only good for performing and administrating ceremonies in communities. The powers to govern and rule remain with the government, and that is also enshrined in the country's Constitution, which is viewed as the supreme law among all the laws of the land. It is, therefore, impossible for traditional leaders to assume their rightful positions, because they have not been assigned the real power and authority to govern, since they are not democratically elected and in no way constitutionally constituted as government (Tlhoale, 2012).

Some General Practical Problems

Although legislation exists to govern the incorporation of traditional leaders into the post-1994 democracy, intense debates on the issue continue. Traditional leaders contribute to several spheres of governance, but their role in crime prevention and the administration of justice is more pronounced. The key question should not be whether traditional leaders should perform such functions but how they can participate in the delivery of local safety (Tshehla, 2006). History teaches that traditional leaders have served in their respective communities in and around Africa with authority over all aspects of communal life. However, in contemporary society and under constitutional democracy, traditional leadership has had many challenges. Some challenges are institutional and governmental, while others are local. A major challenge arises from the recent history of how traditional leadership was perverted and used to support the colonial and apartheid edifice, and how it can now be repurposed. De-politicising it for the service of all citizens regardless of political party affiliation is key to this repurposing.

The need for better service delivery for communities is the first and foremost challenge experienced by traditional leaders. Residents lack consensus on what is most important to develop. An example can be that of people who suggest that schools, health facilities, and all other important infrastructure be developed in an area first, while others argue for neon lights and roads. Traditional leaders do not vote in municipal meetings, and that minimises their chance to advocate for a better budget for development of rural communities. They do not have a fixed budget that they work from but rely on municipalities and councillors to argue for the money and implementation of services. Most policies, legislation, and programs of government departments are developed to suit the government rather than traditional leaders. Such policies often hinder the developmental process and progress of rural communities.

CONCLUSIONS AND RECOMMENDATIONS

Conclusions

The above discussion was based on the role that traditional leadership as an African institution plays in the governance of rural communities. Traditional leaders, and their institution at large, constitute a system that cannot be discarded due to modernity and social urbanisation. Traditional leaders have served local indigenous people for centuries, which makes them and the role they play even more relevant in modern times. Most people residing in rural communities still believe in this system and acknowledge the low level of corruption among traditional leaders, thus making them the most legitimate officers who can foster development in rural areas.

Rural communities and the poor need development, and traditional leaders are the relevant people who can best foster development and progress in their communities. Traditional leaders could be trained, educated, and made aware of the needs of rural communities and how they could go about addressing the needs of their own communities. Natural resources available and accessible to rural communities should be utilised, and IK should be applied to foster development of rural areas. All these need the guidance of traditional leaders who are familiar with the practices of people residing in rural communities.

Laws that governed the institution of traditional leadership during previous governments gave direct power, authority, and functions to traditional leaders. However, the constitutional dispensation changed the position of traditional leaders. The post-apartheid legislation and policies provide no specific powers and functions to traditional leaders. Most development in rural communities is a result of the endeavours of traditional leaders and is carried out with the limited budgets obtained or received from the previous governments. However, the challenges posed by the legal framework have a negative bearing on the role that traditional leaders can play. This framework limits the participation of traditional leaders in governance and development, and that has to be addressed.

Recommendations

Based on the findings of the foregoing study, the following recommendations are advanced:

1. There is a need for capacity-building workshops for traditional leaders to equip them with administrative and legal knowledge and skills. Government needs to establish programmes that involve traditional leaders whereby they would train traditional leaders in the necessary skills to initiate and drive development projects in rural communities.
2. There is a need to organise short courses for traditional leaders on their rights and responsibilities as leaders of local communities. Universities should initiate and implement short courses aimed at equipping traditional leaders with information and management know-how, such as short courses on project management for traditional leaders.
3. There is a need to strengthen the relationship between traditional leaders and government officials in order to ensure proper communication between the two. Communication between traditional leaders and government officials is very poor; if traditional leaders and government officials become united and communicate on a regular basis, much faster development could be achieved.
4. There is a need for revision of policies and legislation on traditional leadership and their role in local governance in order to support traditional leaders and incorporate them in formal government structures. Most policies and legislation

enacted after the transition into democracy only acknowledged traditional leaders and their institution in general but did not afford them real power and authority to lead and effectively participate in governance structures at local and provincial levels.

5. Traditional leaders should be made aware (through short courses and seminars) of the value of incorporating IKS in development projects within rural communities.

Final Remarks

It is clear that traditional leaders could make valuable inputs in the identification of areas for development, as they have very intimate knowledge of the needs of their own communities. Better collaboration between municipal officials and traditional leaders could prove to have very worthy results in the development of rural communities.

Most traditional leaders and councillors are not appreciated by the government council, resulting in a reduction in development projects. Traditional leaders and the institution thereof can contribute greatly to development, and if they work together with elected municipal councillors, much can be achieved. This stems from the fact that traditional leaders have been around from time immemorial. Their understanding of the people and their needs and how they have survived the times of oppression enable them to talk to their people when times are tough and service delivery is not happening as expected. Their contributions could assist municipalities in providing quicker service delivery to the people.

Traditional leaders are important gatekeepers and community assets to the people they serve. They are the centre point for ancestral ceremonies and arbiters of legal and civil disputes. They should be empowered in creative ways so that they serve people and administer government programmes without being embroiled in partisan politics. Their function must be repurposed to deliver the democratic dividend to their communities in post-apartheid South Africa.

REFERENCES

Ayittey GBN (1999) *Africa in Chaos*. Palgrave Macmillan, New York.

Cele SB (2011) Discussion paper on the role of traditional leaders in democratic South Africa. Paper presented to South African Local Government Association (SALGA) conference, Durban, 2011.

CONTRALESA (2011) About us. http://contralesa.org/html/about-us/index.htm.

Craythorn D (2003) *Municipal Administration: A Handbook*. Juta and Co Ltd, Cape Town.

Currie I, de Waal J (2001) *The New Constitutional and Administrative Law*. Juta Academic, Lansdowne.

Davidson B (1992) *Black Man's Burden*. James Currey, London.

Du Plessis W, Scheepers TE (2000) Traditional leaders and development: recognition and the road ahead. Paper delivered at the Colloquium: Constitution and the Law, November 1999, Potchefstroom.

Dusing, S (2002) Traditional Leadership and Democratization in Southern Africa: A Comparative Study of Botswana, Namibia, and Southern Africa. Lit Verlag, Hamburg.

Holomisa P (2011) *A Double-Edged Sword: A Quest for a Place in the African Sun*. Real African Publishers, Houghton.

Hweshe F (2010) Bushmen, Khoisan sue State. http://www.sowetanlive.co.za/news/2010/09/06/bushmen-khoisans-sue-state.

Khan S, Lootvoet B (2001) Tribal authority and service delivery in the Durban uni-city. Paper presented at Gouvernaanceet en Afrique-Australe, November 2001, Lusaka.

Khunou SF (2009) Traditional leadership and independent Bantustans of South Africa: some milestones of transformative constitutionalism beyond apartheid. *Potchefstroomse Elektroniese Regsblad* 12(4):81–122.

Koelble T (2004) Democracy, Traditional Leadership and the International Economy in South Africa. CSSR Working Paper No. 114. University of Cape Town.

Logan C, (2008) Traditional Leaders In Modern Africa: Can Democracy And The Chief Co-Exist? Afro Barometer Working Paper No. 93.

Machingura F (2012) The Judas Iscariot Episode in the Zimbabwean religio-political debate of "selling out." In Gunda MR and Kugler J (eds). *The Bible and Politics of Africa*. University of Bamberg Press, Bamberg. 212–235.

Maubane PP (2007) The Fourth National Annual Local Government Conference of Traditional Leadership and Local Governance in a Democratic South Africa, Southern Sun Hotel-Elangeni, Durban, 30–31 July 2007.

Ntsebeza L (2003)Traditional authorities, local government and land rights. In: Ray DI, Reddy PS (eds). *Grassroots Governance? Chiefs in Africa and the Afro-Caribbean*. University of Calgary Press, Calgary. 173–226.

Olivier NJJ (1969). The governmental institutions of the Bantu peoples of Southern Africa. In *Recueils de la societies Jean Bodin XII*. Foundation Universitaire da Belgique, Bruxelles.

Oomen B (1996) Talking tradition: the position and portrayal of traditional leaders in present-day South Africa. MA Thesis, University of Leiden.

Roodt J, Rusch P, Tandy P (1993) *Rural Local Government*. Rhodes University, Grahamstown.

Rugege S (2003) Traditional leadership and its future role in local governance. *Law, Democracy & Development* 7(2):171–200.

Seiler J (2000) *The Role of Traditional Leadership in Democratic Local Government: A Guide for Councillors, 2000–2001*. CSIR, Pretoria.

Spiegel A, Boonzaier E (1988) *Promoting Tradition: Images of the South African Past*. David Phillip, Johannesburg.

Tlhoale CT (2012) The interface between traditional leadership in shared rural local governance. MA Thesis, University of Johannesburg.

Tshehla B (2006) Here to stay: traditional leaders' role in justice and crime prevention. *SA Crime Quarterly* 11:15–20.

Weber M (2007) Traditional authority in the modern state. *International Journal of Leadership in Public Services* 8(1): 4–20.

CHAPTER 16

A Brief Survey of Early Indigenous Knowledge Which Influenced Modern Agronomic Practices

Estonce T. Gwata

CONTENTS

ABSTRACT

Background Agricultural methods developed by early or prehistoric farmers are shown to provide the foundation for many modern agricultural practices. Important practices that derive from traditional agricultural knowledge include crop domestication and selection for desired characters; seed selection and protection during storage; preparation of land by terracing, irrigation, and mulching; and mixed or rotational cropping systems to reduce disease or preserve fertility.

Relevance The possible benefit and wider application of other inexpensive, sustainable traditional practices should be re-examined. This might result in earth-friendly agricultural practices.

INTRODUCTION

Modern agriculture in many parts of the world is strikingly different from early (prehistoric) forms of agriculture. In particular, the level of sophistication which is involved in modern agronomic practices is overwhelmingly superior in comparison with the simple methods that the early farmers used. However, there are numerous examples in the various disciplines of agriculture where the current sophisticated methods are clearly improved versions of the methods that were employed by prehistoric farmers. (For purposes of this discussion, "prehistoric," refers not only to peoples living in the very distant, mostly non-agricultural past, as is often implied by the common English usage of the word "prehistoric" but to any peoples who did not keep written history and so passed down their knowledge and skills as an oral tradition.) Therefore, this chapter highlights specific examples where early agricultural practices that were based on indigenous knowledge influenced modern agriculture, particularly in crop husbandry. It is important to emphasize now that the modernization of agricultural activities has been necessitated by various factors such as climate change, changing human needs, socio-economic imperatives, and access to advanced agricultural knowledge. The broad term of agricultural activity is used in this chapter to encompass a diverse spectrum of specialized agronomic activities which include crop production, genetic improvement of crops, land and water management, and seed systems. The development of early agricultural practices proceeded in a variety of ways that depended largely on indigenous communities and their knowledge about the existing environmental conditions, ecologies, and plants in their areas.

TERRACING, IRRIGATION, AND MULCHING

Archaeological studies show that inhabitants on hillslope regions in Belize resorted to terracing (Healy 1983) as well as canal construction in parts of northern Belize and Mexico (Pope and Dahlin 1989). Similar activities were carried out by prehistoric farmers in the Inyanga region in eastern Zimbabwe and the Engaruka in the northern region of Tanzania during the Iron Age (Sutton 1984). These activities indicated that the indigenous peoples in these contrasting regions had at least some basic knowledge of hydrology or irrigation, microclimate, and landscape modification to improve crop cultivation. By reducing surface runoff, the flatbeds that were created on terraces improved the conservation of both soil and moisture. The technique of conserving soil moisture using lithic mulch was practiced in geographically diverse dryland areas, including the Israeli Negev, the Peruvian Atacama, New Zealand, and China (Lightfoot 1996).

DOMESTICATION OF WILD CROP RELATIVES

The early domestication of wild relatives of modern crop plants provided mankind with the initial germplasm that was adapted to cultivation and utilization. For example, the domestication and selection of the potato crop (about 10,000 years ago) by the indigenous people of South America from a wild species native to highland regions of Peru, Bolivia, Ecuador, and Peru (Brush et al. 1981, Pearsall 2008) is recognized as one of the earliest forms of crop improvement (or breeding). The potato was a regional staple food, but the indigenous people observed that the wild potato was generally bitter and could even be toxic to humans due to the occurrence of glycoalkaloids (Gregory et al. 1981). The potato plant required significant levels of glycoalkaloids in order to resist insect pests and diseases. The indigenous people selected relatively less bitter types over generations and propagated them. Some of the indigenous people in the mountainous Altiplano region of the Andes also discovered a special type of clay that they would ingest after eating toxic potatoes in order to neutralize the glycoalkaloids (Johns 1986).

The indigenous farmers from the region were able to use their knowledge to correctly distinguish tetraploid (*Solanum tuberosum* ssp. *andigena*) and diploid (*S. goniocalyx, S. phureja*, and *S. stenotomum*) species without any modern methods such as chromosome counting or isozymes (Quiros et al. 1990). This indicated, at least in part, that folkloric methods of selection for this crop were effective. At present, distinguishing between diploid, triploid, and tetraploid forms of the potato is achievable through the use of a range of sophisticated approaches, including electrophoretic (Bauw et al. 2006), flow cytometric (Uijtewaal 1987), and cytogenetic (Dong et al. 2000) methods. DNA marker technologies (Bryan et al. 1999, Nakagawa and Hosaka 2002, Hardigan et al. 2017) are now used for rapid and accurate selections of distinct forms of the potato in genetic improvement programs.

RETENTION OF CROP SEED

Seed systems during the period of early agriculture consisted primarily of retaining part of the harvest for planting in the next season. In practical terms, the preferred seed was identified by farmers based on its outward appearance (phenotype). For instance, maize farmers would retain the large, plump kernels in the middle of the cob, discarding the small or shrivelled (shrunken) kernels elsewhere on the cob. The selection was often carried out after the harvesting but before shelling. In the case of small-grain cereals such as finger millet, the selected seed was stored either in sealed bottles or tins or clay pots, or simply stored as unthreshed panicles (Chigera et al. 2007) but separate from those for home consumption. Farmers evidently knew that the percent germination of the selected kernels was optimum in comparison with that of the discarded kernels. The ultimate goal of the farmers was to optimize germination, hence crop establishment and therefore the grain yield.

At present, commercial seed production involves rigorous steps (starting from the breeder seed, through the foundation, registered, and certified seed phases) that will facilitate the production of high-quality seed. The seed is often tested for germination and contamination by pathogens. These tests are often conducted under well-controlled laboratory conditions. The process of commercial seed production also requires grading of the seed in order to discard any damaged or shrunken seed, thus enhancing the crop quality and yield.

MIXED CROPPING SYSTEMS

Traditionally, prehistoric farmers from different parts of the world were inclined to practice mixed cropping systems in which several different crop species (for example, maize, cowpea, watermelon, and pumpkin) would be planted next to each other in the same field. There is no evidence today that shows unequivocally that farmers from the prehistoric era had an understanding of biological nitrogen fixation that influenced them to plant legumes with cereals, for instance. However, the farmers required fertile and healthy soils and can be presumed to have observed that mixed cropping improved soil health. As another approach to maintaining productivity, crop rotations that were introduced in Europe, for instance, during the Roman Empire (Ralls 2013), catered for both soil health (fallow phase) and improvement of soil fertility through the inclusion of a legume crop. In modern agriculture, cropping systems still generally follow rotations in which leguminous crops are often an integral part of the system.

The early farmers also practiced mixed cropping partly because they intended to safeguard against crop failure caused by a number of factors, such as diseases, insect pest damage, or drought. In some cases, farmers had limited arable land, hence more or less by default they resorted to mixed cropping systems in order to accommodate as many crop species as necessary per unit land area. In other words, the farmers utilized the available land to an optimum level, and often sustainably.

DROUGHT-TOLERANT CROPS

Early farmers domesticated and cultivated a wide range of cereals that are tolerant to drought. The domestication of maize from its wild progenitors is well documented (Benz 2005). The farmers allowed intermating between the selected plants, thus producing open pollinated varieties (OPVs). Although OPVs were not uniform genetically, they offered many advantages, including ecological plasticity, simple improvement procedures, a wide spectrum of disease resistance, and affordable seed production methods. Naturally, the OPVs were adapted to the local agro-ecological conditions, including soil moisture stress.

The quest to improve tolerance to drought in maize is still ongoing (James 2015), more so because of the increase in both the intensity and frequency of drought in many parts of the world. In addition to maize, farmers in some dry regions cultivated finger millet and sorghum, which were considered tolerant to drought. At present, there is still considerable emphasis on small-grain cereals as sources of drought-tolerance genes in breeding activities. The early farmers also selected annual crops that are relatively easier to manage under limited rainfall conditions. At present, many crop breeding programs select for early maturity (or short duration) in crop cultivars which can escape terminal season drought stress.

GRAIN STORAGE

The ability of early farmers to store harvested grain was one of the major developments for sedentism and farming across cultures. The ingenuity underlying the storage structures and methods that were used then clearly influenced modern grain storage facilities and methods. The farmers constructed simple structures using locally available materials such as bamboo sticks or reeds, mud, and thatch (Karthikeyan et al. 2009). Most of the structures were built on raised platforms in order to keep the grain away from moisture in the ground as well as minimize the damage by rodents. The farmers understood the need to dry the grain prior to storage. For example, in maize, the cobs could be dried in the field prior to shelling. Alternatively, drying was facilitated by hanging the cobs over a source of smoke or by placing the cobs in cribs in order to allow natural airflow. At present, large-scale drying of grain utilizes various types of grain dryers, including fluidized-bed dryers (Prachayawarakorn et al. 2004) and solar maize dryers (Gatea 2010). Some of the farmers preferred to use underground storage structures for storage of grain. It is tempting to think that this practice influenced the construction of the Svalbard global seed vault (Wendle 2018), which is the biggest modern underground seed storage facility in the world.

In order to protect the stored grain, some of the ancient farmers utilized botanicals such as neem (*Azadirachta indica*) and fish bean (*Tephrosia vogelii*) among others. There are numerous recent studies (e.g. Rajashekar et al. 2010, Khatun et al. 2011, Rayhan et al. 2014, Sori 2014) that have investigated the efficacy of botanicals in controlling insect pests that attack stored grain. These studies confirm and are based upon the knowledge of early farmers about the insecticidal properties (as repellents or sterilants) of such plants. In contrast, modern grain storage facilities often use harmful pesticides that pollute the environment.

CONCLUDING REMARKS

The selected examples of agronomic practices clearly demonstrate connections between the ingenuity of early and prehistoric farmers and modern practices in agronomic sciences. Indigenous knowledge in agronomy from ancient times evidently paved the way for modern methods that are applied in crop husbandry. In comparison with the methods that were used by early farmers, the current agronomic methods are generally more costly and not readily accessible. Some of the current methods, such as pesticide applications in granaries, are associated with negative effects on the environment, while in contrast, early farmers utilized safe methods to protect the stored grain. Probably there is strong merit in revisiting some of the simple agronomic practices that were employed by prehistoric farmers.

REFERENCES

Bauw G, Nielsen HV, Emmersen J, Nielsen KL, Jørgensen M, Welinder KG (2006) Patatins, Kunitz protease inhibitors and other major proteins in tuber of potato cv. Kuras. *FEBS J* 273(15):3569–3584.

Benz BF (2005) Archaeological evidence of teosinte domestication from Guilá Naquitz, Oaxaca". *Proc Natl Acad Sci USA* 98(4):2104–2106.

Brush SB, Carney HJ, Humán Z (1981) Dynamics of Andean potato agriculture. *Econ Bot* 35:70–88.

Bryan GJ, McNicoll J, Ramsay G, Meyer RC, DeJong WS (1999) Polymorphic simple sequence repeat markers in chloroplast genomes of Solanaceous plants. *Theor Appl Genet* 99:859–867.

Chigora P, Dzinavatonga N, Mutenheri F (2007) Indigenous knowledge systems and the conservation of Sangwe communal lands of Chiredzi in Zimbabwe. *J Sustain Dev Africa* 9(3):146–157.

Dong D, Song J, Naess SK, Helgeson JP, Gebhardt C, Jiang J (2000) Development and application of a set of chromosome-specific cytogenetic DNA markers in potato. *Theor Appl Genet* 101:1001–1007.

Gatea AA (2010) Design, construction and performance evaluation of solar maize dryer. *J Agric Biotech Sustain Dev* 2(3):39–46.

Gregor P, Sinden SL, Osman SF, Tingey WM, Chessin DA (1981) Glycoalkaloids of wild, tuber-bearing *Solanum* species. *J Agric Food Chem* 29:1212–1215.

Hardigan MA, Laimbeer FPE, Newton L, et al. (2017) Genome diversity of tuber-bearing *Solanum* uncovers complex evolutionary history and targets of domestication in the cultivated potato. *Proc Natl Acad Sci USA* 114(46):E9999–E10008.

Healy PF, Lambert JDH, Arnason JT, Hebda RJ (1983) Caracol, Belize: evidence of ancient Maya agricultural terraces. *J Field Archaeol* 10(4):397–410.

James M (2015) Water efficient maize for Africa hits key milestone for drought-tolerant maize. Grain SA. https://www.grainsa.co.za/water-efficient-maize-for-africa-hits-key-milestone-for-drought-tolerant-maize.

Johns TJ (1986) Detoxification function of geophagy and domestication of the potato. *J Chem Ecol* 12(3):635–646.

Karthikeyan C, Veeraragavathatham D, Karpagam D, Firdouse SA (2009) Traditional storage practices. *Indian J Tradit Know* 8(4):564–568.

Khatun A, Kabir G, Bhuiyan MAH, Khanam D (2011) Effect of preserved seeds using different botanicals on seed quality of lentil. *Bangladesh J Agric Res* 36(3):381–387.

Lightfoot DR (1996) The nature, history, and distribution of lithic mulch agriculture: an ancient technique of dryland agriculture. *Agric Hist Rev* 44(2):206–222.

Nakagawa K, Hosaka K (2002) Species relationships between a wild tetraploid potato species, *Solanum acaule* Bitter, and its related species as revealed by RFLPs of chloroplast and nuclear DNA. *Am J Potato Res* 79:85–98.

Pearsall DM (2008) Plant domestication and the shift to agriculture in the Andes. In: Silverman H, Isbell WH (eds), *The Handbook of South American Archaeology*. Springer, New York, pp. 105–112.

Pope KO, Dahlin BH (1989) Ancient Maya wetland agriculture: new insights from ecological and remote sensing research. *J Field Archaeol* 16(1):87–106.

Prachayawarakorn S, Soponronnarit S, Wetchacama S, Chinnabun K (2004) Methodology for enhancing drying rate and improving maize quality in a fluidised-bed dryer. *J Stored Prod Res* 40(4):379–393.

Quiros CF, Brush SB, Douches DS, Zimmerer KS, Huestis G (1990) Biochemical and folk assessment of variability of Andean cultivated potatoes. *Econ Bot* 44(2):254–266.

Rajashekar Y, Gunasekaran N, Shivanandappa T (2010) Insecticidal activity of the root extract of *Decalepis hamiltonii* against stored product insect pests and its application in grain protection. *J Food Sci Technol* 47(3):310–314.

Ralls KM (2013) Crop rotations have been around since Roman times. http://oregonstate.edu/dept/coarc/sites/default/files/may_2013_article.pdf.

Rayhan MZ, Das S, Sarkar R., et al. (2014) Bioefficacy of neem, mahogoni and their mixture to protect seed damage and seed weight loss by rice weevil in storage. *J Biodivers Environ Sci* 5(1):582–589.

Sori W (2014) Effect of selected botanicals and local seed storage practices on maize insect pests and health of maize seeds in Jimma zone. *Singapore J Sci Res* 4(2):19–28.

Sutton J (1984) Irrigation and soil-conservation in African agricultural history: with a reconsideration of the Inyanga Terracing (Zimbabwe) and Engaruka Irrigation Works (Tanzania). *J Afr Hist* 25(1):25–41.

Uijtewaal BA (1987) Ploidy variability in greenhouse cultured and in vitro propagated potato (*Solanum tuberosum*) monohaploids (2n=x=12) as determined by flow cytometry. *Plant Cell Rep* 6(3):252–255.

Wendle J (2018) 'Doomsday vault' protects Earth's food supply—here's how. *National Geographic*. https://www.nationalgeographic.com/environment/future-of-food/norway-svalbard-global-seed-vault.

CHAPTER 17

Applications of Indigenous Knowledges in the 21st Century

Soul Shava and Chamunorwa Togo

CONTENTS

ABSTRACT

Background Indigenous knowledges cut across several multidisciplinary sectors. Although they have made significant contributions in the global era, their role is largely unrecognized and undervalued.

Relevance This chapter reviews the role and applicability of indigenous knowledges in the modern era.

INTRODUCTION

Indigenous knowledges are bodies of knowledge that span different sectors of society, including biology, chemistry, physics, mathematics, ecology, environmental management, agriculture, food processing, medicine and pharmacology, architecture and construction, manufacturing, mining and metallurgy, technology/engineering, astronomy and meteorology, languages, history, philosophy, arts and culture, governance, economics, and trade and commerce. While there is (or despite the) growing recognition of the role of indigenous knowledge across different sectors of society, that recognition is paradoxically juxtaposed against Western narrative ecologies of continuing denigration of indigenous knowledges as primitive, accompanied by their continuing cannibalization into modern knowledge systems, innovations, and practices. Indigenous knowledges have made various significant contributions to the welfare of the global community, which are largely unacknowledged. However, the resistance to recognition and acknowledgement of the positive contributions of indigenous knowledges in present-day contexts can only be thwarted through reclaiming, reconstructing, rewriting, and validating indigenous knowledges (Shizha, 2010). This requires the continued generation of counternarratives and production of a cumulative body of evidence that reveals the continuing role of indigenous knowledges in the present era. This chapter seeks to add voice to the emerging decolonial and anticolonial counternarratives in support of indigenous knowledges by drawing attention to evidence of indigenous innovations, namely, practices, tools, techniques, strategies, and intellectual resources (Emeagwali, 2014) and their contributions (applications) across different sectors in the 21st century.

ENVIRONMENTAL CONSERVATION: INDIGENOUS ENVIRONMENTAL KNOWLEDGE AND ENVIRONMENTAL GOVERNANCE

The 21st century has been characterized by increasing global environmental crises due to the effects of unsustainable Western/Euro-Americentric capitalistic anthropogenic development pathways and untrammeled consumerism. These include climate change effects, pollution, waste accumulation, biodiversity loss, poverty, disease epidemics, and food insecurity, which the global Sustainable Development Goals (SDGs) are trying to address (UNESCO, 2017). Indigenous knowledges have become a subject of renewed attention in global efforts to address these issues against a background of waning confidence in Western/Euro-Americentric sciences as a prime source of solutions to the world's environmental problems, which the West has largely created.

African livelihood systems have a relational foundation with the land. The lived environment is the major resource base for livelihood sustenance for indigenous peoples, and indigenous people have developed holistic ways to sustain the land, water, and biodiversity so that they can continue to sustain their livelihoods. These include the use of taboos, totems, maintenance of sacred forests and water bodies, hunting and gathering restrictions and laws, which are communicated through traditional means of knowledge exchange (stories, myths, narratives, proverbs, observance of ceremonies, and customs). The in-depth knowledge that indigenous people have of their lived environments (indigenous classification systems, knowledge of climate and seasons, and indigenous harvesting strategies) have been proven to be useful and applicable to ecology, botany, zoology, biodiversity conservation, and ecosystem management.

Indigenous knowledges relating to the environment and their potential role in biodiversity conservation have been alluded to by several authors (Gagdil, Berkes & Folke, 1993; Berkes, Folke & Gadgil, 1994; Campbell, 1997; Dold & Cocks, 1999; Nygren, 1999; Byers, Cunliffe & Hudak, 2001; Murombedzi, 2003; Cocks, 2006a, 2006b; O'Donoghue, Shava & Zazu, 2013; Shava & Schudel, 2013). There are many linkages between indigenous knowledge and environmental sustainability emanating from indigenous peoples' ways of living on the land and their maintaining of local biodiversity and ecosystems. This ability to sustain their way of life includes their knowledge of local biodiversity, ecology, indigenous environmental conservation, and sustainable resource use practices.

AGRICULTURE: INDIGENOUS AGROBIODIVERSITY AND INDIGENOUS FARMING SYSTEMS

Indigenous farming systems and agrobiodiversity have contributed in many ways to (1) maintenance of the livelihood of indigenous peoples over time by providing them with food security and sovereignty, and to (2) global food security. The world's major crops have been derived from the crops of various indigenous peoples across the world, with major staple foods coming from Asia, Africa, and Latin America.

Indigenous agrobiodiversity (crops and crop varieties) provide the genetic resource base of many major crops and animal breeds, the basis for food security, and the source of future crops and livestock breeds. Indigenous crop farming systems are agroecologically diverse compared to the crop monocultures of modern farming systems. They are complex and innovative systems integrating several crop species (multicropping/polyculture) and indigenous landraces. These yield a range of products and include the selective conservation of numerous plant species (such as wild and semi-domesticated leafy vegetables, wild fruits, medicinal plants, fuelwood trees, and shade trees) which attract pollinators and are habitats for wild animals, insects, and birds. Related to indigenous crop knowledge and practices are indigenous pest and weed management systems. Many indigenous crops, which may currently be of minor importance to global agriculture, are well adapted to local climates and still continue to play an important role in food security and in meeting dietary and nutritional needs of indigenous rural communities – against the background of a changing environment due to the increasing frequency and intensity of climate change effects (such as drought in southern Africa) (Shava et al., 2009). They also have the potential to become major crops in the future. Likewise, indigenous livestock husbandry involves the keeping of many livestock species and breeds (cattle, sheep, goats, poultry, and other small livestock) which are also important sources of manure required for nutrient cycling and soil enrichment in the indigenous integrated crop and livestock farming systems. Linked to livestock husbandry is indigenous veterinary knowledge and practice for the health and sustenance of livestock. Indigenous people are the repositories and conservators of this agricultural knowledge, agrobiodiversity, agricultural systems, and related practices.

Indigenous farming systems are integrated systems (even on a minute scale) characterized by diversity, dynamism, adaptiveness, experimentation, and innovation (Brookfield & Padoch, 1994; Shava et al., 2009; O'Donoghue, Shava & Zazu, 2013). Most indigenous integrated agricultural systems are relatively environmentally sustainable compared to modern monocultural farming systems. This is due to their species diversity that mimics natural ecosystems, minimal soil disturbance, the selective conservation or inclusion of wild species on cultivated land, and their indigenous agricultural management practices symbolized by a reliance on natural nutrient recycling and soil enrichment as compared to modern agricultural systems which rely on artificial fertilizers and agrochemicals (herbicides and pesticides).

Indigenous agricultural systems and practices with low costs, low inputs, and high diversity are now being more widely adopted, as there is a necessary and growing trend towards sustainable agriculture, particularly with the onset of global climate change and related global environmental crises. This is evidence that agricultural methods that have been tried and tested by indigenous peoples are sustainable (Campbell, Clarke & Gumbo, 1991; Campbell, 1997). The result of such practices is an integrated system for managing the land (soil), water, agrobiodiversity, and natural species diversity in indigenous peoples' lived environments. Such practices have now been repackaged and rebranded - given new Western terminologies such as **agroforestry** (Steppler & Nair, 1987; Buck, Lassoie & Fernandes, 1998; Umrani & Jain, 2010), **organic farming** or **natural farming** (Fukuoka, 1978, 1985), **permaculture** (Mollison, 2003), **ecological agriculture** (Magdoff, 2007), **ecological land use management**, and **organic farming**, often without due acknowledgement of the indigenous originators of such practices. These "emerging" sustainable agriculture systems are embracing indigenous/traditional agricultural farming practices such as multicropping, heirloom seeds, selective conservation of useful trees (e.g. fruit trees and medicinal plants), shifting agriculture, and use of natural manure and bio-pesticides, This appropriation of indigenous agricultural practices and their re-representation as novel or modern sustainable agricultural systems reveals their relevance in the current era.

Reform of global agriculture towards sustainable agricultural systems requires the examination of (and further research into) the different types of agricultural knowledge necessary for sustainability and the need to challenge the monopoly of modern Western agricultural systems (that are input-dependent monocultures) and their negative environmental impacts.

FOOD: INDIGENOUS FOODS AND BEVERAGES

Indigenous people utilize a wide variety of local resources in their lived environment as food. These include food crops (cereal, pulses, vegetables, root and tuber crops, fruits), livestock, wild food plants (grains, vegetables, bulbs, roots, fruits, nuts, and seeds), wildlife (fish, birds, game animals, insects), and mushrooms. A typical indigenous dish consists of a stiff starchy cereal and relish (made from meat and/or vegetables or sour milk).

Fermentation is an economical and viable way used by indigenous communities across the globe for centuries to preserve local perishable foods for a longer time span. Fermentation also improves the nutritional value of the foods, making them more digestible. Traditional indigenous fermented beverages in southern Africa are usually made from cereals, mainly the indigenous cereals sorghum, millet, and finger millet. These include traditional beer brewed from indigenous grains such as sorghum, a fermented non-alcoholic beverage called (a)mar/hewu (mageu) and sour milk. Traditional sorghum beer, (a)mar/hewu and sour milk (Amasi) are now commercialized in Southern Africa (see Haggblade & Holzapfel, 2004; Holzapfel & Taljaard, 2004). Some indigenous people from Southern Africa such as the amaXhosa in the Eastern Cape of South Africa brew mead (iqilika) from honey, while some brew a liquor from fermented fruits, examples being liquor from marula and mangeti (mongongo) nuts. Marula is now made into a commercial liqueur (Amarula cream) and mead is now made as a commercial brand (Iqilika) in South Africa (Cambray, 2005). The commercial appropriations of indigenous foods and beverages are an indication of their commodification potential, which makes them relevant and acceptable in urban society contexts in the modern economy.

In addition to beers and fermented beverages, several indigenous plant species are used to make herbal teas. These include rooibos tea (*Aspalathus linearis*), honeybush (*Cyclopia genistoides, C. intermedia* and *C. subternata*), bush tea (*Athrixia phylicoides*), buchu tea (*Agathosma betulina*), rose pelargonium tea (*Pelargonium capitatum*), and cancer bush/kankerbos tea (*Sutherlandia frutescens*) in Southern Africa (see van Wyk & Gericke, 2003). Most of these teas have health benefits and are now commercialized. The currency and increasing popularity of indigenous teas is proof of their relevance in the 21st century.

Indigenous people have also used sun-drying to preserve fruits, vegetables, roots, and tuber food plants as well as smoking to preserve meat and fish to prevent rapid deterioration in the tropics. Commercially packaged sundried fruits and smoked meats and fish are now popular and widely available in modern food chain stores.

There is a growing "back to nature" trend, particularly in the global north, in addressing human health concerns that has driven the growth in demand of natural/organic foods and traditional/paleo diets (Pollan, 2006, 2009).

MEDICINE AND PHARMACY: INDIGENOUS MEDICINES

Modern/Western medicine and the pharmaceutical industry have benefitted significantly from new drug cures derived from indigenous medicinal knowledge and use of indigenous (mainly plant) species. There are multiple examples of modern Western medicines that have been derived from indigenous medicines, including aspirin (from willow bark), quinine (from cinchona bark), and morphine (from opium poppy) (Vickers, Zollman & Lee, 2001; Dias, Urban & Roessner, 2012). This is despite the fact that most indigenous medicinal knowledge and practices are frequently denigrated as primitive and superstitious. However, the continuing bioprospecting by pharmaceutical companies in search of indigenous medicinal species in indigenous peoples' lands for new pharmaceuticals, nutraceuticals, and cosmeceuticals proves their currency. Consider the recent case of *Hoodia currorii* (ghaap, khobab), a succulent plant whose stems are eaten as raw food by the Khoi San people of Southern Africa as a hunger suppressant during their long treks across the desert. The plant's extract has now been patented and is now being used as an alternative way to slim and reduce obesity (Wynberg, Schroeder & Chennels, 2009).

Besides indigenous medicines contributing to mainstream modern/Western medicine and pharmaceutical practices, there is a growing global trend towards using herbal medicines and so-called "complementary and alternative medicines" in mainstream Western medical practice (Harris & Rees, 2000; Xue et al., 2007). These complementary and alternative medicines include Ayurveda, chiropractic, homeopathy, osteopathy, aromatherapy, naturopathy, nutritional healing, traditional Chinese medicine (including acupuncture), reiki, and spiritual healing (Ernst, 2000). This trend could be against the background of dissatisfaction with modern medicine, ease of access to herbal remedies, and/or modern synthetic pharmaceutical drugs being expensive and having undesirable and sometimes lethal side effects. The sale of herbal medicines is now a trend in most commercial pharmacies.

TECHNOLOGY: INDIGENOUS TECHNOLOGIES AND INNOVATIONS

Indigenous technologies are diverse ranges of practices that span across different sectors of society. They include agricultural, hunting and gathering, food processing, health, mining and metallurgy, building/construction, craftware (weaving, carving, moulding/pottery), and communication technologies. Some of these indigenous technologies are discussed below.

Agricultural Technologies

Several indigenous technologies have been employed in agriculture. These include farming equipment (such as hoes, axes, digging equipment, yokes, and plows), irrigation technologies (canals and water lifting technologies), soil conservation technologies, pest management techniques, and seed breeding and selection.

Indigenous crop cultivation is based on in-depth knowledge of the macro (climate, topology, and soil types) and micro environments and the knowledge of indigenous crop species as well as the process of developing indigenous crop varieties. Cultivation also involves the knowledge of soil and water conservation techniques (terracing, manuring, selective vegetation conservation on cultivated lands, shifting cultivation, coppicing, grazing regimes, seasonal burning of moribund grass, multicropping, and selective cultivation of plants in micro- and macro-environments), and pest management practices (seed conservation through smoking and use of natural pesticides, use of natural pesticides in the field, and companion planting) (O'Donoghue, Shava & Zazu, 2013). Indigenous people have also invented innovations for irrigation and soil conservation, as has been evidenced in Nyanga, Zimbabwe, where evidence of ancient stone terraces and irrigation canals that were constructed by indigenous communities has been found (Sutton, 1984; Soper, 1996; Mupira, 2003; Tempelhoff, 2009).

Indigenous people have also developed methods of weather forecasting to predict seasons and make decisions on the timing of agricultural activities, prediction of drought years and years of heavy rainfall. This includes using local indicators (observing cues) from the natural environment such as the timing of blossoming (flowering), fruiting and leaf production of certain indigenous trees, migration of bird, animal, and insect species, and animal behavior (Acharya, 2011; Maguti & Maphosa, 2012; Makwara, 2013; Zuma-Nethsuikhwi, Stigter & Walker, 2013). Indigenous weather forecasting remains a preferred and reliable decision-making tool for indigenous farmers despite the dominance of modern meteorological technology, and provides an important source of environmental information (see Green, Billy & Tapim, 2010).

Besides crops, animal husbandry is an important agricultural activity for indigenous communities. Livestock herders have in-depth knowledge of good grazing and browsing species and the areas in which they were found, the location of watering holes, tracking of predators to protect the livestock, and knowledge of veterinary medicines and practices to sustain the health of the livestock. They also have knowledge of the indigenous breeds and breed selection.

Many indigenous agricultural technologies have been taken up in many sustainable agriculture systems. Indigenous seed and livestock breeding strategies still play an important role in agrobiodiversity conservation.

Hunting and Gathering

Hunting and gathering is a specialized activity undertaken by people who had the requisite ability and inclination in the community (Shava & O'Donoghue, 2014). Hunting is normally the domain of men, while gathering is the domain of women. Besides the knowledge of the use of hunting and gathering implements (such as spears and arrows) and the technology for making poison for poison-tipped arrows, making animal and bird traps, indigenous hunters and fishermen have in-depth knowledge of natural breeding seasons and animal, bird, and fish migration patterns and routes. Hunting and fishing are seasonal activities governed by observance of breeding seasons, in which hunting is closed. Hunters are also naturally good trackers of animals and therefore have knowledge of different animal habitats, animal tracks, and animal scat. An important aspect of hunting is the sharing of game among the hunters and within the community so that nobody will go hungry.

Modern game-tracking strategies, though now using modern, sophisticated game equipment, still rely on indigenous game trackers and their knowledge of the veld and animal movement, showing that they are still relevant (Rose & Clarke, 1997; Huntington, Suydam & Rosenburg, 2005).

Gathering is the domain of women and children. They gather wild leafy vegetables, edible mushrooms, edible roots, tubers, and fruits. Gathering requires the knowledge of plant habitats, knowledge of the different edible species, and knowledge of the seasons during which the species could be harvested. Wild fruits, vegetables, and mushrooms are now commercially harvested, packaged, and sold across the globe, courtesy of the indigenous peoples who know their food value and locality.

Mining and Metallurgy

Indigenous African people have knowledge of mining and smelting mineral ores such as iron, copper, and gold (Childs, 1991; Hammel et al., 2000; Shava & O'Donoghue, 2014). Indigenous metallurgy includes mining or smelting using traditional clay furnaces and bellows made from animal hide, and metal smithing (using tools for molding and shaping metal into various tools such as hoes, spears, knives, axes, hatchets, adzes, implements, and ornaments (bangles, necklaces, statues). This work is undertaken by specialists, usually males, in this field, who have to undergo relevant intense apprenticeship. Indigenous mining and metallurgy are technologies that can be revamped and modified for local small-scale mining and metal production in the current era.

Construction

Construction is a collective and specialized activity, with various groups undertaking specific functions. For example, in the construction of an indigenous pole, dagga, and thatch hut, some men specialize in the building of the wall, the women cut and collect the grass, and some men construct the roof, while women make the mud walls, smear the dung for the floor, and weave the mats for the door (Shava & O'Donoghue, 2014). Construction using materials such as stone is similarly a very specialized activity.

The thatch, pole, and dagga construction practices have been taken up in modern architecture construction due to their being environmentally sustainable, durable, and possessing good thermal conductivity properties. These are a feature of modern game lodges and gazebos in the African landscape. Likewise, stone masonry is a continuing construction practice.

Craftware

Indigenous people specialize in various crafts for utilitarian and decorative purposes, including weaving, basketry, carving, clay molding/pottery, stone carving, graphic design painting, beading, and the making of musical instruments (such as drums and mbira) (Shava, 2015).

African indigenous pottery is the domain of women, with some designs being plain and utilitarian, while others are more decorative and ornamental. Some indigenous people are well known for their decorative basketry, such as the Tonga of the Zambezi Valley. The Tsonga (Shangaan) and the Ndebele in South Africa are known for their colorful graphic design decorations on their hut walls. The Zulu women are well known for their beadwork, including their famous Zulu love letters. Carving is normally the domain of men. Most African peoples use wood for carving, while some peoples such as the Shona people of Zimbabwe are well known for their stone carvings. Wood carving is used in the making of a range of household items, such as cooking utensils, cups, plates, stools, benches, chairs tables, beds, tool handles, and ornamental objects such as masks and sculptures.

Communication Technologies

Indigenous communication technology includes various means for conveying important messages. For example, the practice of African women sweeping around the homestead is an important means of communication. This is usually done at the beginning or the end of the day. Upon returning from daily family chores or waking up early in the morning, the man of the house would do an inspection around the outside of the home. The man would look for tracks of foreign human and animal footprints, which would then inform him of any dangerous intruders and the direction that they have taken. This information is then used to protect the home, such as by tracking and killing dangerous animals or setting animal traps.

News in the African community is often circulated using indigenous technologies. For example, a common practice is using the talking drum to convey messages (see Mushengyezi, 2003). A particular drum sound is used for conveying important messages for a particular event such as a funeral, attacking invaders, or a ceremony. The drum is sounded at the source of the information and receiving drummers at other sites use the same tone to sound the message. This process is repeated throughout the village and beyond, and people know and understand the meaning, react accordingly, and trace the source of the message. In this way, news is conveyed in the community.

Communication with the ancestors is also a very specialized and intricate process for African peoples. It involves the slaughter of a beast, the brewing of beer, the playing of musical instruments (drums, mbira, rattles, etc.), and dancing.

Aspects of indigenous communication technologies are still evident in modern communication technologies such as mobile phones, army war mobilization strategies, and game-tracking strategies.

Healthcare

Indigenous healthcare includes health practitioners specializing in various healthcare areas, such as herbalists, traditional healers, midwives, and care-givers for the young and the elderly. These practitioners have in-depth knowledge of techniques and practices relevant to their specialization. Such practices are still relevant today in many rural communities, as traditional medicinal practitioners are still consulted by many community members and those from beyond the immediate community (for example, see Gqaleni et al., 2007).

GOVERNANCE AND ORGANIZATION: INDIGENOUS GOVERNANCE SYSTEMS

Indigenous governance systems in Southern Africa have been characterized by collective consultation and collaborative decision-making processes. Indigenous leaders such as kings, chiefs, and village heads have strongly relied on village elders who are a core aspect of the decision-making process. In addition, spirit mediums, many of them women, play an important role in decision-making processes. These elements of indigenous governance are at the core of modern democratic governance systems and processes that are now globally accepted.

In extended African family or village community settings, collaboration is a key strategy. In child raising, for example, not only the immediate family is involved in the raising of a child. The extended family and the village are also involved, hence the common African saying: "It takes a village to raise a child". Likewise, doing labor-intensive chores such as clearing the fields, cultivating the fields, harvesting the crops, herding cattle, and collecting firewood is a collective endeavor accompanied by feasting, singing, dancing, and learning-by-doing to lighten the burden of the chores. While many different chores in indigenous communities are stratified by gender, the element of collaboration is a common thread. These collaborative practices have been core to sustaining a cohesive society, an aspect which even modern social structures such as social welfare have been unable to replace. It should be noted that while indigenous communities emphasize collaboration, they do not take away individuality and individual wealth.

AESTHETICS

African indigenous aesthetics are expressed in various ways, including music (song and dance), orature, hairstyles, cosmetics, ornamentation, and creative artworks (Shava, 2015). Song, dance, poetry, and orature were common means of cultural expression that were used for different cultural occasions such as work, celebration, mourning, and religious worship. African music has distinctive African rhythms and uses African instruments such as drums, rattles, the mbira, marimba, and kora. Likewise, African dances have their own distinctive rhythms, use of particular movements and body parts, as well as use of various accessories, such as masks and rattles. Similarly, African poetry has its own distinctive styles, including love poetry and praise poetry; African orature distinctively involves the use of proverbs and idioms, while storytelling involves the use of African animal characters. These African indigenous arts are now a prominent feature of modern arts and culture.

Indigenous peoples have their own definitions of body beauty, including the use of various chemical products such as saponaceous plants for soap; using plant oils, animal fats, mud, and ochre for enhancing their skin; and using lye for their hair texture. These indigenous natural beauty enhancers are making their way into modern cosmetic products or cosmeceuticals. Indigenous hairstyles, such as braiding and dreadlocks still remain a distinguishing feature of African peoples worldwide.

EDUCATION

Western/Euro-Americentric "formal" education systems, from pre-school to university are an alienating process and experience. Science, technology, engineering, and mathematics (STEM) are a global priority area in education intended to improve competitiveness in science and technology development. However, there is poor performance in these areas across the globe, including in the United States (Kuenzi, 2008; Chen, 2013). Several studies by indigenous scholars are proving that the use of indigenous languages and contextually relevant examples and case studies from indigenous community practices are enabling epistemological access for indigenous learners and helping them gain understanding of abstract scientific concepts (e.g. see Shava, 2016; Shava & Manyike, 2018). There is therefore a need for indigenous scholars to perform research and promote indigenous epistemologies for the benefit of indigenous learners and for epistemological justice through enabling plural representation of knowledges in formal education contexts.

CONCLUSION

This chapter is not intended to provide a comprehensive overview of the roles of indigenous knowledges in the 21st Century but instead to give some highlights of how indigenous knowledges are still current and applicable across different sectors of modern society today. Indigenous people and their knowledges are still largely marginalized, excluded, and devalued by modern Western knowledge discourses and socio-political-economic practices that continue to benefit from them. We need to acknowledge that knowledge is, and has always been a shared resource, and that global knowledge derives from local knowledges. It is the currency (relevance) of indigenous knowledges at the intersection between them and modern/Western dominant knowledge systems that accords them a place for recognition amongst global knowledge systems and socio-economic practices. Indigenous knowledges are an important aspect in the identity of indigenous peoples and crucial to the dismantling of the Western/colonial conceived hierarchy of knowledge systems that prioritizes Western epistemologies. Indigenous scholars and researchers must create enabling platforms for plural and equitable knowledge representation and application. Because of the close relationship indigenous peoples have with their lands, their knowledges are particularly important in sustainable development pathways for the future of humankind and the natural environment.

There is growing interest in the economic potential of indigenous knowledges, particularly on indigenous plants and their medicinal potential. These include application/use in new pharmaceuticals, cosmeceuticals, nutraceuticals, indigenous crafts, and other indigenous technologies (Van Wyk & Gericke, 2000). There is also a growing back-to-nature trend towards natural/organic foods and traditional/paleo diets (Pollan, 2006, 2009). The above examples are evidence that indigenous knowledges continue to possess currency and applicability in addressing issues affecting the world in the present era. We argue that development pathways are more likely to be sustainable if they adopt indigenous knowledges.

Contrary to the belief that indigenous knowledges are primitive and archaic knowledges that are not relevant outside their contexts of generation, the numerous applications of indigenous knowledge in the current era demonstrate their portability and transferability across the world. However, with this transferability comes the problem of appropriation, usually under the guise of indigenous knowledge aspects being "common knowledge" of an indigenous people, and the commericalization and privatization of this knowledge through intellectual property rights. This calls for the need to recognize the originators of the knowledge and to share the accrued benefits of the commercialized use of the knowledge with the indigenous knowledge holders from which it is derived.

REFERENCES

Acharya, S. 2011. Presage biology: lessons from nature in weather forecasting. *Indian Journal of Traditional Knowledge* 10(1): 114–124.

Berkes, F., Folke, C. and Gadgil, M. 1994. Traditional ecological knowledge, biodiversity, resilience and sustainability. In Perrings, C.A., Mäler, KG., Folke, C., Holling, C.S., and Jansson, B.O. (eds.). *Biodiversity Conservation. Ecology, Economy & Environment*, vol. 4. Dordrecht: Springer.

Brookfield, H. and Padoch, C. 1994. Appreciating agrobiodiversity: a look at the dynamism and diversity of indigenous farming practices. *Science and Policy for Sustainable Development* 36(5): 6–45.

Byers, B.A., Cunliffe, R.N. and Hudak, A.T. 2001. Linking the conservation of culture and nature: a case study of sacred forests in Zimbabwe. *Human Ecology* 29(2): 187–121.

Cambray, G.A. 2005. African mead: biotechnology and indigenous knowledge systems in IQhilika process development. Unpublished Doctor of Philosophy thesis, Rhodes University, Grahamstown, South Africa.

Campbell, B.M., Clarke, J.M. and Gumbo, D.J. 1991. Traditional agroforestry practices in Zimbabwe. *Agroforestry Systems* 14: 99–111.

Campbell, R.J. 1997. Innovations in research-based practice. *Personnel Psychology* 50(2): 453.

Chen, X. 2013. *STEM attrition: college students' paths into and out of STEM fields (NCES 2014-001)*. National Center for Education Statistics, Institute of Education Sciences, U.S. Department of Education, Washington, DC.

Childs, S.T. 1991. Style, technology, and iron smelting furnaces in Bantu-speaking Africa. *Journal of Anthropological Archaeology* 10: 332–339.

Cocks, M. 2006a. Biocultural diversity: moving beyond the realm of 'indigenous' and 'local' people. *Human Ecology* 34(2): 185–200.

Cocks, M. 2006b Wild plant resources and cultural practices in rural and urban households in South Africa. Implications for bio-cultural diversity conservation PhD thesis Wageningen Universitty, Wageningen, The Netherlands.

Dias, D.A., Urban, S. and Roessner, U. 2012. A historical overview of natural products in drug discovery. *Metabolites* 2(2): 303–336.

Dold A.P. and Cocks, M. 1999. Preliminary list of Xhosa plant names from Eastern Cape, South Africa. *Bothalia - African Biodiversity and Conservation* 29(2): a601.

Emeagwali, G. 2014. Intersections between Africa's indigenous knowledge systems and history. In Emeagwali, G. and Dei, G.J.S. (eds.). *African Indigenous Knowledge and the Disciplines*. Rotterdam, Boston, Taipei: Sense Publishers (pp. 1–17).

Ernst, E. 2000. The role of complementary and alternative medicine. *BMJ* 321: 1133–1135.

Fukuoka, M. 1978. *The One Straw Revolution: An Introduction to Natural Farming*. Emmaus: Rodale Press.

Fukuoka, M. 1985. *The Natural Way of Farming: The Theory and Practice of Green Philosophy*. Tokyo, New York: Japan Publications, Inc.

Gadgil, M., Berkes, F. and Folke, C. 1993. Indigenous knowledge for biodiversity conservation. *A Journal of the Human Environment* 22: 151–156.

Gqaleni, N., Moodley, I., Kruger, H., Ntuli, A. and McLeod, H. 2007. Traditional and complementary medicine. *South African Health Review* 1: 175–188.

Green, D., Bill, J. and Tapim, A. 2010. Indigenous Australians' knowledge of weather and climate. *Climate Change* 100(2): 337–354.

Haggblade, S. and Holzapfel, W.H. 2004. Industrialization of Africa's indigenous beer brewing. In Steinkraus, K.H. (ed.). *Industrialization of Indigenous Fermented Foods*. New York, Basel: Marcal Dekker, Inc (pp. 271–352).

Hammel, A., White, C., Pfeiffer, S. and Miller, D. 2000. Pre-colonial mining in southern Africa. *Journal of the South African Institute of Mining and Metallurgy* 100(1): 49–56.

Harris, P. and Rees, R. 2000. The prevalence of complementary and alternative medicine use among the general population: a systematic review of the literature. *Complementary Therapies in Medicine* 8: 88–96.

Holzapfel, W.H. and Taljaard, J.L. 2004. Industrialization of mageu fermentation in South Africa. In Steinkraus, K.H. (ed.). *Industrialization of Indigenous Fermented Foods*. New York, Basel: Marcal Dekker, Inc (pp. 363–407).

Huntington, H.P., Suydam, R.S. and Rosenburg, D.H. 2005. Traditional knowledge and satellite tracking as complementary approaches to ecological understanding. *Environmental Conservation* 31(3): 177–180.

Kuenzi, J.J. 2008. Science, Technology, Engineering, and Mathematics (STEM) Education: Background, Federal Policy, and Legislative Action. Congressional Research Service Reports. 35.

Magdoff, F. 2007. Ecological agriculture: principles, practices, and constraints. *Renewable Agriculture and Food Systems* 22(2): 109–117.

Maguti, T. and Maposa, R.S. 2012. Indigenous weather forecasting: a phenomenological study engaging the Shona of Zimbabwe. *Journal of Pan African Studies* 4(9): 102–113.

Makwara, E.C. 2013. Indigenous knowledge systems and modern weather forecasting: exploring linkages. *Journal of Agriculture and Sustainability* 2(1): 98–141.

Mollison, B. 2003. *Permaculture: A Designer's Manual*. Tyalgum: Tagari Publications.

Mupira, Paul 2003. The case of Nyanga cultural landscape, N.E. Zimbabwe. In 14th ICOMOS General Assembly and International Symposium: 'Place, memory, meaning: preserving intangible values in monuments and sites', 27–31 October 2003, Victoria Falls, Zimbabwe.

Murombedzi, J. 2003. Pre-colonial and colonial conservation practices in southern Africa and their legacy today. IUCN Report. http://dss.ucsd.edu/ccgibson/docs/ Murombedzi %20-%20Pre-colonial%20and%20Colonial% 20Origins.pdf.

Mushengyezi, A. 2003. Rethinking indigenous media: rituals, 'talking' drums and orality as forms of public communication in Uganda. *Journal of African Cultural Studies* 16(1): 107–117.

Nuck, L.E., Lassoie, J.P. and Fernandes, E.C.M. (eds.). 1998. *Agroforestry in Sustainable Agricultural Systems*. Boca Raton, London, New York, Washington: CRC Press.

Nygren, A. 1999. Local knowledge in the environment–development discourse. From dichotomies to situated knowledges. *Critique of Anthropology* 19(3): 267–288.

O'Donoghue, R., Shava, S. and Zazu, C. (eds.). 2013. *African Heritage Knowledge in the Context of Social Innovation*. United Nations University – Institute of Advanced Studies (UNU_IAS), Tokyo. http://www.ias.unu.edu/resource_centre/UNU_Booklet_MB2013_FINAL_Links_v12.pdf.

Pollan, M. 2006. *The Omnivore's Dilemma: A Natural History of Four Meals*. New York: Penguin.

Pollan, M. 2009. *Food Rules: An Eater's Manual*. New York: Penguin.

Rose, D.B. and Clarke, A. 1997. *Tracking Knowledge in North Australian Landscapes: Studies in Indigenous and Settler Ecological Knowledge Systems*. Casuarina: North Australian Research Unit.

Shava, S. 2015. African Aesthetic, The. In Shujaa, M.J. and Shujaa, K.J. (eds.). *The SAGE Encyclopedia of African Cultural Heritage in North America*. Thousand Oaks, CA: SAGE Publications.

Shava, S. 2016. The application/role of indigenous knowledges in transforming the formal education curriculum for contextual and epistemological relevance: cases from southern Africa. In Msila, V.T. and Gumbo, M.T. (eds.). *Africanising the Curriculum: Indigenous Perspectives and Theories*. Stellenbosch: Sun Press (pp 121–139).

Shava, S. and Manyike, T.V. 2018. The decolonial role of African indigenous languages and indigenous knowledges in formal education processes. *Indilinga African Journal of Indigenous Knowledge Systems* 17(1): 36–52.

Shava, S. and O'Donoghue, R. 2014. *Teaching Indigenous Knowledge and Technology*. Fundisa for Change Programme. Grahamstown: Environmental Learning Research Centre, Rhodes University.

Shava, S., O'Donoghue, R., Krasny, M.E. and Zazu, C. 2009. Traditional food crops as a source of community resilience in Zimbabwe. *International Journal of African Renaissance Studies – Multi-, Inter- and Transdisciplinary* 4(1): 31–48.

Shava, S., and Schudel, I. 2013. *Teaching biodiversity*. Fundisa for Change Programme. Grahamstown: Environmental Learning and Research Centre, Rhodes University.

Shizha, E. 2010. Rethinking and reconstituting indigenous knowledge and voices in the academy in Zimbabwe: a decolonization process. In Kapoor, D. and Shizha, E. (eds.). *Indigenous Knowledge and Learning in Asia/Pacific and Africa: Perspectives on Development, Education and Culture*. New York: Palgrave Macmillan.

Soper, R. 1996. The Nyanga terrace complex of eastern Zimbabwe: new investigations. *AZANIA: Journal of the British Institute in Eastern Africa* 31(1): 1–35.

Stepller, H.A. and Nair, P.K.R. 1987. *Agroforestry: A Decade of Development*. Nairobi: International Council for Research in Agroforestry (ICRAF).

Sutton, J.E.G. 1984. Irrigation and soil-conservation in African agricultural history with a reconsideration of the Inyanga terracing (Zimbabwe) and Engaruka irrigation works (Tanzania). *Journal of African History* 25: 25–41.

Tempelhoff, J. 2009. Historical perspectives on pre-colonial irrigation in Southern Africa. *African Historical Review* 40(1): 121–160.

Umrani, R. and Jain, C.K. 2010. *Agroforestry: Systems and Practices*. Jaipur: Oxford.

Van Wyk, B.-E. and Gericke, N., 2000. *People's Plants: A Guide to Useful Plants of Southern Africa*. Pretoria: Briza Publications.

Van Wyk, B.E. and Gericke, N. 2003. *People's Plants: A Guide to Useful Plants of Southern Africa*. Pretoria: Briza Publications.

Vickers, A., Zollman, C. and Lee, R. 2001. Herbal medicine. *Western Journal of Medicine* 175(2): 125–128.

Wynberg, R., Schroeder, D. and Chennells, R. (eds.). 2009. *Indigenous Peoples, Consent and Benefit Sharing: Lessons from the San – Hoodia Case*. Dordrecht, Heidelberg, London, New York: Springer.

Xue, C.C.L., Zhang, A.L., Vian Lin, P.H., Da Costa, C. and Story, D.F. 2007. Complementary and alternative medicine use in Australia: a national population-based survey. *The Journal of Alternative and Complementary Medicine* 13(6): 643–650.

Zuma-Netshiukhwi, G., Stigter, K. and Walker, S. 2013. Use of traditional weather/climate knowledge by farmers in south-western free state of South Africa: agrometeorological learning by scientists. *Atmosphere* 4(4): 383–410.

Index

Printed in the United States
by Baker & Taylor Publisher Services